Marine Ecosystems

From the ***Seas and Oceans*** Set
coordinated by
André Mariotti and Jean-Charles Pomerol

Marine Ecosystems

Diversity and Functions

Edited by

André Monaco
Patrick Prouzet

iSTE

WILEY

First published 2015 in Great Britain and the United States by ISTE Ltd and John Wiley & Sons, Inc.

ISTE Ltd
27-37 St George's Road
London SW19 4EU
UK

www.iste.co.uk

John Wiley & Sons, Inc.
111 River Street
Hoboken, NJ 07030
USA

www.wiley.com

Library of Congress Control Number: 2015948076

British Library Cataloguing-in-Publication Data
A CIP record for this book is available from the British Library
ISBN 978-1-84821-782-9

Contents

Foreword

We have been asked by ISTE to stimulate work in the area of the environment. Therefore, we are proud to present the "Seas and Oceans" set of books, edited by André Monaco and Patrick Prouzet.

Both the content and the organization of this collection have largely been inspired by the reflection, initiatives and prospective works of a wide variety of national, European and international organizations in the field of the environment.

The "oceanographic" community, in France and internationally – which is recognized for the academic quality of the work it produces, and is determined that its research should be founded on a solid effort in the area of training and knowledge dissemination – was quick to respond to our call, and now offers this set of books, compiled under the skilled supervision of the two editing authors.

Within this community, there is a consensus about the need to promote an interdisciplinary "science of systems" – specifically in reference to the Earth's own "system" – in an all-encompassing approach, with the aim of providing answers about the planet's state, the way it works and the threats it faces, before going on to construct scenarios and lay down the elementary foundations needed for long-term, sustainable environment management, and for societies to adapt as required. This approach facilitates the shift of attention from this fundamental science of systems (based on the analysis of the processes at play, and the way in which they interact at all levels and between all the constituent parts making up the global system) to a "public" type of science, which is finalizable and participative, open to decision-makers, managers and all those who are interested in the future of our planet.

In this community, terms such as "vulnerability", "adaptation" and "sustainability" are commonly employed. We speak of various concepts, approaches or technologies, such as the value of ecosystems, heritage, "green" technologies, "blue" chemistry and renewable energies. Another foray into the field of civilian science lies in the adaptation of research to scales which are compatible with the societal, economic and legal issues, from global to regional to local.

All these aspects contribute to an in-depth understanding of the concept of an ecosystemic approach, the aim of which is the sustainable usage of natural resources, without affecting the quality, the structure or the function of the ecosystems involved. This concept is akin to the "socio-ecosystem approach" as defined by the Millennium Assessment (http://millenniumassessment.org).

In this context, where the complexity of natural systems is compounded with the complexity of societies, it has been difficult (if only because of how specialized the experts are in fairly reduced fields) to take into account the whole of the terrestrial system. Hence, in this editorial domain, the works in the "Seas and Oceans" set are limited to fluid envelopes and their interfaces. In that context, "sea" must be understood in the generic sense, as a general definition of bodies of salt water, as an environment. This includes epicontinental seas, semi-enclosed seas, enclosed seas, or coastal lakes, all of which are home to significant biodiversity and are highly susceptible to environmental impacts. "Ocean", on the other hand, denotes the environmental system, which has a crucial impact on the physical and biological operation of the terrestrial system – particularly in terms of climate regulation, but also in terms of the enormous reservoir of resources they constitute, covering 71% of the planet's surface, with a volume of 1,370 million km^3 of water.

This set of books covers all of these areas, examined from various aspects by specialists in the field: biological, physical or chemical function, biodiversity, vulnerability to climatic impacts, various uses, etc. The systemic approach and the emphasis placed on the available resources will guide readers to aspects of value-creation, governance and public policy. The long-term observation techniques used, new techniques and modeling are also taken into account; they are indispensable tools for the understanding of the dynamics and the integral functioning of the systems.

Finally, treatises will be included which are devoted to methodological or technical aspects.

The project thus conceived has been well received by numerous scientists renowned for their expertise. They belong to a wide variety of French national and international organizations, focusing on the environment.

These experts deserve our heartfelt thanks for committing to this effort in terms of putting their knowledge across and making it accessible, thus providing current students with the fundaments of knowledge which will help open the door to the broad range of careers that the area of the environment holds. These books are also addressed to a wider audience, including local or national governors, players in the decision-making authorities, or indeed "ordinary" citizens looking to be informed by the most authoritative sources.

Our warmest thanks go to André Monaco and Patrick Prouzet for their devotion and perseverance in service of the success of this enterprise.

Finally, we must thank the CNRS and Ifremer for the interest they have shown in this collection and for their financial aid, and we are very grateful to the numerous universities and other organizations which, through their researchers and engineers, have made the results of their reflections and activities available to this instructional corpus.

André MARIOTTI
Professor Emeritus at University Pierre and Marie Curie
Honorary Member of the Institut Universitaire de France
France

Jean-Charles POMEROL
Professor Emeritus at University Pierre and Marie Curie
France

1

Marine Biosphere, Carbonate Systems and the Carbon Cycle

1.1. Introduction

It is now accepted that the recent increase in the concentration of carbon dioxide (CO_2) in the atmosphere is a consequence of human activities, in particular the combustion of fossil fuels and the production of cement. Approximately 30% of CO_2 emissions are absorbed by the ocean, which clearly indicates the importance of the sea in the regulation of the level of atmospheric CO_2. Once absorbed by the ocean, this carbon becomes an important parameter in biogeochemical cycles.

For several million years, the concentration of CO_2 in the atmosphere has shown cyclic variations directly linked to global changes in volume. These changes occur at regular intervals of 100,000, 40,000 and 23,000 years (Figure 1.1) which represent the Earth's orbit about the Sun; these phases control the insolation on the surface of the Earth (solar forcing) and are responsible for the natural variability of the climate.

COMMENTS ON FIGURE 1.1.– These parameters, each with their own frequency, combine together and cause climatic variations, particularly glacial/interglacial variations. The two curves to the bottom of the figure show: the variations in isotopes of oxygen in the plankton shells (SPECMAP) [MAR 87] which are determined by water temperature and the global volume of ice over the past 400,000 years. Ice cores entrap bubbles

---

Chapter written by Luc BEAUFORT.

containing past atmospheres. The concentration of CO_2 in these bubbles (EPICA) [LUE 08] is very similar to that in the ocean, by identifying glacial (gray areas) and interglacial areas.

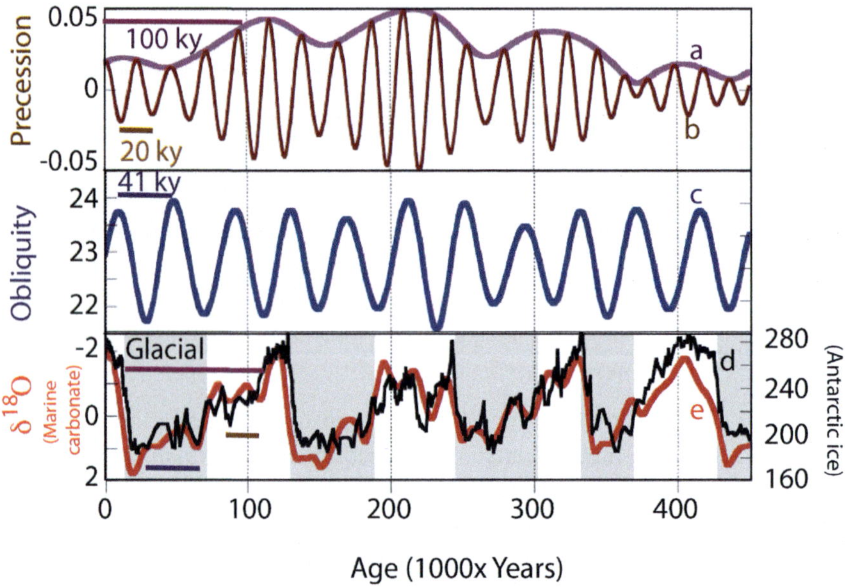

Figure 1.1. *Parameters of the Earth's orbit [LAS 04] molding the global climate (CO_2 concentration in ice core d), oxygen stable isotopes in marine sediments e), precession b) and obliquity c) influence the seasonal contrasts, eccentricity a) modulates (amplifies or not) precession effects (Beaufort, compilation of [LAS 04, LUE 08, MAR 87])*

The reasons why the atmospheric concentration of CO_2 follows climatic variations are still not fully understood, but the ocean could be responsible for their long term natural variations. In fact, the ocean surface layer contains an enormous reservoir of carbon than can react with the atmosphere over these orbital time scales. Marine organisms are of particular importance in these mechanisms as they help incorporate (pump) and transfer a large amount of carbon from the surface of the ocean to the deep sea and into sediments (sinks). Their role becomes crucial in the global carbon cycle.

If we know that human activities, by emitting large quantities of CO_2 into the atmosphere, disturb the biological fluxes of carbon in the ocean, it is extremely difficult to predict the response of marine ecosystems and what the future holds for natural carbon in the biological cycle. Studying marine sediments, which can be considered as historical archives of ocean ecology, allows us to better understand how the ocean participates in the carbon cycle and how marine biodiversity adapts to global changes. In the past, these changes in biodiversity have sometimes had very significant retroactive effects on the environment and climate.

This contribution does not intend to describe the chemistry of oceans or biogeochemical cycles, widely developed in the first volume of this set of books [BER 14, LEG 14]. As stated by Legendre [LEG 14], despite their low biomass, pelagic ecosystems are the driver of biogeochemical cycles in the oceans. In this context, we will focus in particular on calcareous phytoplankton in the dynamics of carbon and its role in the evolution of past and future climates. The complexity of these processes is why this system is often disregarded in experimental research, models and projections.

1.2. Marine organisms and carbon

Marine organisms are clearly adapted to the ocean properties, in which they live, but they also actively contribute to its composition; since organismal biology is based on the chemistry of carbon, this is particularly true for the concentration of carbon dissolved in the sea. In fact, marine organisms use carbon to build their tissues (organic form of carbon), many of which also form solid skeletons, particularly in the form of calcium carbonate (inorganic form of carbon). The concentrations of dissolved and particulate carbon therefore change according to the mass and activity of these organisms. Two equations can express the dual effect of this activity on dissolved carbon: on the one hand, carbon sequestration by photosynthesis into their tissues [1.1] and, on the other hand, the release of CO_2 upon the construction of their carbonaceous skeleton from bicarbonate and dissolved calcium [1.2]:

$$CO_2 + H_2O + \text{light} \rightarrow CH_2O \text{ (organic matter)} + O_2 \qquad [1.1]$$

$$Ca^{2+} + 2HCO_3^- \rightarrow CaCO_3 \text{ (carbonate)} + CO_2 + H_2O \qquad [1.2]$$

These processes can be reversed depending on the environmental conditions. In addition, if the organic matter produced is used by another organism (grazing, predation), this will release carbon dioxide; however, if it is buried in the sediments; the carbon will be trapped, sometimes for very long periods of time (sequestration). With regard to the carbonates that form the skeletons (calcification), they will be either buried in sediment or dissolved in the ocean.

Calcification and biomineralization will depend on the biodiversity of the organisms; different species will not produce the same quantity of matter, skeletons do not have the same degree of calcification as they are produced at different paces. We will therefore see how marine biodiversity impacts the global carbon cycle and therefore climate.

1.3. Variability in the production of organic matter

During photosynthesis, marine algae absorb dissolved CO_2 to produce their biomass, thereby releasing oxygen (equation [1.1]). The algal production is not distributed uniformly throughout the ocean. Most multicellular algae live on the seabed and on a substrate; they are therefore known as benthic; and their distribution is limited to the shallowest depths of the ocean where sunlight can penetrate. On the contrary, algae that constitute the phytoplankton are almost all unicellular and floating; they are widely distributed along oceanic margins, and beyond, over a maximum of approximately 200 m wide in the water column.

Continental margins are generally much more productive in terms of organic matter than zones situated in the center of oceanic basins. In fact, rivers feed the coastal zones with nutritive salts (nitrogen, phosphorus, etc.), required for the growth of phytoplankton. Offshore, in the pelagic zone, this lateral input is increasingly less as one strays from the continents. The wind sometimes brings dust rich in nutritive salts, but these fluxes rarely compensate for this deficit.

Pelagic phytoplankton that lives exclusively in the photic zone (zone exposed to light) depletes the nutrient stores during photosynthesis; the only way to regenerate this stock at the surface is by the vertical pumping of nutritive salts in the deep sea (>200 m). These vertical nutrient transfers, required for pelagic life, are only produced by updrafts caused by gusts of wind in favorable directions (upwelling) or by the deep mixing of surface layers during episodes of strong winds.

One example of this mechanism can be found in the Indian Ocean where monsoon winds are powerful enough to break down the vertical stratification and allow high phytoplanktonic production by fertilization at depth. This mixing acts over a certain depth and is called the mixed layer; when it reaches the thermocline that separates warm surface waters from cold deep waters, the nutritive salts then diffuse towards the surface. We can then easily understand that during periods favored by strong Indian monsoons, or strong trade winds, offshore of West Africa, oceanic primary production (produced by phytoplankton or PP) is reinforced.

Studying sediment cores taken from these zones has revealed that highly biologically productive periods alternate with depletions, and that these changes follow the rhythms of the Earth's orbit. Thus, primary production, expressed in grams of carbon per meter squared per year, has increased from 120 to 200 $gC/m^2/yr$, since the previous precession cycle of the equinoxes in the center of the Indian Ocean [BEA 97] or in the Banda Sea (Figure 1.2). By causing seasonal variations and by heating different tropical zones, these cycles cause the increase or decrease in winds above the Indian Ocean, beginning with oceanic production.

COMMENTS ON FIGURE 1.2.– Top: variations in organic productivity, common (calculated by EOF) to the entire tropical band in the Indian and Pacific Ocean [BEA 01] (continuous line and circles). The organic production varies according to the concentration of CO_2 measured in the core of Antarctic ice [LUE 08] (dotted line and full squares). Bottom: recording of the organic production (dotted line, full circles and vertical error bars) in the Banda Sea. It is highly dependent on the intensity of the Australian monsoon [BEA 10]. The monsoon activity deduced follows the rhythms of insolation which are parameters of the Earth's orbit (continuous line) [LAS 04]. This highlights the difference that can exist between the local and global variations.

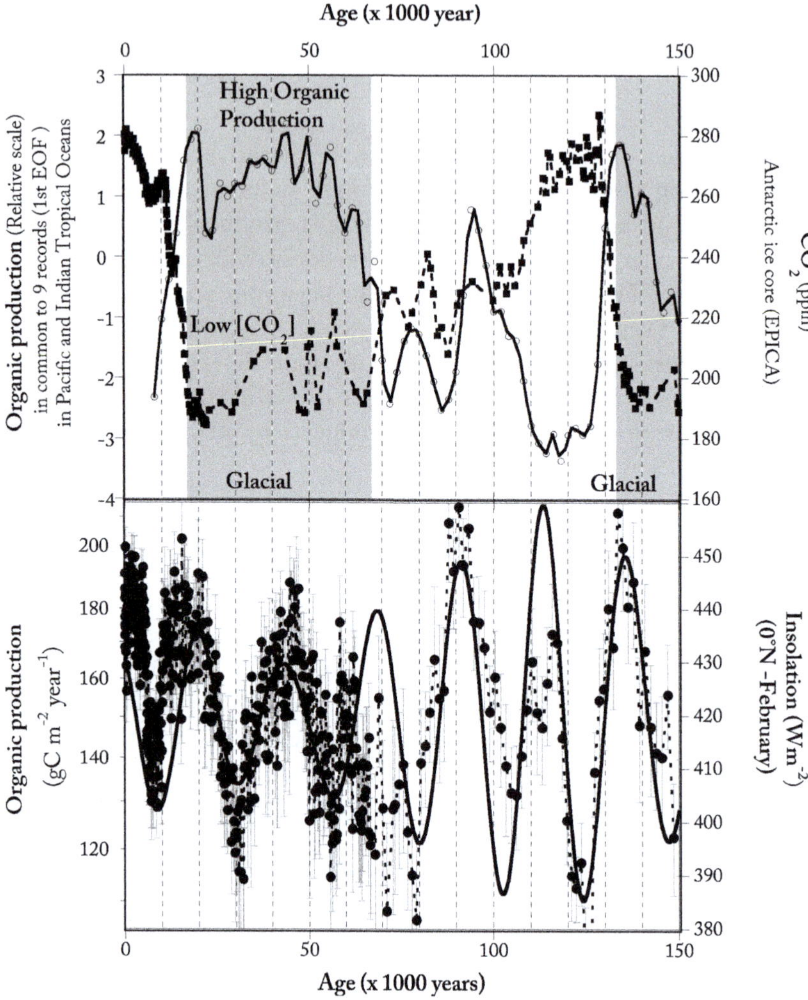

Figure 1.2. *Example of the evolution of organic production in the tropical zone over 150,000 years*

In this example, it is also clear that these changes in primary productivity are accompanied by significant variations in biodiversity. In the sediments deposited in the center of the Indian Ocean 20,000 years ago, 80% of fossil micro flora observed (Figure 1.3) is represented by species known to currently inhabit depths of 100 to 200 m beneath the ocean surface, at the lower limit of the photic zone, in conditions of very low levels of light. However, 10,000 years ago, this microflora represents only 10% of the assembly (assembly of a fossilized community); it has also been replaced by more diverse forms known to live between 0 and 100 m.

We can compare this location in the ocean 20,000 years ago to where we currently find the poorest zones of the central South Pacific (around Easter Island or North Pacific (Hawaii)). These zones, called gyres, represent oceanic deserts (except for areas close to islands) where planktonic organisms live at their deepest, to the extreme limits of light able to access nutritive salts. These zones are occupied by sparse ecological communities, especially at the surface and there are significant vertical differences in the composition of phytoplankton. In the central South Pacific, the specific composition at depth (example 150 m) is more similar to that of the north of the central North Atlantic, at the same depth, than those that live just below the surface. Longhurst [LON 98] defines the base of the mixing zone as the most significant ecological barrier that exists in the ocean. Biogeographical regions or biomes can be defined based on variations in some physicochemical parameters such as the depth of the thermocline and the photic zone, or seasonality.

1.4. From the biosphere to the atmosphere to climate

If climate changes greatly influence phytoplanktonic communities in the sea, in turn, these ecological changes can alter the CO_2 composition in the atmosphere and hence the climate. Similarly, global changes to the wind regime have significant consequences on the carbon cycle and marine biodiversity.

Currently, the global annual primary production is approximately 40–45 gigatons of carbon (45×10^9 g). Most of this biomass is consumed by heterotrophic organisms (planktivores, carnivores and decomposers) and the mineralization of this organic matter will produce CO_2. Ultimately, only a small part falls to the oceanic depths where it is either consumed by heterotrophic benthic organisms, or buried in the sediments. In the former case only, carbon is removed from biogeochemical cycles, often for very long periods of time. We are therefore referring to exported primary production. Although this represents a very small fraction (0.02%) of what is produced at the ocean surface, this fraction of deposited carbon actually represents a significant carbon sink due to the immensity of the ocean surface; it is currently estimated to be more than 10 Mt (10^7 g) of carbon per year. Changes in climate will induce changes in the deposition rate of organic matter via changes to the general circulation speed or degree of oxygenation of the deep ocean which depends on the activity of the benthos; the absence of benthic organisms favors carbon sinks.

The issue that now deserves to be discussed is how a change to the oceanic biological production and its deposition rate as sediments alter the concentration of CO_2 in the atmosphere. This was the first hypothesis proposed by Broecker [BRO 82] to explain the discovery, in 1980, of lower levels of CO_2 in the atmosphere during glacial periods than during interglacial periods. This author proposes an increase in concentration of nutritive salts (essentially phosphate) which would favor organic production. In fact, the decrease in sea levels during the ice age would expose marine sediments rich in phosphate, deposited during the previous interglacial period, and transfer them into the ocean via rivers. This fertilization of the glacial ocean would have induced a greater sequestration of carbon.

Other reasons have been put forth to explain the decrease in concentration of atmospheric CO_2 during the ice age. Note, a better use of nutritive salts by phytoplankton in the Antarctic Ocean [FRA 97]; an increase in strong dust-laden winds bringing iron into zones where this micronutrient is currently not found [MAR 90]; an increase in productivity resulting from a transfer of dissolved silica from the Antarctic Ocean towards the lower latitudes [MAT 02]; a decrease in the general depth of the thermocline at low latitudes which would have fertilized the ocean surface [BEA 01]; or even, an increase in biological production parallel with a decrease in carbonate production [BRO 82]. The latter point will be discussed later on.

Most of these hypotheses themselves cannot completely explain the decrease in concentration of atmospheric CO_2 during ice ages. Moreover, reconstructions of the primary production during these episodes do not always confirm the increases, and large local variations are observed.

At low latitudes, organic production seems to vary independently of global changes in climate [BEA 97] (Figure 1.2). At these latitudes, changes in organic productivity occur every 20,000 years which corresponds to local variations in seasonal insolation, whereas variations in global ice volume occur less frequently (every 41,000 and 100,000 years). Numerical simulations also indicate that local factors, for example wind strength and that of upwelling, are responsible for distribution structures and changes in production of organic matter. Sediments in the tropical zone of the Pacific and Indian Ocean also reveal that the organic production during the last ice age was 50% greater than that produced today (Figure 1.2).

However, many records of organic productivity reveal variations with a dynamic opposite to variations in concentrations of atmospheric CO_2 (Figure 1.2 top); but locally the variations may obey different rules (Figure 1.2 bottom).

At high latitudes (above 40°) and in many tropical regions (for example in the monsoon region), numerical simulations show an increase in PP during the ice age which implies greater sequestration of CO_2 during these periods. A compilation of many data from studies of sediments of the world's oceans indicates that the exported organic production (including sediments) during the last maximum ice age is slightly higher than that of today. However, this variation does not appear to be enough to explain the differences in CO_2 between the glacial and interglacial period. As a result, even if the mechanism proposed by Broecker [BRO 82] cannot fully account for these variations, we know that the primary production has varied greatly in the past and contributed significantly to the atmospheric CO_2 dynamics. So, there is a feedback loop between climate and the marine biosphere, both influencing one another.

1.5. Carbonate production

Calcium carbonate forms the greatest reservoir of carbon on Earth [LEG 14] and many marine species produce shells based on this bio mineral. These secretions, just like organic production, play an important role in the chemistry of the oceans. Just as Broecker stated for organic production, Berger [BER 82] also mentioned changes in the production and conservation of biogenic carbonates to explain variations in the concentration of CO_2 in the atmosphere during glacial and interglacial periods.

The phenomena responsible for this relationship are more difficult to explain since they involve complex chemical processes. So, if we conventionally assume that the dissolution of limestone by acid produces carbon dioxide, in the ocean, this is the opposite reaction: the dissolution of limestone sequesters CO_2 and secretes CO_2. As expressed in equation [1.2], calcification (precipitation of $CaCO_3$) is accompanied by the release of CO_2 from the ocean into the atmosphere. This is what is called the carbonate counter-pump [WES 93].

1.5.1. *Importance of biological carbonate production in the evolution of the planet*

In the ocean, most calcification is of biological origin. Non-biological precipitations of carbonate are only produced in the presence of extreme concentrations of carbonate or bicarbonate ions. These conditions only currently exist in rare environments, at warm shallow depths with high evaporation, but they are responsible for the formation of carbonaceous rocks in the primitive ocean.

1.5.1.1. *The Neoproterozoic*

In the primitive ocean and before the first secretions of biological origin, chemical precipitations only operated in conditions where there was an over saturation of carbonate ions which therefore meant atmospheric CO_2 was not controlled by the ocean. Carbonate ions from the weathering of alumino-siliceous continental rocks are transported towards the ocean. It was shown that the carbon input into the ocean from erosion was greater than that released into the atmosphere by marine calcification; the content of CO_2 in the atmosphere would have progressively declined. For example, if the flow of carbon into the ocean is greater than 25% of the flow of carbon into the atmosphere, in only one million years, the atmosphere would be depleted of CO_2 [BER 97]; the Earth's surface would therefore have been covered in ice, without a greenhouse effect.

The biological carbonate production began 3.5 billion years ago, with stromatholiths from the activity of photosynthetic cyanobacteria. These would have dominated the production of carbonates for a long time but insufficiently so to decrease the saturation of oceans with carbonate ions from the erosion of continental rocks. An increase in continental erosion, sequestration of atmospheric CO_2 and a decrease in the greenhouse effect [DON 04] arose as a result of extreme glacial events. The oceans were certainly completely covered in ice; between 0.8 and 0.6 billion years ago, we find traces of glaciers up to the equator, which was called Snowball Earth by Kirschivink [KIR 92].

It therefore appears that carbonate ions deposited into the ocean were not used quickly enough by stromatoliths and the biological release of oceanic carbon into the atmosphere was too limited to prevent these extreme conditions from occurring.

1.5.1.2. *The Paleozoic*

At the beginning of the Paleozoic (primary era), approximately 550,000,000 years ago, several different groups of organisms, mollusks, sponges, corals, arthropods, etc., began to produce carbonaceous skeletons in shallow environments. This revolution of marine biodiversity would have had consequences on atmospheric CO_2 concentrations.

The rate of carbonate production depended on the relative sea level height and the usable area for calcification, during periods of high sea levels (submerged continental shelves), the total carbonate production will be significant and will cause an increase in CO_2 in the atmosphere; this relates to relatively warm periods. During periods of low sea levels, there are fewer calcifying organisms in shallow zones, which will cause a decrease in the total marine carbonate production which will not be able to compensate for the carbonate input from the continents. In particular, this is the case of periods of tectonic activity and uplift of mountain ranges where erosion is great. If the glaciations were severe, they would not have had the same extensions and consequences as before this first revolution, as during Snowball Earth.

1.5.1.3. *Mesozoic revolution*

During the Triassic (220 million years ago), a group of unicellular algae, the coccolithophores, began to produce a carbonaceous skeleton made of microscopic plates (coccoliths) external to their cells. These microalgae can develop offshore due to their floating ability and rapidly colonize the oceans. From the Jurassic onwards, they are accompanied by planktonic for aminifera, which, despite also being unicellular, belong to the zooplankton. Most current coccolithophores have a diameter of between 2 and 15 µm whereas for aminifera usually have a diameter of up to several hundred micrometers (Figure 1.3). Despite their tiny size and their low weight (less than a nanogram), coccolithophores can produce very dense blooms extending over thousands of kilometers squared. The density of calcareous scales is so great that they can be detected by satellite images. Found at the bottom of oceans, remains of coccolithophores can contribute to the formation of huge quantities of light carbonate-rich sediments; the extreme example is chalk from the Cretaceous (from the Latin *Creta*) uniquely composed of coccolithophore remains (Figure 1.4).

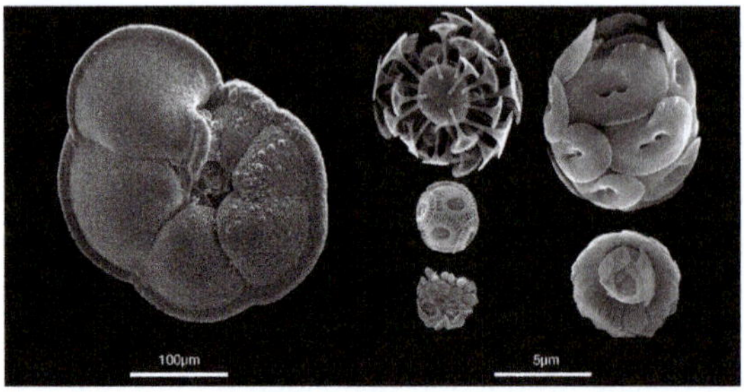

Figure 1.3. *Photographs taken using a scanning electron microscope of several representatives of the two main groups of pelagic calcifying organisms*

COMMENTS ON FIGURE 1.3.– To the left, planktonic for aminifera *Globorotalia menardii*; to the right, five coccolithophores (from left to right and top to bottom): *Discosphaera tubifera*, *Helicosphaera carteri*, *Emiliania huxleyi*, *Solisphaera galbula* and *Florisphaera profunda*. The two last species at the bottom left are typical of the deepest parts of the photic zone. Note the difference in scale (horizontal bars) for the foraminifera and for coccolithophores (author's photos).

The colonization of the entire surface of the ocean by calcite-producing organisms was a major evolutionary event in the history of the climate. It was called the "mid-Mesozoic revolution" [RID 05] due to its high impact on marine chemistry. Pelagic calcification helped better stabilize the climate; even during glacial episodes of the quaternary period, it helped maintain an equilibrium between the production of marine carbonates and atmospheric carbon.

Numerical models have shown that carbonate concentrations have remained relatively stable over the past 100 million years, whereas they were two to three times greater and highly variable before the coccolithophores appeared (more than four times greater before the Paleozoic [RID 05]). This shows how important the role the diversity of the biosphere plays in the stability of the Earth's climate.

1.5.2. *Carbonate compensation depth*

Currently, coccolithophores and planktonic foraminifera produce the largest portion of calcium carbonate in the pelagic domain which represents

90% of the sea surface. However, if the open ocean contributes to more than half of the marine production, only a third approximately is accumulated in the form of $CaCO_3$. This difference is due to the fact that bio-mineral particles produced in shallow zones have a much greater chance of being buried than those produced offshore. So, aragonite crystals (one of two forms of $CaCO_3$) secreted by coralin tahiti certainly contribute to reef growth whereas calcite crystals (the other form of calcium carbonate), secreted at the surface by a coccolithophore tens of kilometers offshore of the reef, will not be buried in the ocean floor if they are too deep. In fact, beyond a depth of 4,000 m, calcite dissolves. It is estimated that 50–80% of carbonates that are secreted on continental margins are stored in coastal sediments, against 45% of carbonates produced in the open ocean. In comparison, each year, 11×10^{12} moles of $CaCO_3$ accumulate in the pelagic zone and 7×10^{12} moles of $CaCO_3$ in coral reefs [MIL 93]. Despite their better efficiency in storing calcium carbonate, coral reefs still accumulate less due to a relatively low useable area 0.6 million km^2 versus 106 million km^2 for the pelagic zone whose depth is less than 4,000 m. Carbonate particles dissolve below a certain depth mainly due to the high pressure; the solubilities of calcite and aragonite strongly depend on pressure, and at great depths, the pressure is great enough that $CaCO_3$ dissolves; this limited depth is called the lysocline.

First note, the transfer of particulate carbon towards the sea floor due to the aggregation of diverse particles produced on the surface and the ballast effect of coccoliths; since this falling of particles is referred to as "marine snow", the deep sea is often represented as an alpine landscape with snow-capped peaks (white calcite and aragonite), and the deeper snow-free parts (without carbonates) (Figure 1.4). This physiographic limit corresponds to the base of the lysocline, carbonate compensation depth (CCD) below which there are no carbonate deposits. On the map of the central Pacific (Figure 1.4(c)) the white/black limit which corresponds to the mean depth of the lysocline (4 000 m +/– 900) represents the contrast between oceanic basalt (black) and the carbonates (light). A photo of the depths (Figure 1.3(b)), in this zone, at –2,000 m, shows that the sediments are whitish; a core sampled at this depth (Figure 1.4(c)) allows us to verify, from assessing samples using an electronic microscope, that the sediment is almost entirely composed of planktonic foraminifera (Figure 1.4(d)) and calcareous scales from coccolithophores (Figure 1.4(e)). In the same zone, but at greater depths, basalt occurs without being covered by sedimentary rock.

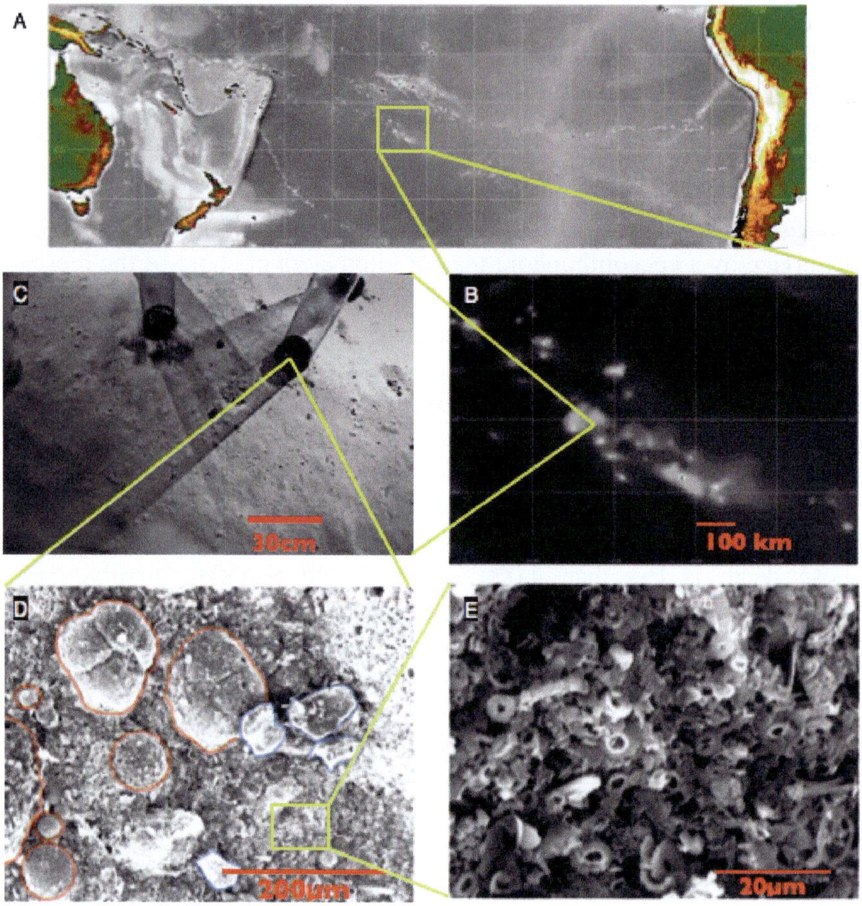

Figure 1.4. *Images of the deep sea in the Southern islands at a depth of 2,500 m (images from the oceanographic campaign SO2O4 Pacenpal by Beaufort, 2010)*

COMMENTS ON FIGURE 1.4.– (a) General map. (b) Detailed map; light areas are depths of less than 4,000 m occupied by carbonaceous sediments and the shaded areas are depths greater than 4,000 m where basalticrocks are found. (c) Photograph of the deep sea during the sampling of white calcareous sediment cores. (d) Photograph of the sediment viewed under a scanning electron microscope, showing the foraminifera (the largest particles) and their fragments (smaller identifiable elements). (e) Detailed image of (d) showing that the cement surrounding the foraminifera is exclusively composed of coccolithophore skeletons.

Carbonate compensation does not always occur at the same depth since the solubilities of calcite and aragonite also depend on the concentration of carbonate and calcium. So, a high productivity of $CaCO_3$ at the surface, during a coccolithophore bloom, will cause a high vertical flux of calcareous skeletons which, upon arrival at the CCD, will dissolve and release calcium and carbonate ions into the water; the increase in concentration of these ions will tend to displace the lysocline at depth. However, a zone of low carbonate production at the surface will have alysocline situated at middle depth. As long as there is a pelagic production of carbonate, the overall chemical concentration of Ca and carbonate ions in the ocean remains fairly constant and is redistributed by the currents.

The overall mechanism that transfers the sum of all chemical species of inorganic carbon or $\sum CO_2$ from the surface towards the sea floor and that maintains the high concentrations at depth is called the "carbonate-counter pump" and plays a key role in the regulation of the climate; Legendre [LEG 14] describes four types: physical, organic (or biological pump), carbonate-counter pump and the microbial pump.

1.5.3. *Carbonates and climate*

Very soon after the discovery of low concentrations of atmospheric CO_2 during ice ages, Berger [BER 82] stated that the possession of continental margins by coral reefs during deglaciation could explain a substantial part of the post-glacial increase in CO_2. During the last major ice age which occurred between 70,000 and 15,000 years ago, the volume of ice distributed over the continents was so large that it caused a decrease in the sea level by 120 m. Organic and carbonate production was limited to a small coastal fringe and the global production of $CaCO_3$ decreased to almost half [BER 82]. This decrease was slightly compensated for by the more constant production of pelagic carbonate following a deepening of the lysocline by approximately 1,000 m, as in the Pacific [FAR 89]. During deglaciation, the continental margins became gradually submerged and benthic calcification caused a rapid increase in reefs and therefore a gradual increase in CO_2 in the atmosphere.

The numerical modeling of this phenomenon shows that this amount of carbon accounts for a large part of the difference in concentration of atmospheric CO_2 between a glacial and an interglacial period. However, this scenario is often overlooked in current studies on the marine carbon cycle. The reason for the infrequent consideration of this phenomenon is due to the fact that it occurs as a result of climate change and therefore cannot initiate it. In fact, it requires that the sea levels rise as a result of warming. This initial warming cannot therefore be attributed to this phenomenon. Nevertheless, changes in the habitat area of marine organisms and ecological systems have certainly had a non-negligible retroactive effect on the climate variations during deglaciation.

Various experimental data show that during ice ages, the concentration of carbonate ions is greater, and hence the pelagic carbonate production is probably greater:

1) the planktonic foraminifera shells are then thicker and coccolithophores produce more calcified plates;

2) the pelagic production at low latitudes and particularly that of coccolithophorids is greater than that during interglacial periods;

3) as we have seen, the lysocline retreats.

So, a greater production of carbonate and a better conservation of carbonates will cause an increase in the concentration of atmospheric CO_2 which will mitigate the effects of glaciation. However, a precise quantification of the pelagic carbonate production and its conservation has rarely been carried out because of the complexity of its implementation.

1.6. The coupling of carbonaceous and organic productions

We have seen that the marine organic production as well as the carbonate production interfere, at different scales of time and space, with the concentration of CO_2. Some models play with this combination.

In this category, the most elegant is definitely that of the rain ratio by Archer and Maier-Reimer [ARC 94]. During the ice age, in the pelagic zone, there appears to have been an increase in organic productivity coupled with a decrease in carbonate productivity. So, carbon has a better chance of being sequestered in the sediments causing the observed decrease in carbon during this period. For these authors, a 40% decrease in the ratio of $CaCO_3$ in organic matter in sediment particles would be sufficient to explain the decrease in concentration of atmospheric CO_2 recorded during ice ages.

Several reasons could explain the decrease in this ratio; they are all linked to a major ecological change in the pelagic zone: the replacement of coccolithophores by diatoms, phytoplanktonic organisms by a siliceous skeleton. Diatoms would outcompete coccolithophores due to an influx of silicic acid whose sources, according to hypotheses, would be: wind (silica-rich dust), rivers from exposed margins, or even an oceanic source of silica from the Antarctic.

This theory, still a very topical issue, has been faced with observations that contradict it:

1) as we saw above, the coccolithophores do not appear to have decreased in abundance nor in calcification; rather the opposite is observed;

2) in the cores from low and middle latitudes, we do not find a synchronous or large increase in the number of diatom skeletons.

We know that the process for storing marine organic matter is much more efficient than the aggregation of matter in suspension (marine snow). In this sense, calcareous coccolithophorid scales are the best vectors of carbon towards the sea floor and sedimentary sinks. In these conditions, if the coccolithophores were outcompeted by diatoms, the hypothesis of organic carbon sequestration in sediments would not be valid.

1.7. Modification of equilibria and consequences on marine life

Time constants play an important role in the carbon cycle; as described above, the marine carbon cycle follows the rhythms of global climate variations (Figure 1.1), but greater and more prolonged alterations, often of geodynamic origin (plate tectonics, uplift and erosion of mountain chains, large volcanic eruptions), greatly alter these rhythms. Note, in particular, the two best documented acidification events in the history of the Earth: one at the Permian-Triassic boundary, 250 million years ago, and the other at the Triassic-Jurassic boundary, 200 million years ago. These are also two of the most dramatic mass extinction events, with periods of volcanic activity where large amounts of CO_2 were released into the atmosphere, over a relatively long scale of time (several hundred thousand years) [WIG 01].

The effects of this were particularly felt by calcifying marine organisms, since most of them disappeared, as well as most species of coral, at the

Triassic-Jurassic boundary. It took several millions years for new coral reef species to emerge from the few that survived [KIE 09]. Annual levels of CO_2 emissions during these events were much lower than today, but they lasted for a very long time; the chemical and ecological effects during these periods were certainly very different to what the ocean would encounter in the future [RID 05]. In fact, in the current conditions of high CO_2 emissions occurring over a relatively short period of time, it is difficult to predict the consequences on the marine biosphere and biogeochemical cycles.

Since the beginning of the Industrial Revolution, human activities have released huge quantities of CO_2 into the atmosphere, which is added to the natural carbon cycle. We have come to know the effects on the climate quite well and we are beginning to see the effects on the ocean via acidification. In fact, an increase in the CO_2 (partial pressure) in the atmosphere causes CO_2 to dissolve more readily in seawater, which consequently increases the H^+ ions and therefore decreases pH.

pH measurements over the past decades correspond exactly to numerical models, and emission projections allow us to predict ocean pH in the future. What we do not know is how to predict the consequences of this phenomenon on the marine biosphere. Many studies are currently being carried out on different groups of organisms with different methods (crop cultivation, *in situ* observation, etc.). One chapter in this set of books dedicated to the ocean covers the state of current research in this domain [GAT 14].

Logically, a decrease in pH would have consequences first and foremost on calcifying organisms and the disappearance of carbonate production would have severe consequences on the carbon cycle; in the future, the Earth could be heading towards a new ice age, worse than that two/three centuries ago with the little ice age.

Nevertheless, given the current trends, it is unlikely that this scenario will occur. Results from experiments on the possible effects of acidification on pelagic organisms, threatened in zones of high pH and different carbonate concentrations, are mixed [BEA 11]; they show a worrying general trend to decrease the degree of calcification of coccolithophore skeletons (Figure 1.5), but contrarily, in the most acidic waters of the world's oceans, in the coastal resurgence of Chili and Peru, we find coccolithophores with more calcified skeletons than anywhere else. The photo in Figure 1.6 (left) shows a highly calcified *Emiliania huxleyi*, collected in water with a pH of

7.66 (global mean surface pH is 8.1) offshore of Peru in April 2014. Despite the low pH values in this resurgence high in CO_2, coccolithophores have managed to adapt possibly through physiological and energetic mechanisms.

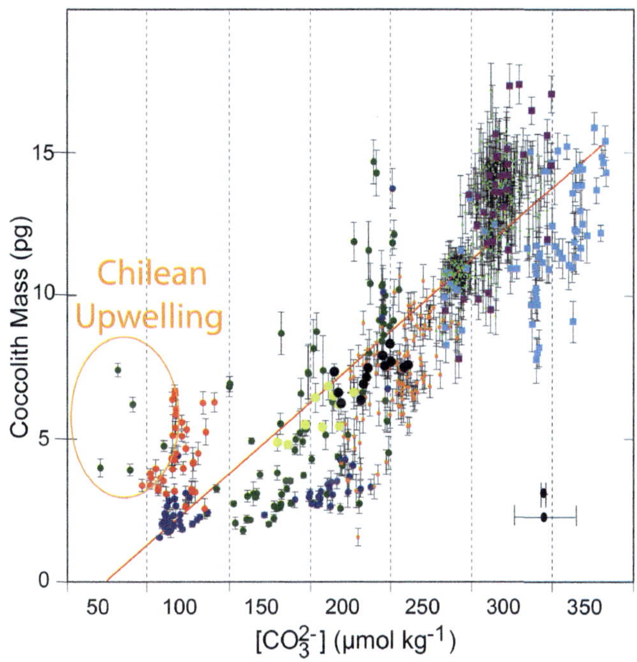

Figure 1.5. *Correlation between mass and calcified plates of coccolithophores (coccoliths) and the concentration of dissolved carbonate in seawater, in different regions of the current and past oceans (0 – 40,000 years ago) [BEA 11]*

COMMENTS ON FIGURE 1.5.– There is a positive relationship between mass (calcification of coccoliths) and carbonate concentration. In one of the most corrosive regions of the current ocean, upwelling from Chile (zone circled), we observe the opposite trend, with highly calcified coccoliths which should have disappeared due to the low carbonate concentration consistent with the trend indicated by regression line.

The models cannot predict that the ocean surface pH could drop to values lower than 7.6 over the next century, which makes the disappearance of these organisms possible and jeopardizes their role in the pelagic biosphere.

We should, however, note that in zones where the pH has slightly decreased to values close to 8.0, skeletons become malformed and partially dissolved (Figure 1.6 right). Changes in the composition of assemblies are expected, without really knowing if the species will adapt or if the non-calcified will ultimately dominate.

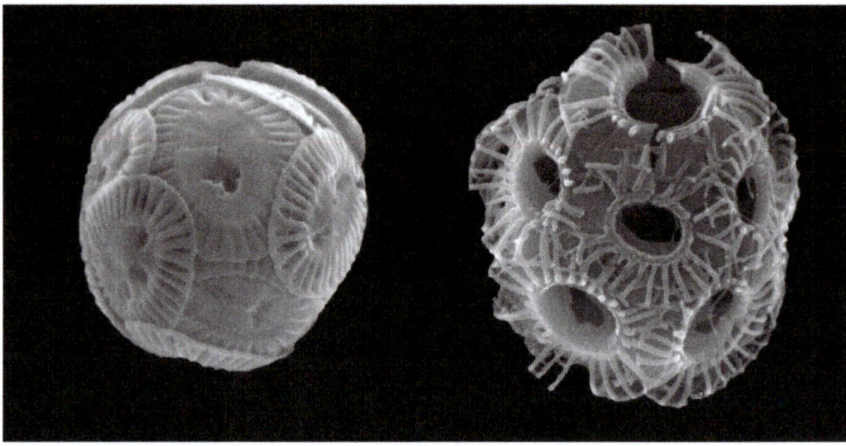

Figure 1.6. *Two images taken using a scanning electron microscope of the coccolithophore Emiliania huxleyi. Left: collected in April 2014 offshore of Pisco (Peru) in water with a pH of 7.66. Right: this coccolithophore was collected in February 2009 between Cap Horn and Antarctica in water with a pH of 8.05*

1.8. Conclusion

The marine biosphere marine plays a decisive role on the global carbon cycle and the diversity of ecosystems. It intervenes via two antagonistic processes: one, photosynthesis which sequesters carbon dioxide; another, calcification which is a source of CO_2. Both phenomena are in equilibria and have maintained a stable climate for millions of years.

Large bio-calcification first occurred on the continental margins 550 million years ago and spread by pelagic calcification, also known as oceanic calcification, 220 million years ago, mitigating the effects of ice ages.

For almost one million years, the Earth has experienced ice ages every hundred thousand years, cycles which follow the concentration of atmospheric carbon dioxide. The dynamic of the marine biosphere causes changes in the carbon cycle, with a retroactive effect on the climate.

Currently, one of the major risks to the marine biosphere is ocean acidification. Even if we are quite confident that pelagic calcification will not immediately disappear, it is still probable that significant disruptions to the diversity of communities of calcifying marine organisms are to be expected.

Another effect, often not included, is the impact of acidification on calcareous phytoplankton, the quantity and quality of coccolithophores; with regard to the production of organic matter at the surface, these organisms play a ballast role in the vertical transfer and the burying of organic particulate matter in sedimentary sinks. Any alteration to this mechanism, essential in the carbon cycle, would have consequences on the biological pump, the removal of atmospheric CO_2 and therefore on climate warming and acidification.

1.9. Bibliography

[ARC 94] ARCHER D., MAIER-REIMER E., "Effect of Deep-Sea sedimentary calcite preservation on atmospheric CO_2 concentration", *Nature*, vol. 367, pp. 260–263, 1994.

[BEA 97] BEAUFORT L., LANCELOT Y., CAMBERLIN P. *et al.*, "Insolation cycles as a major control of the Equatorial Indian Ocean primary production", *Science*, vol. 278, pp. 1451–1454, 1997.

[BEA 01] BEAUFORT L., DE GARIDEL THORON T., MIX A.C. *et al.*, "ENSO-like forcing on Oceanic Primary Production during the late Pleistocene", *Science*, vol. 293, pp. 2440–2444, 2001.

[BEA 10] BEAUFORT L., VAN DER KAARS S., BASSINOT F.C., *et al.*, "Past dynamics of the Australian monsoon: precession, phase and links to the global monsoon concept", *Climate of the Past*, vol. 6, pp. 695–706, 2010.

[BEA 11] BEAUFORT L., PROBERT I., DE GARIDEL-THORON T *et al.*, "Sensitivity of coccolithophores to carbonate chemistry and ocean acidification", *Nature*, vol. 476, pp. 80–84, 2011.

[BER 82] BERGER W.H., "Increase of carbon dioxide in the atmosphere during deglaciation – The coral-reef hypothesis", *Naturwissenschaften*, vol. 69, pp. 87–88, 1982.

[BER 97] BERNER R.A., CALDEIRA K., "The need for mass balance and feedback in the geochemical carbon cycle", *Geology*, vol. 25, pp. 955–956, 1997.

[BER 14] BERTRAND P., "The ocean in the Earth System: Evolution and Regulation", in MONACO A., PROUZET P. (eds), *Ocean in the Earth System*, ISTE Ltd, London and John Wiley & Sons, New York, 2014.

[BRO 82] BROECKER W.S., "Glacial to interglacial changes in the ocean chemistry", *Progress in Oceanography*, vol. 11, pp. 151–197, 1982.

[DON 04] DONNADIEU Y., GODDERIS Y., RAMSTEIN G. *et al.*, "A "snowball Earth" climate triggered by continental break-up through changes in runoff", *Nature*, vol. 428, pp. 303–306, 2004.

[FAR 89] FARRELL J., PRELL W.L., "Climatic change and $CaCO_3$ preservation: an 800,000 year bathymetric reconstruction from the central Equatorial Pacific ocean", *Paleoceanography*, vol. 4, pp. 447–466, 1989.

[FRA 97] FRANÇOIS R., ALTABET M.A., YU E.F. *et al.*, "Contribution of Southern Ocean surface-water stratification to low atmospheric CO_2 concentrations during the last glacial period", *Nature*, vol. 389, pp. 929–935, 1997.

[GAT 14] GATTUSO J.P., HANSSON L., GAZEAU F., "Ocean Acidification and its Consequences", in MONACO A., PROUZET P. (eds), *Ocean in the Earth System*, ISTE Ltd, London and John Wiley & Sons, New York, 2014.

[KIE 09] KIESSLING W., RONIEWICZ E., VILLIER L. *et al.*, "An early Hettangian coral reef in Southern France: implication for the End-Triassic reef crisis", *Palaios*, vol. 24, pp. 657–671, 2009.

[KIR 92] KIRSCHIVINK J.L., "Late Proterozoic Low-Latitude Global Glaciation: the Snowball Earth", in SCHOPF J.W., KLEIN C. (eds), *The Proterozoic Biosphere: A Multidisciplinary Study*, Cambridge University Press, Cambridge, 1992.

[LAS 04] LASKAR J., ROBUTEL P., JOUTEL F. *et al.*, "A long-term numerical solution for the insolation quantities of the Earth", *Astronomy & Astrophysics*, vol. 428, pp. 261–285, 2004.

[LEG 14a] LEGENDRE L., "Marine biogeochemical cycles", in MONACO A., PROUZET P. (eds), *Ocean in the Earth System*, ISTE Ltd, London and John Wiley & Sons, New York, 2014.

[LEG 14b] LEGENDRE L., "Pelagic marine ecosystems and biogeochemical cycles", in MONACO A., PROUZET P. (eds), *Ecosystem Sustainability and Global Change*, ISTE Ltd, London and John Wiley & Sons, New York, 2014.

[LON 98] LONGHURST A., *Ecological Geography of the Sea*, Academic Press, San Diego, 1998.

[LUE 08] LUETHI D., LE FLOCH M., BEREITER B. *et al.*, "High-resolution carbon dioxide concentration record 650,000-800,000 years before present", *Nature*, vol. 453, pp. 379–382, 2008.

[MAR 90] MARTIN J.H., GORDON R.M., FITZWATER S.E., "Iron in Antarctic water", *Nature*, vol. 345, pp. 156–158, 1990.

[MAR 87] MARTINSON D.G., PISIAS N.G., HAYS J.D. *et al.*, "Age dating and the orbital theory of the ice ages: development of a high-resolution 0 to 300,000 year chronostratigraphy", *Quaternary Research*, vol. 27, pp. 11–9, 1987.

[MAT 02] MATSUMOTO K., SARMIENTO J.L., BRZEZINSKI M.A., "Silicic acid leakage from the Southern Ocean: a possible explanation for glacial atmospheric pCO(2)", *Global Biogeochemical Cycles*, vol. 16, 2002.

[MIL 93] MILLIMAN J.D., "Production and accumulation of calcium carbonate in the ocean–Budget of a nonsteady state", *Global Biogeochemical Cycles*, vol. 7, pp. 927–957, 1993.

[RID 05a] RIDGWELL A., "A mid mesozoic revolution in the regulation of ocean chemistry", *Marine Geology*, vol. 217, pp. 339–357, 2005.

[RID 05b] RIDGWELL A., ZEEBE R.E., "The role of the global carbonate cycle in the regulation and evolution of the Earth system", *Earth and Planetary Science Letters*, vol. 234, pp. 299–315, 2005.

[WES 93] WESTBROEK P. *et al.*, "A model system approach to biological climate forcing – the example of Emiliania huxleyi", *Global and Planetary Change*, vol. 8, nos. 1–2, pp. 27–46, 1993.

[WIG 01] WIGNALL P.B., "Large igneous provinces and mass extinctions", *Earth-Science Reviews*, vol. 53, pp. 1–33, 2001.

2

Biodiversity of Phytoplankton: Responses to Environmental Changes in Coastal Zones

2.1. Introduction

The term "plankton" refers to "all organisms adapted to living, for part or all their life, in suspension in the sea, lakes, ponds and rivers" [REY 06]. Therefore, "phytoplankton" refers to all unicellular microalgae living in suspension in the water column in aquatic environments. Phytoplankton communities dominate pelagic ecosystems, where they account for approximately 50% of the annual primary production on Earth and yet represent only 1% of the global ocean biomass [FIE 98]. This primary production is at the base of trophic networks, but the efflorescences associated with microalgae can have negative consequences on organisms and ecosystems, through the production of toxins (for example: neurological, paralytic and diarrhetic toxins) or red tides responsible in some cases for the mass death of farm fish[1].

Changes to phytoplankton may affect the overall biomass [BOY 10], its diversity [PEN 05], the biogeographic distribution of species [LET 05], their phenology [EDW 04] or the functional diversity [DAV 12]. These responses are complicated by the involvement of multiple stress factors and ecosystem

Chapter written by Tania HERNÁNDEZ-FARIÑAS and Cédric BACHER.

1 For additional information, refer to *Vulnerability of Coastal Ecosystems and Adaptation* [MON 14a] from the "Seas and Oceans" Set for additional information on the toxicity of phytoplanktonic communities.

characteristics. They are observed at all levels: from individuals (for example: physiology, cell size) to communities (for example: structure, phenology). Analyzing the responses to environmental changes and gradients must be done over several taxonomic and functional levels, and ecological concepts (phenology, adaptive strategies, ecological niche and biological traits) must be associated with observations *in situ*.

In this chapter, we will first describe the ecology of phytoplankton, before examining the different studies that have assessed the changes in coastal zones with regard to anthropogenic pressures. We will then clarify how observation systems can help analyze the changes and structuring of phytoplankton communities.

2.2. Phytoplankton ecology

2.2.1. *General characteristics*

Marine phytoplankton are a polyphylogenetic group composed of many diverse species. Sournia *et al.* [SOU 91] state that there are approximately 4,000 species described. Other studies have stated this figure to be 5,000 [TET 95], whereas [FAL 07] identify 6,800 species for the diatoms and dinoflagellates groups only. Therefore, the number of species identified is constantly evolving, and new species have been described due to genetic analyses and electron microscopy. These figures represent not only the number of different taxa, but also the huge morphological heterogeneity, as well as their phylogenetic divergence (Table 2.1). The taxonomy of the phytoplankton has been based on morphological criteria for a long time, but genetics and electron microscopy have also allowed a better characterization of species. As with estimating the total number of species, the phytoplankton systematics is constantly evolving and the detailed classification of these organisms is often revised.

Marine microalgae are represented by organisms belonging to the bacteria and eukarya. Cyanobacteria are prokaryotic organisms that are able to photosynthesize and have played an important role over the course of evolution. In fact, the acquisition of photosynthesis by eukaryotic algae is based on endosymbiosis[2] involving cyanobacteria and eukaryotic protists

2 Beneficial cooperation between two living organisms where one lives inside the other.

[REY 06]. Eukaryotic microalgae are classed in several kingdoms (*Chromista*, *Plantae*, *Protozoa*; Table 2.2 and Figure 2.1). The following groups: diatoms (class *Bacillariophyceae*), dinoflagellates (class *Dinophyceae*) and the Prymnesiophyceae contribute significantly to the diversity of phytoplankton communities, respectively representing 40%, 40% and 10% of the number of eukaryotic species described [SIM 09]. Other groups such as the Chlorophyta (kingdom *Plantae*), the Euglenoidea (kingdom *Protozoa*) or even Cryptophyta compose the marine phytoplankton.

Kingdom	Phylum	Class	Examples of representative taxa
Bacteria			
	Cyanobacteria	Cyanophyceae	*Synechococcus*
Chromista			
	Ochrophyta		
		Bacillariophyceae	*Guinardia, Rhizosolenia, Pseudo-nitzschia*
		Chrysophyceae	*Dinobryon*
		Dictyochophyceae	*Dictyocha*
		Raphidophyceae	*Chattonella*
	Cryptophyta	Cryptophyceae	*Cryptomonas*
	Haptophyta	Prymnesiophyceae	*Phaeocystis*
	Myzozoa	Dinophyceae	*Prorocentrum, Dinophysis, Ceratium*
Plantae	Chlorophyta	Chlorophyceae	*Scenedesmus*
Protozoa	Euglenozoa	Euglenoidea	*Eutreptiella*

Table 2.1. *Systematics of several groups within the marine phytoplankton. The classification is based on the taxonomic database WoRMS [WOR 14]*

The heterogeneity of shapes and sizes of phytoplankton highlights the sheer diversity of these primary producers. Cell size varies from less than 2 µm to more than 200 µm (Figure 2.2, Table 2.2) and these cells may either be single or in colonies. With the exception of some extremely large species, the phytoplankton cannot be seen with the naked eye. However, phytoplankton efflorescence sometimes produces visible phenomena called colored "algal blooms". This morphological variation reflects their ecophysiological diversity because their size, cell shape and capacity to form colonies affect the metabolic activity (for example: growth, respiration, photosynthesis), absorption of nutrients, sedimentation rate and the herbivore selective pressure. The plasticity of these traits is great and varies between taxa, greatly influencing the specific composition of the phytoplankton community [LIT 08]. It has also been observed that genera with greater plasticity appear in a wider variety of environments [DUA 90].

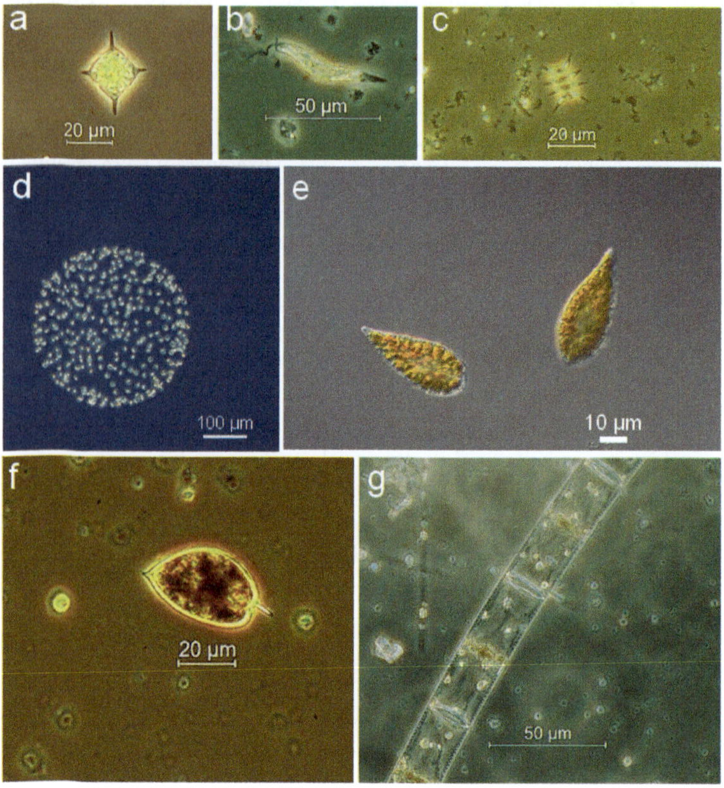

Figure 2.1. *Representative species of certain groups within the marine phytoplankton. a) Dictyocha (Dictyochophyceae), b) Eutreptiella (Euglenoidea), c) Scenedesmus (Chlorophyceae), d) Phaeocystis (colonial form, Prymnesiophyceae), e) Chattonella (Raphidophyceae), f) Prorocentrum (Dinophyceae) and g) Guinardia (chain, Bacillariophyceae) (photo credits a), b), c), e), f) and g) Ifremer LER Pertuis Charentais, d) Robert Andersen)*

2.2.2. *Lifecycle*

The lifecycle of organisms that form the phytoplankton depends on several abiotic and biotic factors, variable in time and which are in constant interaction. These factors favor the growth and reproduction of microalgae

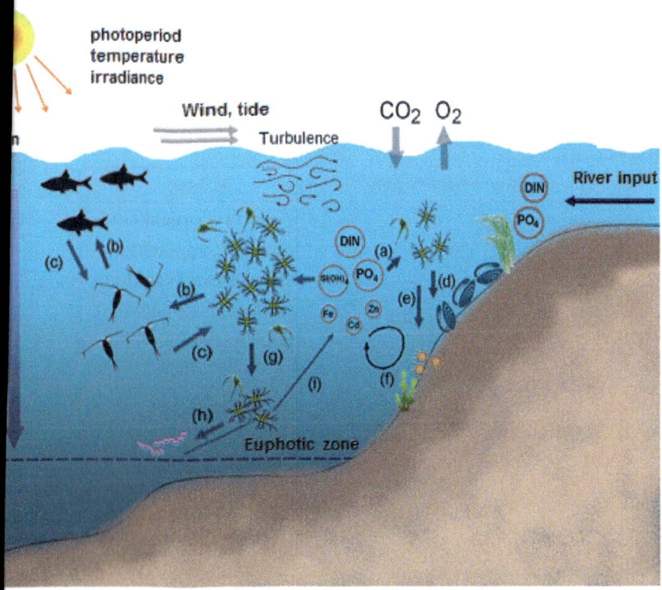

gure 2.3. *Diagram showing the physical and biological factors that influence the phytoplankton*

ON FIGURE 2.3.– CO_2 and light form the basis of photosynthesis. s spatially (attenuation according to depth) and temporally d). (a) Nutrients such as nitrogen, phosphate, silicate and others l for phytoplankton (bottom-up control). In coastal ecosystems, ide nutrients and the turbulence favors the transport of nutrients photic zone. (b) This primary production is at the base of the phic web. (c) The trophic web controls the concentration of ton (top-down control). (d) In coastal zones, phytoplankton is med by benthic filter feeders. (e) In the absence of favorable some species may form cysts that sediment out. (f) When conditions for growth return, the cysts germinate, and they are d in the water column. (g) Other processes such as ion contribute the decrease in concentration of phytoplankton in the mn. (h) Finally, this organic matter is used by microorganisms (i) Bacterial activity favors the regeneration of nutrients in the water

ts such as phosphate, nitrogen, iron, zinc and others are essential plankton, hence the importance of physical processes such as or, conversely, decrease phytoplankt sedimentation, herbivory[3], parasitism or in phytoplankton abundance. These pro abundance of a given species increases c

Linear dimensions	
< 2 µm	p
2–20 µm	n
20–200 µm	m
200 µm–2 mm	m
>2 mm	ma

Table 2.2. *Phytoplankton classification a*

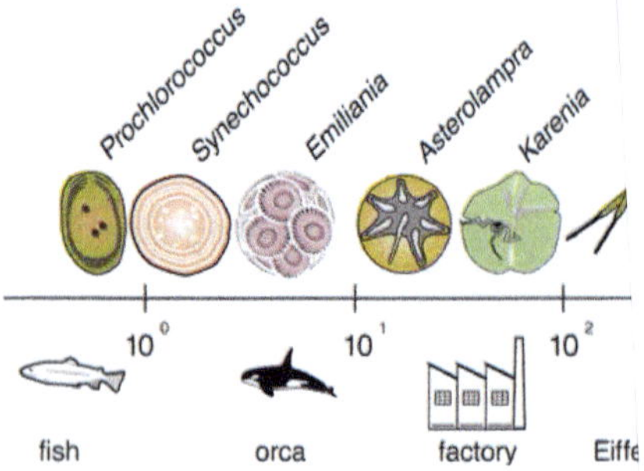

Figure 2.2. *Different size scales of macroscopic organisms and ob*

Energy provided by the sun and C mainstarting elements for photosynthesi organisms require light energy and nutrien reproduction (Figure 2.3). As the light tr phytoplankton must primarily maximize the euphotic zone (Figure 2.3, [REY 06]). This the turbidity of water, extends from the su intensity corresponds to 1% of light at the sur to time, for example circadian rhythms. I variability of the photo period during the ye of these organisms.

3 An animal that feeds on plants.

Fi

COMMENT Light vari (photoperi are essenti rivers prov into the e pelagic tr phytoplan also cons conditions favorable resuspend sedimenta water col (bacteria). column.

Nutrie for phyto

turbulence in the mixing of the water column and transport of nutrients into the euphotic zone[4]. Diatoms and silicoflagellates also require silica. These nutrients present in the water column are either transported by rivers, by regeneration (bacterial activity) or due to nitrogen fixation (cyanobacteria). The energy produced by phytoplankton through the process of photosynthesis supports secondary heterotrophic production and, consequently, higher trophic levels. However, distinguishing between autotrophic and heterotrophic microalgae is not always clear. In fact, some species can capture particles or prey (heterotrophy) [REY 06]. The fact that there are groups that nourish themselves using strategies other than autotrophy shows that the phytoplankton are not all the same from a functional point of view [CLO 05]. Thus, mixotrophy[5] is considered advantageous when the availability of nutrients is limited. Currently, this is primarily found in dinoflagellates and some types of chrysophytes. Finally, heterotrophy in microalgae plays an important role in the pelagic carbon cycle [REY 06].

In the coastal environment, organic matter produced by phytoplankton is also consumed by benthic filter feeders. Moreover, the resuspension of benthic microalgae in the water column revealed the occurrence of benthos-pelagos coupling. In addition, some species of phytoplankton produce cysts that form sediment when essential resources become limited, and germinate when conditions in the water column are favorable once again.

Lifecycles determine the main interactions between phytoplankton and the environment. These include processes that control the phytoplankton biomass via the assimilation of nutrients (bottom-up regulation) or herbivory (top-down regulation) [CLO 05, MET 96, REY 06]. This lifecycle and the diverse factors that influence it can be used as a general schematic, since the physiological and morphological diversity of these organisms reflects their many survival strategies and adaptation to their environment. Cell shape and size, buoyancy (for example: presence of gas vacuoles), the formation of chains or colonies or even vertical migration (motility of some species), are some of these diverse adaptation strategies and greatly influence the spatial and temporal dynamics of these organisms. This diversity is consistent with the complexity of interactions between species, which define the community structure of the community.

4 For additional information, we can refer to *Ocean in the Earth System* [MON 14b] from the "Seas and Oceans" Set.

5 Capacity to feed either by autotrophy or heterotrophy.

2.2.3. *Spatial distribution*

The spatial and temporal distribution of phytoplankton is one of the main lines of research carried out by ecologists. The marine environment is heterogeneous due to the spatial differences in temperature, salinity, turbidity, etc., each of these factors being subject to continuous variations [REY 06]. Consequently, the spatial distribution of phytoplankton is heterogeneous, which is either in a vertical or horizontal plane. This spatial variability is highly influenced by hydrodynamic processes and shaped by turbulence phenomena, for example under the effect of wind. Turbulence is not a stable phenomenon over time, and hence the spatial organization of the phytoplankton changes over the hours, days or seasons [CLO 96].

Despite the contiguity of the marine water masses on the planet, variations in temperature, irradiance, nutrients, stability of the water column, currents and geographic variations lead to differences in phytoplankton composition and abundance. In the subtropical zone of the North Pacific, the circulation maintains a coherent mass of water and separates it from adjacent habitats. These waters are characterized by limitations in nutrients and the high transparency of the water. In this zone, approximately 50% of the phytoplankton biomass comes from prokaryotic organisms (*Prochlorococcus* and *Synechococcus*) [REY 06]. The groups of diatoms (e.g. *Hemiaulus*, *Rhizosolenia*) and dinoflagellates (e.g. *Prorocentrum*, *Neoeratiumtripos*) only represent a small fraction of the plankton. Along the coast of Peru (South America), upwellings increase the levels of nutrients. Under these conditions, the diatoms, primarily *Rhizosolenia delicatula*, *Thalassiosira subtilis*, *Skeletonema costatum* and *Chaetoceros debilis* dominate the phytoplanktonic fauna. Dinoflagellates and coccolithophorids also represent a small fraction of the biomass [REY 06].

Phytoplankton dynamics are not identical in all types of ecosystems. Coastal zones, at the land-sea interface, receive nutrients transported by rivers, and experience exchanges with the offshore water mass and the seabed in shallow zones. These highly dynamic zones offer many possibilities to the phytoplankton for growth and production of biomass. This variability in environmental conditions favors the species diversity and biomass, and, at the same time, a greater variability in abundance [REY 06]. We find many examples of this in studies on the spatial variability of phytoplankton diversity along the coasts of France. Gailhard *et al.* [GAI 02] studied the microphytoplankton diversity in 17 sites distributed over the three coastlines of mainland France between 1992 and 2000. The authors highlight the particularities at the different sampled sites, from which three geographic zones were distinguished (Western Channel, Bay of Biscay and the Mediterranean). Recently, David *et al.* [DAV 12] analyzed the spatiotemporal variations in the specific and functional diversity along the Atlantic coast and the western English Channel between 1993 and 2010. The variations in diversity observed corresponded to the geographic distribution driven by hydrodynamic characteristics and the latitudinal gradient.

The heterogeneity in the water column can be represented by vertical gradients in light, temperature and concentration of nutrients. Species may occupy different spaces of this vertical gradient to optimize their growth and multiplication, which adds to the diversity at the community level. The ability of some phytoplankton groups (e.g. dinoflagellates) to control their vertical position due to motility is considered to be a competitive advantage, allowing them to move toward light and nutrients [ROS 07]. Many researchers are interested in the vertical distribution of the abundance of different species of phytoplankton. For example, in an experiment that monitored the Antifer port in Normandy for 48 h (English Channel), Lassus *et al.* [LAS 91] observed a significant increase in *Dinophysis* in the subsurface layer (5 m) as well as vertical movements (phototactic migration).

2.2.4. *Temporal variability and successions*

The temporal distribution of species occurs due to some factors that have been previously mentioned (irradiance, nutritive elements, temperature). The vertical migrations of some organisms alternate from day to night. During the day, these organisms exploit light at the surface and during the night, the nutrients of deeper waters [REY 06]. The different levels of variability in phytoplankton biomass also depend on the level of variation in environmental factors. Episodic phenomena such as storms increase short-term fluctuations in phytoplankton. For example, the arrival of freshwater rich in nutrients following a typhoon in the Hiroshima bay favored the development of *Chattonella marina*, a species responsible for mass deaths of fish [KIM 73].

Phytoplankton is also characterized by seasonal cycles. In this context, phenology is defined as the study of periodic variations in the lifecycle of a species according to seasonal climate variations. Phytoplankton experience great seasonal variations characterized by increased growth rates in favorable conditions, followed by a maximum abundance and a decrease. On this basis, Beliaeff *et al.* [BEL 01] define a "phytoplankton event" to describe the successions of species on an intra-annual scale along the French coasts. These seasonal variations differ according to the ecosystems. Three types of seasonal production have been described by Cushing [CUS 96] (Figure 2.4): tropical, temperate and arctic (or Antarctic). In the Polar Regions, solar radiation and the ice layers are the main factors limiting primary production during the winter. From March onwards (in the Arctic), the irradiance increases from zero to its maximum in June with 24 h of light, favoring annual blooms. In temperate zones, two annual blooms are observed, between the end of winter and the beginning of spring and in autumn. In winter, conditions are still unfavorable as light levels are low, wind is strong and microalgae do not remain in the euphotic zone for long enough. At the end of winter and in spring, the temperature begins to increase, nutrients are sufficient and light is no longer a limiting factor, which favors a bloom dominated by diatoms. In summer, after the depletion of nutriments and the grazing pressure by zooplankton, biomass decreases. This period is characterized by the presence of dinoflagellates, better adapted to these conditions. In autumn, further nutrient input causes a second bloom. In tropical zones, [CUS 96] mention a continuous primary production process with almost stable production rates where the population growth of microalgae is always regulated by herbivory.

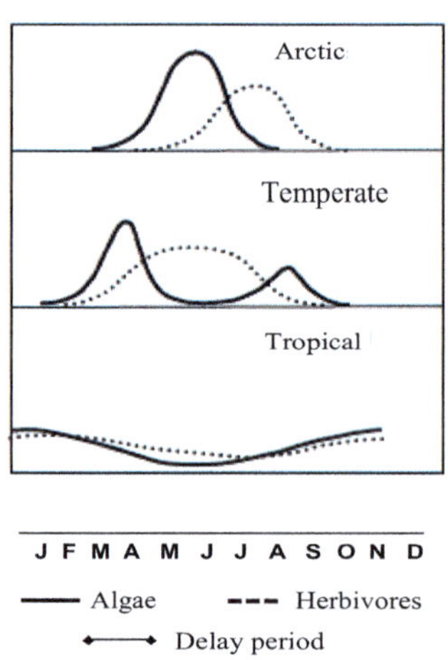

Figure 2.4. *Representation of seasonal cycles of phytoplankton (continuous line) and herbivores (dotted line) at different latitudes according to Cushing [CUS 96]*

COMMENTS ON FIGURE 2.4.– The term "delay period" refers to the grazing capacity of herbivores. Egg production by herbivores depends on the level of available food, and the egg then takes time to develop into a copepodite stage and ultimately have a significant grazing capacity. Over this period, phytoplankton biomass continues to increase.

2.2.5. *Adaptive strategies*

The seasonal variability of physical parameters (e.g. temperature, concentration of nutrients) establishes a cycle of species successions where each of them exploits favorable conditions for growth and reproduction. Therefore, the temporal dynamics of phytoplankton also depends on their survival strategies under variable environmental conditions. For example, it is assumed that diatoms would be favored by turbulence in the water column and high concentrations of nutrients, whereas stratification would be favorable to dinoflagellates [MAR 78]. This species succession model is called "*Mandala*" (Figure 2.5(a), [MAR 97]). It represents the distribution of different phytoplankton groups under turbulence and nutrient concentration gradients. The combination of these two factors would act in the temporal regulation of different morphotypes of phytoplankton, with physiologically similar forms and characteristics. Margalef defines these groups as r-strategists versus K-strategists: r-strategists involve species with rapid reproduction (e.g. diatoms) favored by unstable turbulent environments and relatively high concentrations of nutrients, whereas K-strategists have a slower reproduction and dominate more stable environments (e.g. dinoflagellates).

Later studies developed this theory and provided new elements ([CUL 02], Figure 2.5(b)). Nevertheless, situations with high turbulence and low concentrations of nutrients, considered by Margalef as being "gaps" in his model, appear to exist in waters of the Southern Ocean. The short window of favorable conditions for blooms supports the modest growth of phytoplankton, before the primary production is limited by the iron concentration. The main species found in these zones include the diatoms *Chaetoceros neglectus*, *C. atlanticus*, species belonging to the genus *Fragilariopsis*, *Nitzschia* and *Rhizosolenia* and the haptophyte *Phaeocystis antarctica* [REY 06].

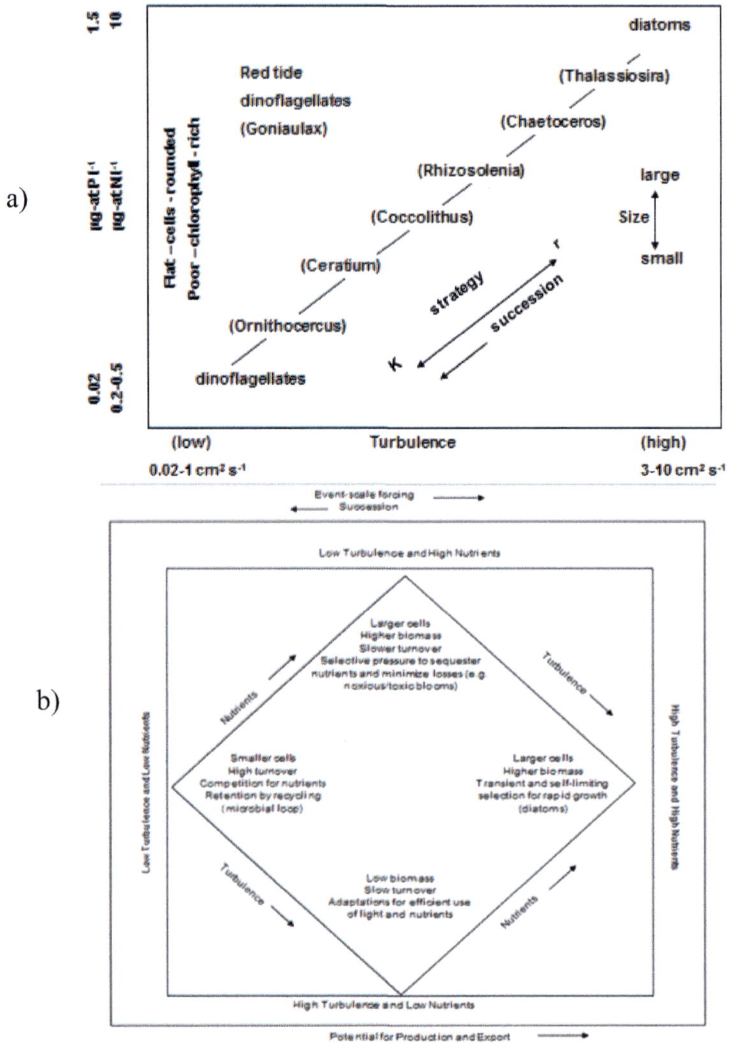

Figure 2.5. *Models of phytoplankton organization*

COMMENTS ON FIGURE 2.5. – (a) The "Mandala" model developed by Margalef explains the dominance of different phytoplankton groups under different conditions of environmental factors (according to Margalef [MAR 97]). (b) The modification of the "*Mandala*" model proposed by Cullen *et al.* [CUL 02] considers a broader range of relationships between physical processes, nutrient inputs, and phytoplankton assemblages structure.

The CSR model, that is to say colonizing-invasive species (C), tolerant to stress (S) and disturbance-tolerant ruderal species (R), has also been used to classify phytoplankton species based on their morphology and physiological traits [REY 88, REY 06]. Species with strategy (C) have small cells, a high surface/volume ratio (S/V), generally low sedimentation rates and are very exposed to herbivory. Species with strategy (S) have a low S/V ratio and are present when the mixing in the water column favors vertical gradients of nutrient concentrations and light. These conditions are easily exploited by organisms with alternative means of acquiring nutrients, for example by nitrogen fixation, phagotrophy[6] or vertical migrations. Species with strategy (R) have large cells (generally elongated) and, despite their dimensions, have a high S/V ratio. These species are favored by turbulence and nutrient concentration. This association in organisms was initially developed in terrestrial plants, then adapted to phytoplankton in fresh water environments. The spatial and temporal distribution of some types and species of phytoplankton is shown to be directly correlated with strategies C, S or R (Figure 2.6). Recently, it has been used to explain species associations along the French Atlantic coast [DAV 12]. Moreover, the seasonal variations observed in the English Channel are associated with these strategies: diatoms thought to have strategy (R) dominate during the spring, then the assemblies change toward a dominance by diatoms-dinoflagellates (*Rhizosolenia spp.*, *Gyrodinium*, *Heterocapsa*, *Prorocentrum*) in the beginning of summer, and by the genus *Ceratium* (strategy S) toward the end of summer [REY 06].

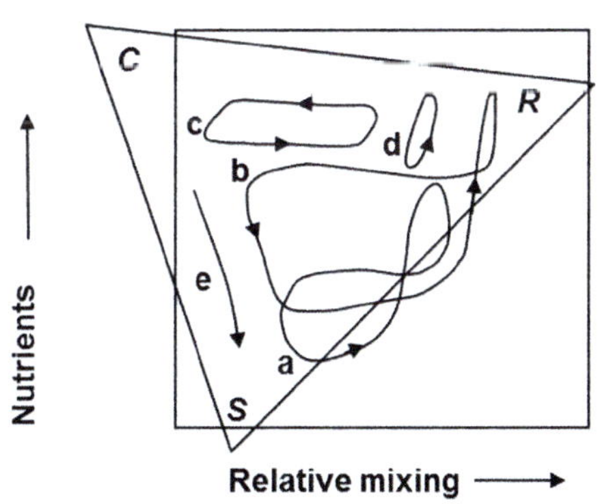

Figure 2.6. *Representation of the CSR organization with regard to turbulence gradients and nutrient concentration. The loops represent temporal successions over one year in a) oligotrophic lakes, b) eutrophic lakes, c) and d) small and rich systems and e) autogenic succession (figure according to Reynolds [REY 06])*

6 Feeding mode involves feeding on particulate organic matter.

Different scales of variability have been observed with regard to phytoplankton and their abundance. Temporally, the continuous variation of environmental conditions prompts phytoplankton responses on a timescale of days or even hours. The following figure shows a temporal series of chlorophyll *a* at two points of time in the San Francisco bay. The measurements were taken every hour for 14 days. During the first week, the chlorophyll *a* data show semi-diurnal periodic fluctuations. This is related to the spatial gradients of chlorophyll *a* oscillating on the sensor with the tides. During the second week, a significant increase in phytoplankton biomass was observed. In this context, semi-diurnal fluctuations in biomass, even if they are still observed, are partly masked by the rapid growth of phytoplankton. This daily variation in biomass characterizes a phytoplankton bloom. Fluctuations over a larger time scale are also visible. These include seasonal, inter-annual and long-term variations (for example: abrupt changes). The seasonal peak in chlorophyll *a* is observed in summer (low flow rate). The inter-annual variability is marked by the summer blooms, lessened over the years due to an extremely high fresh water input and short residence time in water bodies. The apparently persistent change in chlorophyll *a* and the disappearance of the summer bloom following the introduction of *Potamocorbula amurensis* (family *Corbulidae*) in 1987.

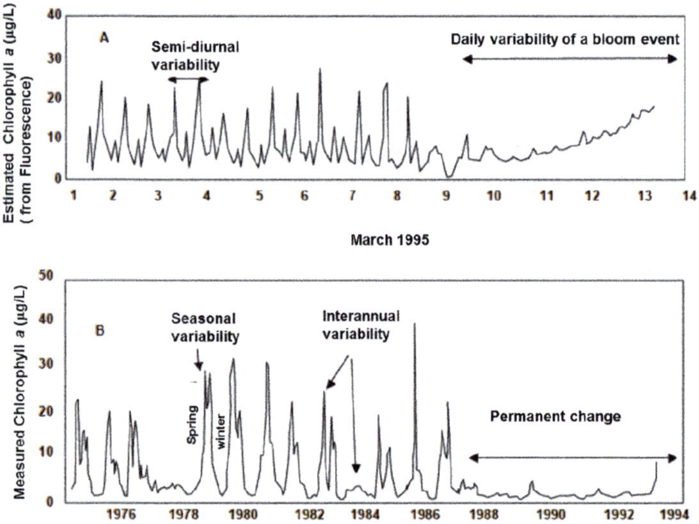

Temporal and spatial variability of phytoplankton expressed on different scales. Chlorophyll a measurements taken hourly in the San Francisco bay for 14 days a) and monthly for 21 years b) (according to [CLO 96])

Remote sensing (figure below) allows us to study the spatial variability of phytoplanktonic primary production by eco-region but also over time. This involves the direct observation of primary production, phytoplankton blooms (*Karenia mikimotoi*) and the monitoring of the water quality [GOH 11].

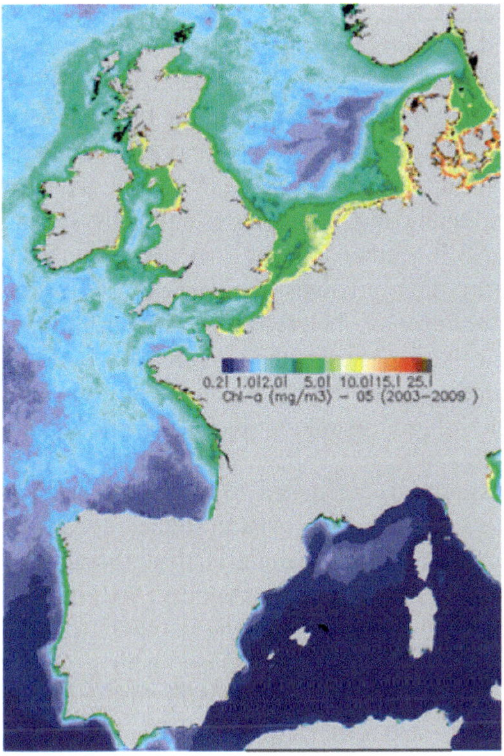

Spatial variability in chlorophyll a in the north-east Atlantic ocean, in May, measured using satellite images (according to [GOH 11]). For a color version of the figure, see color section

Box 2.1. *Illustration of the spatiotemporal dynamics of the phytoplankton*

Although seasonal variations in the phytoplankton can be identified, this temporal variation is very variable, especially in coastal ecosystems experiencing irregular and punctual variations [CLO 08]. Therefore, inter-annual fluctuations in phytoplankton biomass reflect the variability of hydroclimatic conditions (Figure 2.1).

2.2.6. *Functional diversity*

To measure the biological diversity, Clarke and Warwick proposed several indices accounting for the taxonomic differences between species [CLA 98, CLA 01, WAR 95]. These indices integrate not only the distribution in abundance between species, but also their taxonomical relationships. Taking into account the taxonomy is a first step towards studying the functional diversity. The latter has many definitions, such as "the variety of functional traits that influence how the ecosystem functions" [TIL 01] or even "the number, type and distribution of functions performed by organisms within an ecosystem" [DIA 01]. This generally involves understanding and analyzing communities and ecosystems based on the idea of what organisms do, rather than on their evolutionary history [PET 06]. For phytoplankton, this involves the capacity to produce biomass and feeding modes (autotrophy, heterotrophy, mixotrophy), the production of toxins, pumping of CO_2. These functions involve biological processes that depend on the lifecycle and the physiological and biological characteristics of different species of phytoplankton (biological and functional traits).

Functional diversity is supposed to favor ecosystem productivity and increase resistance to invasions. Many studies focus on the relationship between the degradation of the habitat quality, the anthropogenic pressures and the loss of functions in these ecosystems [MIC 05, VIL 10]. Within the phytoplankton, not all species have the same functional role, hence the concept of functional group which groups species with similar functions. Therefore, the specific composition greatly influences the pathways and efficiency of energy transfer to the trophic levels [CLO 05]. The diatoms and silicoflagellates influence the silica cycle. Some cyanobacteria are capable of fixing atmospheric nitrogen. Since 1978, Margalef has been studying the impact of environmental variables on the dominance of different phytoplankton morphotypes with physiologically similar forms and characteristics [MAR 78, MAR 97].

Different taxonomic groups of phytoplankton are often used as different functional groups [MIE 08]. Taxonomic classification is based on the pigment composition of the cell wall, the motility, the physiology and the morphology. The advantage of regrouping taxonomic groups is that most of the species that have been identified can be associated to a group. This approach assumes that all species of phytoplankton belonging to the same group respond similarly under similar conditions. However, organisms in different taxonomic groups often have similar function, which suggests that the strict taxonomic separation may prove to be an inefficient strategy for distinguishing functional groups [MIE 08].

Other clustering methods are based on the adaptive strategies of different organisms (for example: r-K, CSR). One of the challenges of this approach is associating species to one of these groups. As it is incredibly difficult to establish limits for each group, choosing to which group to allocate an organism may prove to be problematic. Regrouping by functional traits is also used. These include using light, acquiring resources (e.g. nutrients) or even the capacity to avoid grazing (Figure 2.7, [LIT 08]). However, it is difficult to obtain this information for each species. Kruk *et al.* [KRU 10] proposed that regrouping species based on morphological traits characterizes the variability in the functional properties of phytoplankton. To characterize the functional aspects of phytoplankton, Weithoff [WEI 03] proposes traits such as cell size, the potential to fix nitrogen, requirement for silicates, phagotrophy, mobility and cell shape. These analyses require exhaustive documentation on organisms and deciding to which functional group organisms belong, which sometimes may require compromise. Problems identifying the taxonomy exist in many groups of phytoplankton, hence the challenge when basing functional classification on taxonomy.

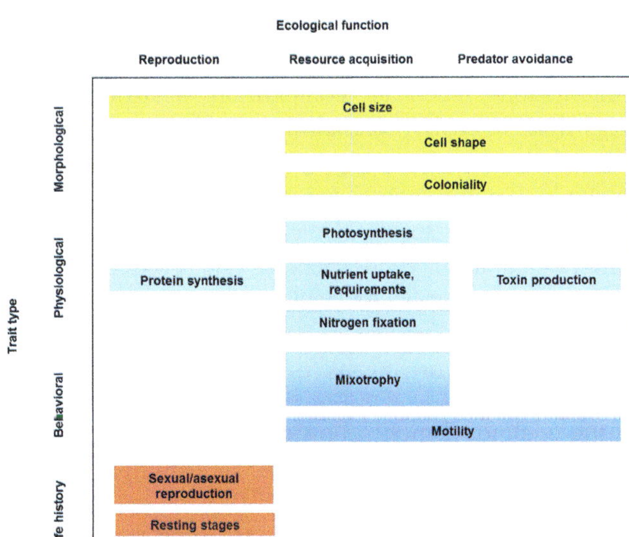

Figure 2.7. *Typology of phytoplankton traits according to Litchman and Klausmeier [LIT 08])*

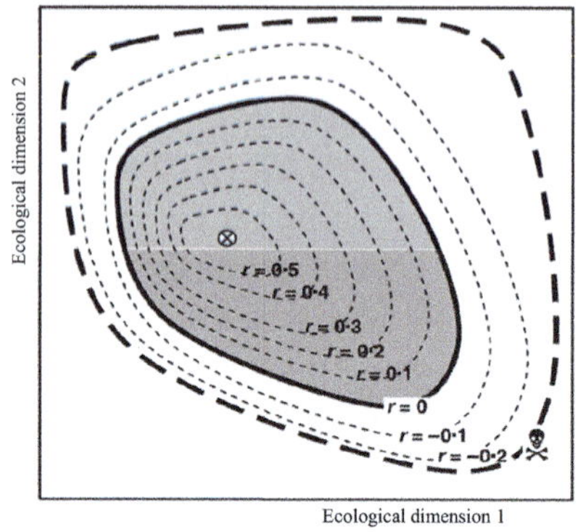

Figure 2.8. *Schematic representation of an ecological niche in a two-dimensional environment. The grey zone represents the space where growth (r) is positive (from Hirzel and Le Lay [HIR 08])*

2.2.7. *Ecological niche*

Hutchinson [HUT 57] defines the niche of a species as an environmental space that allows growth (Figure 2.8). This space is defined as the set of biotic and abiotic environmental conditions in which the species can thrive. The environmental factors essential for the existence of a species form the fundamental niche. The realized niche is defined as the niche that is actually occupied, sub-unit of the fundamental niche resulting from biological interactions, such as competition. Considering these source gradients (e.g. light intensity, availability of nutrients), species evolve by utilizing different levels of these gradients [WHI 72]. The concept of an ecological niche is therefore associated with diversity and allows us to study the distribution patterns of species. Diversity in the community will evolve by the niche differentiation of each species, and consequently, the position occupied in their environment. As well as describing the ecological niche, studies on species distribution also associate global changes to modifications in the distribution range of organisms (expansion or decrease).

Studying the ecological niche allows us to analyze different topics such as the distribution of diversity [JET 12], the future impact of climate change [THO 04], species invasion [BRO 07, PET 03], or even the distribution of disease vectors [PET 09]. Three hypotheses are often put forth:

– the distribution of each species is uniquely determined by its abiotic environment;

– dispersion is not limited;

– there is no potential for adaptation.

Nevertheless, the realized niche is limited to the fundamental niche of a species because of biological interactions such as competition.

The concept of ecological niche offers a relatively simple framework to quantify the space offered to a species. Litchman *et al.* [LIT 12] believe that the concept of ecological niche will help to predict how species of phytoplankton respond to diverse stress factors that act simultaneously. These authors highlight that the phytoplankton confirms most of the hypotheses from niche models, since they are highly influenced by abiotic factors, unlimited dispersion and their behavior does not allow them to modify the environmental influences. Despite the large number of studies that refer to the concept of niche in ecology, this approach is less frequently used for phytoplankton. Two well-known data sets (Continuous Plankton Recorder (CPR) and Helgoland Roads) have been recently analyzed to estimate the niche of marine phytoplankton species [GEB 09, GRU 11, IRW 12]. The position of the niche and its width with regard to the environmental gradients are often measured. This was the case for the diatoms *Paralia sulcata* and *Odontella regia* and the dinoflagellate *Neoceratium fusus* [GEB 09, GRU 11]. These measurements were used to study the temporal variation of the realized niche of these species, and in particular show that the niche of *Paralia sulcata* has been modified over time [FRE 12, GEB 09]. We must also wonder about the calculation of these parameters on an annual basis, that is to say position and width. A modification in these parameters leads us to ask: has the niche really changed? Does this reflect the environmental variability? The presence of one species in very different environmental conditions highlights the occurrence of strains with variable lifecycles, which represents the resilience of the ecosystem in the face of environmental variability [CLO 05].

Studies on the ecological niche also allow groups of species to be compared and gain more insight into the community organization. With this intention, Irwin *et al.* [IRW 12] studied the realized niche of phytoplankton in the North Atlantic Ocean. The position occupied by 119 species in environmental gradients (temperature, salinity, irradiance, nutrients) revealed a high variability in responses, despite finding a strong signal separating the diatoms and dinoflagellates.

2.3. Phytoplankton responses to anthropogenic pressures

According to Litchman *et al.* [LIT 12], the phytoplankton can respond to environmental variations in three ways:

– due to their phenotypic plasticity, the organisms are capable of thriving in this environment;

– when reaching the limits of their phenotypic plasticity, some species poorly adapted to new environmental conditions are replaced by better adapted ones;

– species can adapt through the selection of new genotypes (for example, mutation).

Therefore, the modifications observed in the environment are the result of this processes acting simultaneously or sequentially (Figure 2.9, [LIT 12]). Research has shown the direct or indirect impact of environmental variations on organisms, populations and communities, from field observations, mesocosm experiments or modeling results. In the following sections, we will describe the environmental variations linked to anthropogenic pressures that can cause phytoplankton to respond in terms of biomass, biodiversity, spatial distribution, phenology, or biological traits.

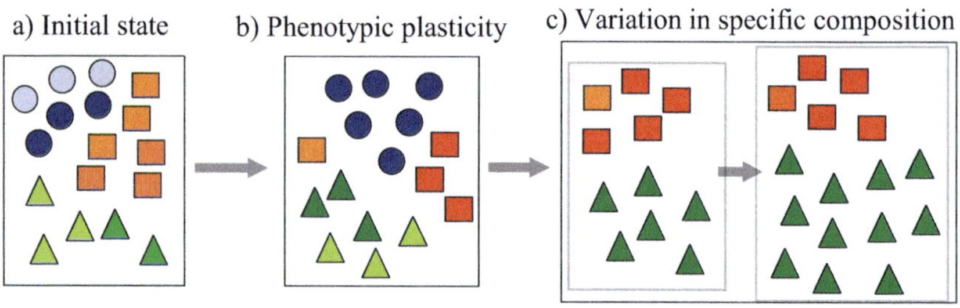

Figure 2.9. *Possible responses of three species of phytoplankton to changes in their environment. Each species is represented by a shape (circle, triangle or rectangle) and each phenotype (and potentially a genotype) by a given color*

COMMENTS ON FIGURE 2.9.– (a) The community in its initial state consists of three species; (b) following a change in their environment, the species respond by varying their phenotype (phenotypic plasticity); (c) a poorly adapted species (circle) and does not thrive in the community whereas the better adapted species (triangle) increases in abundance. Some phenotypes (darker shade), and potentially genotypes, are better adapted and increase in frequency through natural selection (adapted from Litchman *et al.* [LIT 12]).

2.3.1. *Main sources of anthropogenic variability*

One of the main problems caused by human activity is the increase in the concentration of atmospheric carbon dioxide (CO_2). This has increased by 40% since the beginning of the industrial revolution, reaching the most unprecedented levels for 800,000 years [IPC 13]. One direct effect of this increase is the increase in the temperature of the oceans. This causes icebergs to melt, sea level to rise, stratification of the water column, acidification and alteration of ocean circulation currents, precipitations and affects freshwater inputs[7].

An increase in sea surface temperature (SST) is found at all latitudes despite regional variations also being observed [GOH 09, TRE 07]. The mean global temperature has increased by 0.74°C ± 0.18°C over the previous century and an increase of 1.8–4°C is predicted for the 21st century [TRE 07]. In most cases, different scenarios of such a change will have severe consequences on diversity, such as a higher risk of species extinction, changes in the functioning of ecosystems, their stability as well as the ecological benefits ecosystems provide [DON 03, IPC 07, WAR 11]. An increase in SST will increase vertical stratification of the water column and cause a decrease in nutrient availability in the photic zone with consequences on pelagic primary producers (Figure 2.3, [ITO 10]).

Temperature is one of the most influential abiotic factors on all life on Earth. Biological processes such as enzymatic reactions, photosynthesis, respiration, growth or even physical processes, such as the rate of sedimentation or stratification of the water column, depend on temperature. Each organism has an optimum that determines its survival and reproduction. These optima vary from one species to another, thereby defining the distribution range of organisms and largely their lifecycle (Figure 2.3). Temperature variations may modify the size structure of communities, the spatial distribution of organisms and the abundance of populations throughout the seasons. In turn, these changes alter, at the

7 Refer to *Vulnerability of Coastal Ecosystems and Adaptation* [MON 14a] from the "Seas and Oceans" Set for more information.

ecosystem level, by means of bottom-up or top-down processes, interactions between species and the trophic web (Figure 2.3).

An increase in the concentration of CO_2 also causes the gradual acidification of the ocean. Since the industrial revolution, the pH has decreased by 0.1 unit, which represents a 26% increase in the H^+ concentration [IPC 13]. Another consequence is the decreased availability of carbonate CO_3^{2-}. It is predicted that ocean acidification will affect marine ecosystems, in particular organisms with calcareous structures such as scleractinian corals, some mollusks and coccolithophorids.

The oxygen content in the oceans is also expected to decrease through the combined effect of lower oxygen solubility due to higher temperatures and less ventilation due to stratification. In coastal zones, the number of hypoxic regions has increased considerably [DIA 08], largely due to human activities and increased nutrient input. In these zones, hypoxia will be modulated by climate change due to alterations in mixing of the water column and wind regimes [RAB 10].

Using fertilizers in agriculture or husbandry and the release of domestic waste water have increased the input of nutrients such as nitrogen (present in ammonium, nitrates and nitrites) and phosphorus (present in phosphates) [MEN 01, RAB 09][8]. This phenomenon has become a concern due to ecological and economic consequences. In the marine environment, this excess of nutrient inputs causes the excessive multiplication of phytoplankton and other algae, and disrupts the equilibria between the production and metabolism of organic matter in these systems [CLO 01]. The negative effect arises when the concentration of nutrients is greater than the ecosystem can utilize this primary production [RAB 09]. The microbial decomposition of organic matter also gradually consumes dissolved oxygen and hypoxic and anoxic events may take place, with harmful consequences on the diversity of benthic fauna [CLO 01, DIA 08]. A second consequence of microbial degradation of organic matter is the production of CO_2 and a decrease in the pH of the water [WAL 14].

The effects of an increase in CO_2 are not an isolated case. Along with these impacts, there are other pressures that vary from one region to another, in

8 Also refer to Chapter 4 of *Vulnerability of Coastal Ecosystems and Adaptation* [MEN 14] from the "Seas and Oceans" Set.

particular: the degradation of habitats, the intensive use of fertilizer, the over exploitation of fish stocks, more intense aquaculture and the introduction of non-indigenous species. Thus, marine ecosystems experience multiple stress factors (Figure 2.10). In particular, coastal ecosystems are exposed to the both climate and anthropogenic impacts. This dual effect causes modifications to biological communities at all levels, form individuals (physiology, size) to communities (for example: structure, phenology, distribution). However, biological communities do not respond equally to these variations. Depending on the system (ocean, coastal, estuary, etc.) and their inherent properties (for example: tides, residence times in water bodies), different responses can be expected. The simultaneous effect of multiple factors and interactions between organisms makes it more complex to study the responses of biological communities. Despite this complexity, it is important to understand the variations in the diversity of phytoplankton communities in the long term, to analyze the responses of biological systems when faced with environmental change.

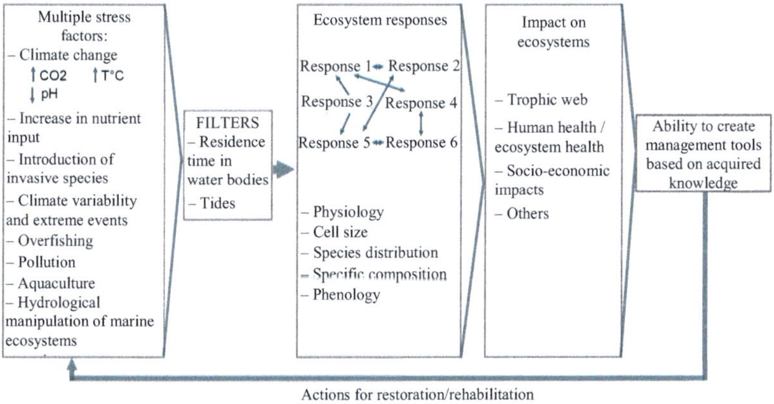

Figure 2.10. *Conceptual diagram of responses and impacts on ecosystems under multiple stress factors. The effects these factors have on ecosystems depend on the inherent properties of systems, which act as filters, and mitigate or exacerbate these effects*

COMMENTS ON FIGURE 2.10.– Within biological communities, these responses are observed at all levels, from individuals (for example: physiology, size) to communities (for example: structure, phenology, distribution). The modification of biological communities has large-scale impacts on the trophic web or on human health. Understanding these mechanisms allows us to design management strategies to restore the functions of ecosystems and biological communities (adapted from Cloern [CLO 01]).

2.3.2. Responses with regard to biomass, abundance and species composition

Over several decades, many studies have focused on the effects of increasing SST on the physiological and ecological aspects of phytoplankton. Boyce *et al.* [BOY 10] indicate that there is a global decline in phytoplankton biomass (chlorophyll *a* measurements) by approximately 1% per year for the past century. This decline in phytoplankton biomass has been linked to changes in ocean conditions particularly the increase in sea surface temperature. However, these trends do not appear to be uniform in all oceans. Moreover, these authors documented an increase in the concentration of chlorophyll *a* in the southern Indian Ocean. The changes in phytoplankton biomass are the result of two opposite mechanisms caused by the increased stratification of the water column: on the one hand, a lower input of nutrients essential for phytoplankton and on the other hand, a more efficient use of light, allowing a longer growth period (Figure 2.11, [BOP 01]). It is predicted that, in higher latitudes (colder), phytoplankton biomass will increase, favored by the extended growth period and a sufficient input of nutrients [BOP 01, DON 06, RIC 04]. Nevertheless, this mechanism does not explain the decrease in chlorophyll *a* observed in polar zones [BOY 10].

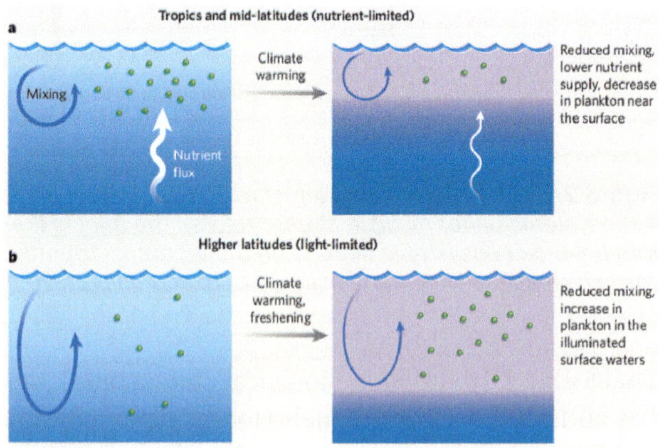

Figure 2.11. *Predicted response of phytoplankton as the ocean temperature increases (according to Doney [DON 06])*

COMMENTS ON FIGURE 2.11.– (a) In zones at lower latitudes, phytoplankton is typically limited by nutrients. An increased temperature and stratification of the water column will cause a decrease in the nutrient input and consequently a decrease in phytoplankton biomass. (b) In polar and subpolar latitudes, phytoplankton is limited by light. Decreased mixing should increase phytoplankton biomass due to an increased availability of light.

These variations in phytoplankton biomass imply that the main groups forming the phytoplankton are also affected by climate variations. The diatoms represent a large proportion of the phytoplankton and often dominate communities, primarily when there are high nutrient concentrations [CLO 05]. Bopp *et al.* [BOP 05] predict a 10% decrease in the relative abundance of this group on a global scale and approximately a 60% decrease in the north Atlantic. Other studies also confirm a decrease in diatoms in the Atlantic Ocean, the English Channel and the Baltic Sea over the past four decades [KLA 11, LET 05, WID 10, ZHA 13]. This decrease will be accompanied by an increase in other groups such as dinoflagellates. However, these results differ to those obtained by Hinder *et al.* [HIN 12] in the North-East Atlantic (between 1960 and 2009), who reported an increase in diatoms such as *Pseudonitzschia spp.* and *Thalassiosira spp.*, whereas the abundance of the dinoflagellates *Neoceratium furca* and *Prorocentrum spp.* decreased. These alterations were attributed to the combined action of an increased SST and summer breezes, favoring conditions that are more turbulent. The effect of a change in diatom/dinoflagellate ratio on the ecosystem is unknown. Typically, diatom blooms are considered to sustain secondary production and the development of copepods, and consequently, the growth of fish larvae, which depend on them for food [IRI 02]. However, a recent study conducted by Vehmaa *et al.* [VEH 11] concluded that a change in this ratio in favor of dinoflagellates would have a positive effect on the reproduction of copepods.

The impact of climate variations on an inter-annual scale on the structure of phytoplankton communities has been observed in several littoral zones [BRE 06, DAV 12, IRI 00]. For example, the north-Atlantic oscillation (NAO) (Figure 2.12), defined as the difference in atmospheric pressure between the Azores anticyclone and the Iceland depression, has been found to have a positive correlation with the abundance of diatoms and a negative correlation to that of dinoflagellates [IRI 00, ZHA 13]. The positive phase of the NAO is associated with a predominance of westerly winds and mild and rainy winters in Northern Europe, but dryer around the Mediterranean. Low

NAO index values, associated with weaker westerly winds, cause a decreased vertical mixing in the water column and a decrease in turbulence, thereby favoring the dinoflagellates [ZHA 13]. The Atlantic Multidecadal Oscillation (AMO) has also been correlated with long-term plankton variations (Figure 2.12 [DAV 12, EDW 13]). The AMO is an indicator of climate variability with regard to the sea surface temperature of the North Atlantic [DIJ 06]. Several studies show the relationship between AMO and sea surface temperature, precipitation, hurricanes in the Atlantic and the summer climate in North America and Europe [ENF 01, GOL 01, SUT 05].

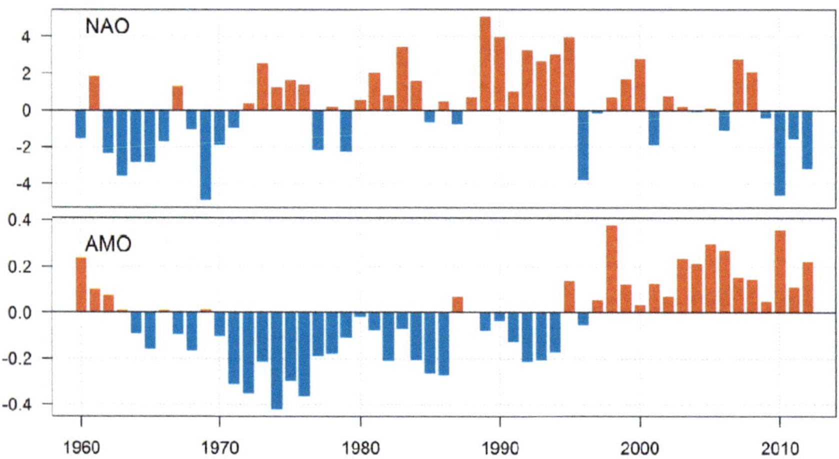

Figure 2.12. *Time series of the winter mean (from January to March) of the climate index north Atlantic oscillation (NAO) and the annual mean of the index Atlantic multidecadal oscillation (AMO)*

The effect of an increase in CO_2 concentration and ocean acidification has been studied in phytoplankton particularly on those with calcareous structures such as coccolithophorids. The response by this group of phytoplankton, and more specifically by the species *Emiliania huxleyi*, has been studied previously. Although the diversity and abundance of these organisms being significantly affected [ZIV 14], other studies suggest that the adaptation processes will help maintain the functionality of these groups when faced with change [LOH 12].

In coastal zones, one concern is the greater input of nutrients such as nitrogen and phosphorus. This phenomenon results in a greater production of phytoplankton biomass and a loss of equilibria with regard to biodiversity.

Within marine ecosystems, some algae blooms have been attributed to eutrophication [AND 02]. Therefore, causal links exist between the increase in chlorophyll *a* concentration and the increase in nutrient input in coastal zones and estuaries [CLO 01]. In the North-East Atlantic, 104 zones, including the Bay of the Seine and the Bay of the Vilaine, are considered to be particularly sensitive [OSP 08]. In July 1982, the Bay of the Vilaine experienced anoxic conditions causing the mass death of fish and benthic invertebrates. This phenomenon was linked to two successive stormy episodes in the river basin, enriching the surface waters and causing haline stratification of the water column. The phytoplankton continued to develop due to calm weather and sunshine and its subsequent degradation caused significant oxygen depletion. Stratification of the water column and weak currents favored the oxygen deficit [CHA 91]. Other examples reveal a relationship between the enrichment of marine zones and an increase in phytoplankton biomass. For example, in the summer of 1997, anoxia occurred in the Mariager fjord in Denmark [FAL 00]. This phenomenon was the result of a huge dinoflagellate bloom, by the species *Prorocentrum minimum*, after meteorologically favorable conditions (particularly hot summer, sunny and calm). The northern part of the Gulf of Mexico often experiences severe hypoxia. This is due to the high nutrient input into the Gulf from anthropogenic activities (in particular from agriculture) in the basin of the Mississippi river, which favors an increased phytoplankton production.

The correlation between an increased nutrient input and an increase in phytoplankton biomass is not always clear. This was found to be strongly dependent on the system and its unique processes, which explains why some marine ecosystems are more sensitive to nutrient input than others. Therefore, some processes such as tides, winds or residence time in water bodies act as filters that determine the type and size of the response by ecosystems [CLO 01]. Moreover, high nutrient input is not an isolated process, and it is becoming increasingly more evident than the interaction with multiple factors such as climate can lessen or exacerbate the consequences on ecosystems [CLO 01, RAB 09].

With regard to the economic and environmental challenge caused by the high nutrient input into coastal zones, the concentrations of chlorophyll *a* are often used as the assessment criteria to measure eutrophication levels and the ecological state of ecosystems, for example in Marine Strategy Framework Directive [FER 11, LEF 12, MEN 01]. For the Water Framework Directive

(WFD), chlorophyll *a* and the abundance and diversity of phytoplankton are parameters that help assess coastal water bodies [AND 06].

Over the past decades, Harmful Algal Blooms (HAB) have often been attributed to eutrophication[9]. In the "Mandala" phytoplankton organization model by Margalef [MAR 97], dinoflagellates that form red tides are associated with conditions of high nutrient input and low turbulence. Some studies are based on measuring dinoflagellate cysts in sediments over the past 100 years to study the link between high nutrient input into marine waters from human activities and HAB (Oslo fjord, Norway – [DAL 01]). According to this author, high nutrient input into temperate systems with limited nutrients would cause an increase in HAB, particularly during the summer. The species concerned is *Lingulodinium polyedricum*, producer of diarrheic toxins. We cannot generalize the impacts of aquaculture on HAB. In zones of intensive aquaculture, the cumulative effect of high nutrient input may have an impact on phytoplankton composition and abundance. However, the intensity of this impact largely depends on the toxic species present, the relative abundance of nutrients, the hydrodynamic characteristics of the zone as well as the level of herbivory and availability of light [AND 02]. Another aspect studied is the effect of the synergic action of climate and human activities on the occurrence and growth of HAB [HAL 10].

At high concentrations, some elements such as cadmium, zinc, and mercury are toxic to phytoplankton and the same goes for pollutants such as PCB and DDT. The contamination of coastal zones by these elements may cause variations in species dominance, affecting the community structure and composition [FIS 73, MOS 72]. Moreover, microalgae accumulate some of these contaminants and some of them are not, or only slightly, broken down. Given that these organisms are primary producers, the phytoplankton are a key link in the transfer of pollutants to higher trophic levels including humans.

Finally, in the literature, phytoplankton has been shown to have highly variable responses to changes in environmental conditions, and therefore it is difficult to state a single response by these organisms (Table 2.3). These different variations highlight the complexity of responses by this biological

9 For more information refer to Chapter 4 of *Vulnerability of Coastal Ecosystems and Adaptation* [MEN 14] from the "Seas and Oceans" Set.

group. Moreover, the multiple factors acting on the phytoplankton explain why it is difficult to state the predominant effect of one factor or another.

Zone	Frequency of observations	Variations observed	Reference
Regions of the Baltic Sea	Variable according to sampling site	Phytoplankton biomass increased in the central region of the Baltic Sea (Baltic Proper) but decreases in the Mecklenburg bay. The decrease in diatom biomass in spring in some regions of the Baltic Sea is accompanied by an increase in dinoflagellates. Environmental variations observed include an increase in temperature, decrease in salinity and concentration of nutrients. In the Kiel bay, autumn and summer blooms of Chaetoceros and Skeletonema observed at the beginning of the 20th Century have been substituted by Cerataulina pelagica, Dactyliosolen fragilissimus, Proboscia alata and Pseudo-nitzschia spp. New proliferating species are potentially toxic (e.g. Prorocentrum minimum, Pseudo-nitzschia spp.). This study highlights the existence of local trends in the Baltic Sea.	[WAS 98, WAS 03, WAS 08, WAS 11]
Regions of the Baltic Sea	Variable according to sampling site	The relative proportion of dinoflagellates compared to diatoms has increased during winter-spring blooms, particularly in regions of the Northern Baltic Sea (Gulf of Finland and Gulf of Bothnia). The authors hypothesize that mixing of the water column in winter and the re-suspension of benthic cysts, followed by stratification of water under a thin layer of melting ice favors the accumulation of dinoflagellates. Milder winters and a thin layer of melting ice (10–20 cm) favor complex blooms of Biecheleria baltica (B. baltica, Gymnodinium corollarium and Scrippsiella hangoei).	[KLA 11, KLA 13]
Gulf of Finland and the northern region of the Baltic Sea	Approximately twice in summer primarily in August, between 1979 and 2003	This study shows a significant increase in the biomass of Chrysophyceae Chlorophyceae. The biomass of dinoflagellates and cyanobacteria increases in the north region of the Baltic Sea and the Gulf of Finland, respectively. The biomass of Euglenophyceae and Cryptophyceae decrease in these zones, respectively. The variations in environmental conditions include the increase in summer temperature and concentration of dissolved inorganic nitrogen (DIN) and the decrease in salinity. These changes to the specific composition of communities has been linked to the hydroclimatic variations and continuous eutrophication.	[SUI 07]
Central reigon of the North Atlantic	Data obtained monthly	The analyses carried out between 1991 and 2009 reveal an increase in the proportion of dinoflagellates and a decrease in diatoms. This study found a link between the climate index NAO, SST and change in structure of the phytoplankton community.	[ZHA 13]
North-east region of the North Atlantic and the North Sea	Monthly	The abundance of dinoflagelles such as Prorocentrum spp. and Neoceratium furca decreased and that of the diatoms increased (e.g. Pseudo-nitzschia spp., Thalassiosira spp.). These variations caused a significant icnrease in the relative abundance of diatoms with regard to dinoflagellates. The combined effect of an increased SST and winds, which favor more turbulent conditions, was also associated with these variations.	[HIN 12]

Western region of the English Channel	Weekly	Between 1992 and 2007, a significant decrease in the abundance of diatoms and Phaeocystis was observed whereas that of coccolithophorids and the dinoflagellate Prorocentrum minimum increased.	[WID 10]
North Atlantic and adjacent seas	Monthly	This study described an abrupt change in several biological components of the ecosystem (e.g. phyto-zooplankton) during the 1980s. In particular, the phytoplankton color index increased (indicator of phytoplankton biomass using CPR data). There was an association between the increase in SST and a low herbivore pressure by Calanus finmarchicus (its abundance decreased).	[BEA 09]
Calvi Bay	Five periods between winter-spring between 1979 and 1998	Limited silica between 1986 and 1988 favored the decrease in abundance of diatoms, accompanied by an increase in the abundance of non-siliceous phytoplankton. After this period, the decrease in the availability of nitrates explains the gradual decrease in abundance of non-siliceous phytoplankton.	[GOF 02]
Narragansett Bay (United States)	Weekly	This study shows a 45% decline in the abundance of Skeletonema between 1959 and 1997. It also highlights the relationship between this trend and climate (NAO index, temperature and irradiance). Mild winters are associated with summer blooms of Skeletonema and a decrease in annual abundance.	[BOR 09]
Atlantic coast (France)	Bi-monthly	Study of temporal and spatial variations in the specific and functional diversity of phytoplankton. Despite the temporal variation not being able to explain the variation in diversity, the main factors that explained the inter-annual variations were large-scale climate indices, mainly the AMO.	[DAV 12]
Belgian coast	Twice a week to bi-monthly between 1988 and 2001	The diatom/Phaeocystis ratio is determined by the combined effect of the north-atlantic oscillation (NAO) and freshwater and nitrate input from the river Escaut.	[BRE 06]
Bay of the Seine (Normandy, France)	Bi-monthly between 2001 and 2011	There was a decrease in chlorophyll a in summer but a slight increase in winter. An increase in the abundance of dinoflagellates was observed in summer. The abundance of diatoms did not vary significantly. The decrease in phosphorus corresponds to an overall decrease in phytoplankton biomass, but the higher concentrations of nitrogen favor the occurrence of harmful microalgal blooms.	[ROM 13]

Table 2.3. *Examples of studies on temporal variations of biomass and the abundance of phytoplankton, based on observations in different marine regions*

2.3.3. *Changes in spatial distribution*

Species distribution is sensitive to global changes [HAY 05]. In particular, the impact of an increase in sea temperature on the spatial distribution of phytoplankton has been the subject of several studies. Thomas *et al.* [THO 12] hypothesized that the diversity in tropical zones decreases as phytoplankton are more sensitive to an increase in temperature. These authors also state a possible increase in diversity in zones at higher latitudes due to the displacement of species toward the poles.

CPR data reveal several changes in the distribution range of phytoplankton. In particular, the dinoflagellate *Neoceratium trichoceros*, identified only in the south of the United Kingdom before 1970, is currently observed on the west coast of Scotland and north of the North Sea [HAY 05]. In the North Sea, the establishment of species considered as thermophiles (e.g. *Alexandrium minutum*, *Gymnodinium catenatum*, *Proboscia indica*) would have been favored by higher temperatures [NEH 98].

Another example of a change in distribution range involves the diatom *Neodenticula seminae*, previously observed in the subarctic North Pacific. Using the CPR data and analysis of sediment cores, this species was found in the 20th century in the North Atlantic, after an 800,000 year absence. This appearance has been associated with an increased influx of Pacific waters toward the Atlantic via the Arctic Ocean, due to the melting of sea ice and/or variations in the Arctic Ocean circulation [MIE 13, REI 07].

The geographic expansion of phytoplankton has occasionally been attributed to the release and dispersion of ballast waters of ships. This method of introducing species alters their distribution and has been suspected for several taxa such as *Coscinodiscus wailesii* in the English Channel and North Sea [EDW 01] or *Alexandrium catenella* in the Thau Lake, in France [LIL 02]. These introductions often cause unexpected blooms and are sometimes accompanied by the occurrence of toxins (e.g. Paralytic Shellfish Poisoning, a paralyzing toxin). How these species come to settle and even proliferate is still difficult to explain [NEH 98]. These species can exploit previously unoccupied habitats. Moreover, ecosystems disturbed by the combined effects of climate change and human activities are more subject to invasion [STA 02].

In the context of global change, predicting the distribution range of species, their extinction risk as well as underlying alterations to the community is still a challenge. With regard to available data, some models account for only temperature and do not consider the possible combined effect of temperature and other parameters (e.g. inter specific interaction). Moreover, laboratory experiments are often based on the effect of a single factor on organisms, which only allows the study of physiological plasticity in the short term. The possible adaptation of these organisms is still not widely studied, but some studies have begun to provide evidence for adaptation phenomena [LOH 12]. Kremp *et al.* [KRE 12] show that depending on the strain of *Skeletonema marinoi*, the effects of temperature on growth may be either positive, negative or negligible.

The impact of these variations on the primary production of the ecosystem is still poorly understood. The importance of maintaining diversity for the productivity of ecosystems has been widely discussed beyond the context of climate change [CAR 09]. Productivity is favored by diversity. We must therefore expect the variations in phytoplankton diversity alter the productivity and stability of ecosystems.

The complexity of these interactions reveals a challenge when forming future scenarios. Predicting the changes in diversity and distribution of species may be partly mitigated by the capacity of species to adapt, especially those with fast population turnover time, which is the case for phytoplankton [THO 12]. In addition, more laboratory studies are required to account for the adaptive response of phytoplankton in models that predict species distribution.

2.3.4. *Changes in phenology and amplitude of blooms*

Phenology is often expressed by the date the algal blooms occur, their amplitude and duration. It is directly linked with climate and fluctuations in climate form an integral part of the phenological variability of primary producers. In terrestrial ecology, the literature on phenological variations with regard to climate change is abundant. A number of these studies have found a correlation between some parameters such as leaf coloration in autumn and the date of the bloom and increase in temperature, while highlighting the complexity of responses when accounting for the interaction with other factors such as precipitation [CLE 06, CLE 07, MEN 03]. These studies have inspired research into similar patterns in phytoplankton phenology, from total biomass or estimated for different taxonomic levels.

The increasingly early spring blooms of phytoplankton due to the increase in temperature is still one of the most common hypotheses [HAS 07, LEW 10, WIN 04a]. This phenomenon has been observed in aquatic environments, linked to the stratification of the water column or break up of lake ice [THA 08, WAL 13, WEY 01, WIN 04a]. Land observations are consistent with this hypothesis [EDW 04]. However, the response by phytoplankton is far from being unique, and the interaction with other parameters as well as inter specific variations lead to diverse responses (Table 2.4).

Remote sensing methods are widely used to estimate the start date and maximum peak of phytoplankton biomass [KAH 11, RAC 12, SIE 02, ZHA 11]. This method has the advantage of providing a wide spatial overview and a refined temporal resolution (of approximately one day) of primary production processes and can address these with regard to climate variations. A study that lasted over a decade (1997-2009) revealed a significant trend in the occurrence of earlier blooms (approximately 50 days) in the northern Pacific and Arctic Oceans, and later blooms in some parts of the North-Atlantic [KAH 11]. The annual and decadal variability in chlorophyll *a* was also attributed to climate variations such as the North Atlantic Oscillation (NAO), the Pacific Decadal Oscillation (PDO) or even El Niño Southern Oscillation (ENSO) [CHI 12, KAH 11, RAC 12, ZHA 11, ZHA 13].

Walters *et al.* [WAL 13] highlight the importance of taking into account refined taxonomic resolution levels since the responses to climate change can vary according to species and affect the composition of communities [ANN 02, GUI 10]. Thus, observation systems *in situ* allow the study of interannual phenological variations, often at the species level. CPR data reveal earlier blooms (by approximately 23 days) of dinoflagellates *Ceratium*, *Protoperidinium* and *Dinophysis* in the North Sea [EDW 04]. On the other hand, the diatoms did not experience modifications to their spring blooms. This diversity in responses is associated with factors that strongly influence the seasonal cycle of different taxonomic groups. The phenological response by dinoflagellates to the increase in temperature is not solely physiological: it can also be attributed to the earlier stratification of the water column. The diatoms are probably more dependent on day length or light intensity [EDW 04, EIL 95, WIL 08].

Marine ecosystems can be particularly vulnerable to these changes. The variation in phytoplankton phenology can lead to a desynchronization between primary and secondary producers, as well as a bottom-up cascade toward higher trophic levels with negative consequences, such as for example, on there productive success of marine birds [DUR 03] or on human activities such as fishing [CUR 08, EDW 04].

Taxa	Methodology	Zone	Metrics	Phenological variations	Observation frequency	Reference
Diatoms and dinoflagellates at the species level	Field observations–CPR	Center of North Sea	Date of seasonal peak	Earlier blooms (approximately 23 days) of dinoflagellates *Ceratium*, *Protoperidinium* and *Dinophysis*. No change in diatoms.	Monthly	[EDW 04]
Diatoms	Field observations	Heligoland Roads, North Sea	Date of start and exponential growth of bloom	No trend in start date of bloom despite significant alterations to abiotic conditions.	Daily data (workable)	[WIL 08]
Chlorophyll *a* and phytoplankton at the species level	Field observations	Bahía Blanca Estuary (Argentina)	Date of start and decline of bloom	Typical winter bloom and early spring bloom replaced by relatively short blooms throughout the year. *Thalassiosira curviseriata* was replaced by other species (e.g. *T. minima* and *Cyclotella*). The dates of growth and decline are earlier. Changes are attributed to milder winter and low precipitation.	Twice a week to one every 15 days	[GUI 10]
Chlorophyll *a* and phytoplankton count	Field observations	Narragansett Bay, United States	Date of maximum bloom and abundance	Typical winter-spring blooms regularly disappear over the previous years. Relatively short diatom blooms in spring, summer and autumn remain, replacing the summer flagellate blooms. Changes are attributed to milder winters and a decrease in wind velocity and increased cloud cover.	Weekly	[NIX 09]
Diatom group, but also at the species level	Field observations	Lake Washington	Date of maximum biomass in spring, center of gravity of spring-autumn peak	Earlier blooms (approximately 19 days) linked to stratification, but specific response to taxa.	Weekly to monthly	[WAL 13, WIN 04b]
Biomass of phytoplankton	Field observations	Swedish lakes	Spring decline in start date of cyanobacterial blooms in summer	Earlier decline in biomass following a particularly mild winter, earlier summer blooms (cyanobacteria) and no response y dinoflagellates, chrysophytes and cryptophytes.	Irregular frequency, often four samples per year	[WEY 01]

Measurements of primary production and biomass of phytoplankton (e.g. picoplankton, diatoms, dinoflagellates)	Mesocosm	Conditions in the Kiel Bay (Baltic Sea)	Start date, peak and end of spring blooms	Start date, peak and end of earlier blooms when the temperature is higher ($\Delta T = 6°C$), the duration does not vary.	Three times a week	[LEW 10]
Concentration of chlorophyll *a*, total biomass of phytoplankton and abundance of diatoms and dinoflagellates	Field observations–CPR and satellite	North Atlantic Ocean	Start date of spring bloom	Interannual variation in start date of spring bloom and negative correlation with NAO index and positive with SST.	Satellite data collected over eight days	[ZHA 13]
Concentration of chlorophyll *a*	Satellite	Arctic Ocean and adjacent boreal seas	Date of annual peak in chlorophyll *a*	In 11% of the Arctic Ocean, the peak in phytoplankton biomass is earlier, by approximately 50 days in some cases. In the North Atlantic, the peak appears to be later.	Daily data collected over five days	[KAH 11]

Table 2.4. *Examples of studies on the phenological variations in phytoplankton. The results were obtained using data from observations on land, satellites and mesocosm experiments*

2.3.5. *Size spectrum*

Over the past few years, several authors have been interested in the effect of climate change and increase in sea temperature on the structure of phytoplankton communities by size class, favoring smaller species [LEW 10, SOM 08]. Cell size is directly linked to processes such as nutrient absorption and the cell sedimentation rate [LIT 08]. Smaller sized species have a greater surface/volume ratio, which leads to lower sedimentation rates and more efficient nutrient absorption. This is advantageous in a scenario where the increase in temperature increases stratification of the water column and depletion of nutrients in the euphotic zone. These conditions favor an abundance of small diatoms such as *Cyclotella* and *Cymbella* and chlorophyceae such as *Gloeotila* and *Botryococcu*, to the detriment of larger diatoms (e.g. *Fragilaria*, *Synedra*) [WIN 08].

Similarly, the size structure of phytoplankton in some estuaries has been modified, favoring smaller diatoms, e.g. Thalassiosira minima and Cyclotella [GUI 10].

Recently, mesocosm studies have shown a relationship between an increase in temperature and an increase in picophytoplankton biomass, small dinoflagellates and decrease in the mean size of phytoplankton cells [LEW 10, SOM 11]. These results suggest a greater effect on the grazing of larger cells by zooplankton [LEW 10].

2.4. Observation systems for the identification of phytoplankton

2.4.1. *Detecting changes in time series*

Observation systems are essential to understand the impacts of global changes on marine ecosystems (Table 2.5). Time series from these observation systems are used to identify temporal trends and relationships between biological variables and environmental factors. Finally, these studies highlight the wide array of data analysis methods. The development and application of new data analysis methods allow us to better understand the set of questions to which data from observation systems can answer, while respecting the specificities of these data.

Identifier	Location	Description	Temporal coverage	Website
Narragansett Bay, station II	East coast of North America (41.56 N, 71.38 W)	Physicochemical (temperature, salinity, nutrients) and biological parameters (Chl *a*, phytoplankton abundance and composition, zooplankton)	1959– weekly	www.gso.uri.edu/ phytoplankton/
BATS (*Bermuda Atlantic Time-Series Study*)	Northwest Atlantic (31.66 N, 64.16 W)	Physicochemical (temperature, salinity, dissolved oxygen, nutrients) and biological parameters (primary production, abundance of bacteria, zooplankton biomass)	1988–	http://bats.bios.edu/
CPR (*Continuous Plankton Recorder*)	North Atlantic Ocean	Phytoplankton and zooplankton data	1931– monthly	www.sahfos.ac.uk/
DYFAMED (*dynamics of flow of matter into the Mediterranean*)	Central zone of Ligurian sea (43.41 N, 7.86 E)	Physicochemical (temperature, salinity, dissolved oxygen, nutrients) biological (primary production, zooplankton biomass) and meteorological parameters	1991– monthly, but also over a limited period	www.obs-vlfr.fr/sodyf/
ESTOC (*European Station for Time-series in the Ocean, Canary Islands*)	North of Canary Islands (29.167 N, 15.50 W)	Physicochemical (temperature, salinity, oxygen, pH, nutrients) and biological parameters (Chl *a*, zooplankton, ichthyoplankton)	1994–2004	www.eurosites.info/ estoc.php

Identifier	Location	Description	Temporal coverage	Website
HELCOM (*Baltic Monitoring Program of Helsinki Commission*) – coordinator of monitoring programs in the Baltic Sea	Baltic Sea	Physicochemical (temperature, salinity, pH, nutrients), and biological parameters (Chl *a*, primary production, abundance of phytoplankton and mesozooplankton)	1979– varied according to sampling site	http://helcom.fi/action-areas/monitoring-and-assessment
Helgoland Roads	North Sea (54.18 N, 7.9 E)	Physicochemical (temperature, salinity, nutrients), and biological parameters (abundance of phytoplankton and zooplankton)	Working days	www.awi.de/en/research/research_divisions/biosciences/shelf_sea_ecology/long_term_studies/helgoland_roads_long_term_data_series/
HOT (*Hawaii Ocean Time series station*)	Subtropical North Pacific (22.75 N, 158 W)	A number of physicochemical and biological parameters including Chl *a*, *b*, *c*, pigments, bacteria, nano-microphytoplankton, mesozooplankton	1987– weekly, monthly	http://hahana.soest.hawaii.edu/hot/hot_jgofs.html
REPHY	Coast of English Channel, Atlantic, Mediterranean, French Antilles, Guyana and La Réunion	Physicochemical (temperature, salinity, turbidity, dissolved oxygen, nutrients) together with regional networks and biological parameters (Chl *a*, abundance of phytoplankton and composition)	1987– weekly, monthly	http://envlit.ifremer.fr/surveillance/phytoplancton_phycotoxines/presentation
Si.Di.Mar (*Sistema Difesa del Mare*)	Italian coast (Mediterranean Sea)	Physicochemical (temperature, salinity, pH, nutrients) and biological parameters including Chl *a*, phytoplankton and mesozooplankton	2001–2007	www.sidimar.tutelamare.it/
SOMLIT (*service of observation in coastal environment*)	Coast of English Channel, Atlantic and Mediterranean	Physicochemical (temperature, salinity, pH, nutrients) and biological parameters including Chl *a*	Varies according to sampling sites	http://somlit.epoc.u-bordeaux1.fr/fr/
WGPME (*Working Group on Phytoplankton and Microbial Ecology*)	North Atlantic Ocean	Physicochemical and biological parameters (phytoplankton, bacteria)	–	www.wgpme.net/

Table 2.5. *Main observation networks of phytoplankton and the network characteristics*

By definition, a time series is a sequence of chronologically ordered observations where the number N of observations defines the length of the series. Many analysis techniques are available to study this. For example, segmentation methods allow a time series to be divided into "contiguous blocks of data" [KEH 04]. When analyzing the series, the issue is to not only determine the trends, but also to detect abrupt changes. Currently, several methods to detect points of change are available[10].

In the marine environment, time series from observation and surveillance systems are often subject to irregular sampling frequencies. This is often associated with difficult navigation conditions, especially during the winter. While the analysis of time series often requires regularly obtained data, an irregular frequency often provides a time series with missing data. In these cases, there are models that can "fill in" these data and imputation techniques are often used (e.g. cubic spline interpolation [DAV 12]).

The non-stationary character of time series, which can often result in a variable mean, the presence of an irregular frequency, missing data, outliers data and seasonal components, requires the use of adapted methodologies. Some methods are based on descriptive approaches such as the moving average, and others are not specific to time series, e.g. regression (Table 2.6). A simple method is the use of cumulative sums [IBA 93]. Methods based on smoothing have been widely used to determine and describe the long-term trends in phytoplankton groups as well as environmental variables [KLA 11, SUI 07, WAS 03]. These smoothing methods have been considered as flexible tools since no previous information of the shape or trend is required, but they do not provide an equation for the evolution. Moreover, to study long-term trends, some methods isolate a seasonal component, which is fixed for all years. However, properties in these cyclic variations (e.g. bloom duration, amplitude) can change from one year to another. If this is the case, this aspect must be taken into account to allow the seasonal component to evolve over time. Recently, Winder and Cloern [WIN 10] used spectral analyses to quantify the temporal evolution of chlorophyll *a*, and detect periodic fluctuations and phenological changes. This type of approach faces the problem of missing data, and imputation methods should be applied to rectify the time series. In this context, dynamic linear models allow the long-term study of the evolution of biological and

10 An exhaustive list is available online at: www.beringclimate.noaa.gov/regimes/Regime_shift_methods_list.html.

environmental parameters. This tool has been previously used to analyze time series in marine ecology, in particular in the analysis of population dynamics or even changes in phytoplankton phenology [COT 98, SOU 97, WIN 04b].

Studying the responses of biological communities involves the simultaneous assessment of several variables. For example, within a community characterized by *S* species, it is possible to study the responses of a set of populations to determine whether all species have similar tendencies or not. In this respect, the most widely used methods in ecology are multivariate descriptive analyses. Recently, principal component analyses (PCA) have allowed the identification of patterns in variability between several biological time series. Schlüter *et al.* [SCH 08] studied the variations associated with forty time series in Heligoland Bay and Bay using a set of chemical, climatic and biological variables to identify regime shift. These authors concluded that an abrupt change between 1987 and 1988 gave rise to a new regime that persisted for more than a decade. Other methods including hierarchical cluster analyses have also been used to study discontinuities in time series.

Studying regime shifts often involve the analysis of time series. Despite there being discrepancies in the definition of regime shift, these changes are often linked to the sudden or abrupt modifications in physical and biological components of the system at several trophic levels [LEE 06]. PCA is the favored method for the analysis of these multidimensional changes as it allows the number if dimensions to be reduced (analysis space). Additional analyses are often used to assess the statistically significant character of the changes. For example, the sequential algorithm proposed by Rodionov [ROD 04] can also detect regime shifts [NIC 11]. Other methods such as multi-table analyses allow the study of the data cube, i.e. three-dimensional data: variables measurements, space and time. This methodology was used to study the spatiotemporal dynamics of phytoplankton populations along the French coast [DAV 12, GAI 02].

2.4.2. *Taxonomic aggregation levels*

The "species" level is often used to study the responses of biological systems when faced with environmental disruptions. Organisms belonging to the same species share similar physiological and morphological characteristics and consequently, are expected to respond in a similar manner when faced with variations in their environment. Other studies place species in *ad hoc* groups to study these variations. It is common to use diatoms and dinoflagellates to describe changes in phytoplankton [KLA 11, ZHA 13]. Species belonging to each group also share common morphological and physiological characteristics (e.g. requirement for silica in diatoms). Each group is supposed to respond in the same way when faced with a change. Moreover, organisms can be arranged into functional groups. Regrouping organisms not by their taxonomic classification, but by their functions in the ecosystem form the basis for the functional approach mentioned previously.

Method	Applications	Reference
Principal component analysis (PCA)	This method is frequently used to identify patterns of variability for a set of TS. Methods to detect abrupt changes are often applied. PCA has been used to detect regime shifts in the North Pacific Ocean using 100 climate and biological series and in the North Sea.	[BEA 04, HAR 00]
Spectral analysis	This analysis identifies periodicities and variations in TS. This approach has been used to analyze TS.	[FER 13, WIN 10]
Sequential segmentation algorithms	Determining points of change in TS and identifying regime shifts. In particular, the algorithm developed by Rodionov [RAD 04] was used to detect abrupt changes in zooplankton abundance and a regime shift in the Bay of Quinte (Canada).	[NIC 11, VAN 11]
Chronological classification	The method identifies discontinuity in TS by considering the time sequence of measurements, Legendre *et al.* [LEG 85]. This method has been applied to TS in the North Sea and Wadden Sea to identify regime shift.	[WEI 13]
Dynamic linear models (DLM)	These models specific for the analyses of time series have several advantages: the dynamic aspect accounts for the non-stationnarity; the components of the model (e.g. seasonality) evolve over time; the model generates missing and unique data and accounts for exogenous information (variation in measurement methods). DLM have been used to study the temporal dynamics of phytoplankton populations.	[COT 98, SOU 97]
Cumulative sums	This simple and descriptive method is applied to detect trends in the TS mean. Thus, several studies reveal the main variations nits (e.g. phytoplankton) and identify transition phases.	[BRI 10, EDW 06, LEF 11]
Smoothing techniques (LOESS, LOWESS, GAM)	Studying the trends in parameters such as biomass, abundance and phenological parameters such as date of phytoplankton blooms.	[KLA 11, WAL 13, WAS 03]

Table 2.6. *Examples of analyses applied to time series (TS) in ecology*

In some biological groups, identification at the species level is incredibly complex, and often requires the use of techniques such as genetics or scanning electron microscopy, which are currently tricky to integrate in the context of observation networks. Organization into higher taxonomic levels than the species level is one alternative, measuring other than at the species level and reducing the noise linked to an erroneous taxonomic identification. This approach is justified when the use of these levels does not result in a significant loss of information. In this respect, some studies highlight that disturbances in biological communities can also be detected with a low taxonomic resolution [FER 95, WAR 88]. With regard to phytoplankton, some studies show that little information is lost when using the genus level [CAR 10, COT 98].

2.5. Conclusion

Presenting environmental pressures and their impacts on phytoplankton reflect the range of studies undertaken on this topic, but also the complexity of interactions. In fact, phytoplankton are influenced by a set of environmental variables that determine the lifecycle, their spatiotemporal dynamics and at the level of the community, the structure and composition. In particular, in coastal zones, these pressures act synergistically and the consequences on phytoplankton are difficult to separate.

An important consequence is that the study of the specific diversity of phytoplankton involves characterizing the spatial, temporal and biological scales (e.g. species, functional group). Considering these scales when studying long-term evolutions allows the application of different concepts in ecology: diversity (species richness, distribution of abundance, biological and functional traits), ecological niche of taxa (i.e. the environmental space that they occupy) or even phenology.

In this context, observation systems are essential to analyze and understand the impacts of global changes on marine ecosystems, and in particular, on phytoplankton. Time series from these observation systems are used to identify temporal trends and relationships between the biological variables and environmental forcing using various types of data analysis methods. However, new additional approaches to plankton diversity are required to better understand the structure and functional characteristics of these communities.

2.6. Bibliography

[ALE 10] ALEKSANDRA L., SOMMER U., "Climate change and the spring bloom: a mesocosm study on the influence of light and temperature on phytoplankton and mesozooplankton", *Marine Ecology Progress Series*, vol. 405, pp. 101–111, 2010.

[AND 02] ANDERSON D., GLIBERT P., BURKHOLDER J., "Harmful algal blooms and eutrophication: nutrient sources, composition, and consequences", *Estuaries*, vol. 25, pp. 704–726, 2002.

[ANN 02] ANNEVILLE O., GINOT V., DRUART J.-C. *et al.*, "Long-term study (1974–1998) of seasonal changes in the phytoplankton in Lake Geneva: a multi-table approach", *Journal of Plankton Research*, vol. 24, pp. 993–1008, 2002.

[BEA 04] BEAUGRAND G., "The North Sea regime shift: evidence, causes, mechanisms and consequences", *Progress in Oceanography*, vol. 60, pp. 245–262, 2004.

[BEA 09] BEAUGRAND G., "Decadal changes in climate and ecosystems in the North Atlantic Ocean and adjacent seas", *Deep Sea Research*, vol. 2, pp. 656–673, 2009.

[BEL 01] BELIAEFF B., GROS P., BELIN C. *et al.*, "'Phytoplankton events' in French coastal waters during 1987–1997", *Oceanologica Acta*, vol. 24, pp. 425–433, 2001.

[BOE 10] BOEUF G., BÉNÉ C., CARVAHLO G. *et al.*, Biodiversité en environnement marin, Synthèse et recommandations en sciences environnementale et sociale, Report IFREMER, p. 139, 2010.

[BOP 01] BOPP L., MONFRAY P., AUMONT O. *et al.*, "Potential impact of climate change on marine export production", *Global Biogeochemical Cycles*, vol. 15, pp. 81–99, 2001.

[BOP 05] BOPP L., AUMONT O., CADULE P. *et al.*, "Response of diatoms distribution to global warming and potential implications: a global model study", *Geophysical Research Letters*, vol. 32, pp. 1–4, 2005.

[BOR 09] BORKMAN D.G., SMAYDA T., "Multidecadal (1959–1997) changes in Skeletonema abundance and seasonal bloom patterns in Narragansett Bay, Rhode Island, USA", *Journal of Sea Research*, vol. 61, pp. 84–94, 2009.

[BOY 10] BOYCE D.G., LEWIS M.R., WORM B., "Global phytoplankton decline over the past century", *Nature*, vol. 466, pp. 591–596, 2010.

[BRE 06] BRETON E., ROUSSEAU V., PARENT J.-Y. *et al.*, "Hydroclimatic modulation of diatom/*Phaeocystis* blooms in nutrient-enriched Belgian coastal waters (North Sea)", *Limnology and Oceanography*, vol. 51, pp. 1401–1409, 2006.

[BRI 10] BRICEÑO H., BOYER J., "Climatic controls on phytoplankton biomass in a sub-tropical estuary, Florida Bay, USA", *Estuaries and Coasts*, vol. 33, pp. 541–553, 2010.

[BRO 07] BROENNIMANN O., TREIER U.A., MÜLLER-SCHARER H. *et al.*, "Evidence of climatic niche shift during biological invasion", *Ecology Letters*, vol. 10, pp. 701–709, 2007.

[CAR 09] CARDINALE B.J., BENNETT D.M., NELSON C.E. *et al.*, "Does productivity drive diversity or vice versa? A test of the multivariate productivity: diversity hypothesis in streams", *Ecological Society of America*, vol. 90, pp. 1227–1241, 2009.

[CAR 10] CARNEIRO F.M., BINI L.M., RODRIGUES L.C., "Influence of taxonomic and numerical resolution on the analysis of temporal changes in phytoplankton communities", *Ecological Indicators*, vol. 10, pp. 249–255, 2010.

[CHI 12] CHIBA S., BATTEN S., SASAOKA K. *et al.*, "Influence of the Pacific Decadal Oscillation on phytoplankton phenology and community structure in the western North Pacific", *Geophysical Research Letters*, vol. 39, pp. 1–6, 2012.

[CLA 98] CLARKE K.R., WARWICK R.M., "A taxonomic distinctness index and its statistical properties", *Journal of Applied Ecology*, vol. 35, pp. 523–531, 1998.

[CLA 01] CLARKE K.R., WARWICK R.M., "A further biodiversity index applicable to species lists: variation in taxonomic distinctness", *Marine Ecology Progress Series*, vol. 216, pp. 265–278, 2001.

[CLE 06] CLELAND E.E., CHIARIELLO N.R., LOARIE S.R. *et al.*, "Diverse responses of phenology to global changes in a grassland ecosystem", *Proceedings of the National Academy of Sciences*, vol. 103, pp. 13740–13744, 2006.

[CLE 07] CLELAND E.E., CHUINE I., MENZEL A. *et al.*, "Shifting plant phenology in response to global change", *Trends in Ecology Evolution*, vol. 22, pp. 357–365, 2007.

[CLO 96] CLOERN J., "Phytoplankton bloom dynamics in coastal ecosystems: a review with some general lessons from sustained investigation of San Francisco Bay", California, *Reviews of Geophysics*, vol. 34, pp. 127–168, 1996.

[CLO 01] CLOERN J., "Our evolving conceptual model of the coastal eutrophication problem", *Marine Ecology Progress Series*, vol. 210, pp. 223–253, 2001.

[CLO 05] CLOERN J.E., DUFFORD R., "Phytoplankton community ecology: principles applied in San Francisco Bay", *Marine Ecology Progress Series*, vol. 285, pp. 11–28, 2005.

[CLO 08] CLOERN J.E., JASSBY A.D., "Complex seasonal patterns of primary producers at the land: sea interface", *Ecology Letters*, vol. 11, pp. 1294–1303, 2008.

[COL 09] COLWELL R.K., "Biodiversity: concepts, patterns and measurement", in LEVIN S.A., CARPENTER S.R., GODFRAY H.C.J. *et al.* (eds), *The Princeton Guide to Ecology*, Princeton University Press, 2009.

[COT 98] COTTINGHAM K.L., CARPENTER S.R., "Population, community and ecosystem variates as ecological indicators: phytoplankton responses to whole-lake enrichment", *Ecological Applications*, vol. 8, pp. 508–530, 1998.

[CUL 02] CULLEN J.J., FRANKS P.J., KARL D.M. *et al.*, "Physical influences on marine ecosystem dynamics", in ROBINSON A.R., MCCARTHY J.J., ROTHSCHILD B.J. (eds), *The Sea*, John Wiley & Sons, New York, 2002.

[CUR 08] CURY P.M., SHIN Y.-J., PLANQUE B., *et al.*, "Ecosystem oceanography for global change in fisheries", *Trends in Ecology and Evolution*, vol. 23, pp. 338–346, 2008.

[CUS 59] CUSHING D.H., "The seasonal variation in oceanic production as a problem in population dynamics", *Journal du Conseil*, vol. 24, pp. 455–464, 1959.

[DAL 01] DALE B., "The sedimentary record of dinoflagellate cysts: looking back into the future of phytoplankton blooms", *Scientia Marina*, vol. 65, pp. 257–272, 2001.

[DAV 12] DAVID V., RYCKAERT M., KARPYTCHEV M. *et al.*, "Spatial and long-term changes in the functional and structural phytoplankton communities along the French Atlantic coast", *Estuarine, Coastal and Shelf Science*, vol. 108, pp. 37–51, 2012.

[DÍA 01] DÍAZ S., CABIDO M., "Vive la différence: plant functional diversity matters to ecosystem processes", *Trends in Ecology and Evolution*, vol. 16, pp. 646–655, 2001.

[DIA 08] DIAZ R.J., ROSENBERG R., "Spreading dead zones and consequences for marine ecosystems", *Science*, vol. 321, pp. 926–929, 2008.

[DIJ 06] DIJKSTRA H.A., RAA L.T., SCHMEITS M. *et al.*, "On the physics of the Atlantic Multidecadal Oscillation", *Ocean Dynamics*, vol. 56, pp. 36–50, 2006.

[DON 03] DONE T., WHETTON P., JONES R. *et al.*, Global climate change and coral bleaching on the Great Barrier Reef, Final Report to the State of Queensland Greenhouse Taskforce, Department of Natural Resources & Mines, Townsville, Australia, p. 49, 2003.

[DON 06] DONEY S.C., "Oceanography: plankton in a warmer world", *Nature*, vol. 444, pp. 695–696, 2006.

[DUA 90] DUARTE C., AUGUSTI S., CANFIELD JR D.E., "Size plasticity of freshwater phytoplankton: implications for community structure", *Limnology and Oceanography*, vol. 35, pp. 1846–1851, 1990.

[DUF 97] DUFRÊNE M., LEGENDRE P., "Species assemblages and indicator species: the need for a flexible asymmetrical approach", *Ecological Monographs*, vol. 67, pp. 345–366, 1997.

[DUR 03] DURANT J.M., ANKER-NILSSEN T., STENSETH N.C., "Trophic interactions under climate fluctuations: the Atlantic puffin as an example", *Proceedings of the Royal Society of London. Series B: Biological Sciences*, vol. 270, pp. 1461–1466, 2003.

[EDW 01] EDWARDS M., JOHN A., JOHNS D. *et al.*, "Case history and persistence of the non-indigenous diatom *Coscinodiscus wailesii* in the north-east Atlantic", *Journal of the Marine Biological Association of the United Kingdom*, vol. 81, pp. 207–211, 2001.

[EDW 04] EDWARDS M., RICHARDSON A.J., "Impact of climate change on marine pelagic phenology and trophic mismatch", *Nature*, vol. 430, pp. 881–884, 2004.

[EDW 06] EDWARDS M., LETERME S.C., SVENDSEN E. *et al.*, "Regional climate change and harmful algal blooms in the northeast Atlantic", *Limnology and Oceanography*, vol. 51, pp. 820–829, 2006.

[EDW 13] EDWARDS M., BEAUGRAND G., HELAOUËT P. *et al.*, "Marine ecosystem response to the Atlantic Multidecadal Oscillations", *PLoS ONE*, vol. 8, p. e57212, 2013.

[EIL 95] EILERTSEN H.C., SANDBERG S., TOLLEFSEN H., "Photoperiodic control of diatom spore growth: a theory to explain the onset of phytoplankton blooms", *Marine Ecology Progress Series*, vol. 116, pp. 303–307, 1995.

[ENF 01] ENFIELD D.B., MESTAS-NUÑEZ A.M., TRIMBLE P.J., "The Atlantic Multidecadal Oscillation and its relation to rainfall and river flows in the continental U.S", *Geophysical Research Letters*, vol. 28, pp. 2077–2080, 2001.

[FAL 07] FALKOWSKI P.G., RAVEN J.A., *Aquatic Photosynthesis*, 2nd ed., Princeton University Press, 2007.

[FER 95] FERRARO S.P., COLE F.A., "Taxonomic level sufficient for assessing pollution impacts on the southern California Bight macrobenthos-revisited", *Environmental Toxicology and Chemistry*, vol. 14, pp. 1031–1040, 1995.

[FER 13] FERNÁNDEZ DE PUELLES M.L.F. DE MOLINERO J.C., "Increasing zooplankton variance in the late 1990s unveils hydroclimate modifications in the Balearic Sea, Western Mediterranean", *Marine Environmental Research*, vol. 86, pp. 76–80, 2013.

[FIE 98] FIELD C.B., BEHRENFELD M.J., RANDERSON J.T. *et al.*, "Primary production of the biosphere: integrating terrestrial and oceanic components", *Science*, vol. 281, pp. 237–240, 1998.

[FIN 09] FINKEL Z.V., BEARDALL J., FLYNN K.J. *et al.*, "Phytoplankton in a changing world: cell size and elemental stoichiometry", *Journal of Plankton Research*, vol. 1, p. 19, 2009.

[FIS 43] FISHER R.A., CORBET A.S., WILLIAMS C.B., "The relation between the number of species and the number of individuals in a random sample of an animal population", *The Journal of Animal Ecology*, vol. 12, pp. 42–58, 1943.

[FIS 73] FISHER N.S., GRAHAM L.B., CARPENTER E.J. *et al.*, "Geographic differences in phytoplankton sensitivity to PCBs", *Nature*, vol. 241, pp. 548–549, 1973.

[FRE 12] FREUND J., GRÜNER N., BRÜSE S. *et al.*, "Changes in the phytoplankton community at Helgoland, North Sea: lessons from single spot time series analyses", *Marine Biology*, vol. 159, pp. 2561–2571, 2012.

[GAI 02] GAILHARD I., GROS P., DURBEC J. *et al.*, "Variability patterns of microphytoplankton communities along the French coasts", *Marine Ecology Progress Series*, vol. 242, pp. 39–50, 2002.

[GEB 09] GEBÜHR C., WILTSHIRE K.H., ABERLE N. *et al.*, "Influence of nutrients, temperature, light and salinity on the occurrence of Paralia sulcata at Helgoland Roads, North Sea", *Aquatic Biology*, vol. 7, pp. 185–197, 2009.

[GOF 02] GOFFART A., HECQ J.-H., LEGENDRE L., "Changes in the development of the winter-spring phytoplankton bloom in the Bay of Calvi (NW Mediterranean) over the last two decades: a response to changing climate?", *Marine Ecology Progress Series*, vol. 236, pp. 45–60, 2002.

[GOH 09] GOHIN F., OGER-JEANNERET H., Directive Cadre sur l'Eau. Réseaux de surveillance dans les masses d'eau côtières des littoraux Seine-Normandie et Loire-Bretagne: optimisation de la stratégie d'échantillonnage par utilisation des images satellites, Report IFREMER, p. 66, 2009.

[GOH 11] GOHIN F., "Annual cycles of chlorophyll-a, non-algal suspended particulate matter, and turbidity observed from space and in-situ in coastal waters", *Ocean Science*, vol. 7, pp. 705–732, 2011.

[GOL 01] GOLDENBERG S.B., LANDSEA C.W., MESTAS-NUÑEZ A.M. *et al.*, "The recent increase in Atlantic hurricane activity: causes and implications", *Science*, vol. 293, pp. 474–479, 2001.

[GOT 96] GOTELLI N., GRAVES G., *Null Models in Ecology*, Smithsonian Institution Press, Washington D.C., 1996.

[GRE 99] GREENSTREET S.P., SPENCE F.E., MCMILLAN J.A., "Fishing effects in northeast Atlantic shelf seas: patterns in fishing effort, diversity and community structure. V. Changes in structure of the North Sea groundfish species assemblage between 1925 and 1996", *Fisheries Research*, vol. 40, pp. 153–183, 1999.

[GRÜ 11] GRÜNER N., GEBÜHR C., BOERSMA M. *et al.*, "Reconstructing the realized niche of phytoplankton species from environmental data: fitness versus abundance approach", *Limnology Oceanography Methods*, vol. 9, pp. 432–442, 2011.

[GUI 10] GUINDER V., POPOVICH C., MOLINERO J. *et al.*, "Long-term changes in phytoplankton phenology and community structure in the Bahía Blanca Estuary, Argentina", *Marine Biology*, vol. 157, pp. 2703–2716, 2010.

[HAL 10] HALLEGRAEFF G.M., "Ocean climate change, phytoplankton community responses and harmful algal blooms: a formidable predictive challenge", *Journal of Phycology*, vol. 46, pp. 220–235, 2010.

[HAR 00] HARE S.R., MANTUA N.J., "Empirical evidence for North Pacific regime shifts in 1977 and 1989", *Progress in Oceanography*, vol. 47, pp. 103–145, 2000.

[HAS 07] HASHIOKA T., YAMANAKA Y., "Ecosystem change in the western North Pacific associated with global warming using 3D-NEMURO", *Ecological Modeling*, vol. 202, pp. 95–104, 2007.

[HAY 05] HAYS G.C., RICHARDSON A.J., ROBINSON C., "Climate change and marine plankton", *Trends in Ecology & Evolution*, vol. 20, pp. 337–344, 2005.

[HIN 12] HINDER S.L., HAYS G.C., EDWARDS M. *et al.*, "Changes in marine dinoflagellate and diatom abundance under climate change", *Nature Climate Change*, vol. 2, pp. 271–275, 2012.

[HIR 08] HIRZEL A.H., LE LAY G., "Habitat suitability modeling and niche theory", *Journal of Applied Ecology*, vol. 45, pp. 1372–1381, 2008.

[HUR 71] HURLBERT S.H., "The non-concept of species diversity: a critique and alternative parameters", *Ecology*, vol. 52, pp. 577–586, 1971.

[HUT 57] HUTCHINSON G.E., "Concluding remarks", *Cold Spring Harbor Symposia on Quantitative Biology*, vol. 22, pp. 415–427, 1957.

[IGN 94] IGNATIADES L., "Species dominance and niche breadth in bloom and non-bloom phytoplankton populations", *Oceanologica Acta*, vol. 17, pp. 89–96, 1994.

[IPC 07] INTERGOVERNMENTAL PANEL ON CLIMATE CHANGE (IPCC), "Impacts, adaptation and vulnerability", in PARRY M., CANZIANI O., PALUTIKOF J. *et al.* (eds), *Contribution of Working Group II to the Fourth Assessment Report of the Intergovernmental Panel on Climate Change*, Cambridge University Press, Cambridge, UK, 2007.

[IPC 13] INTERGOVERNMENTAL PANEL ON CLIMATE CHANGE (IPCC), "The physical science basis", in STOCKER T.F., QIN D., PLATTNER G.-K. *et al.* (eds), *Contribution of Working Group I to the Fifth Assessment Report of the Intergovernmental Panel on Climate Change*, Cambridge University Press, Cambridge, UK and New York, 2013.

[IRI 00] IRIGOIEN X., HARRIS R.P., HEAD R.N. *et al.*, "North Atlantic Oscillation and spring bloom phytoplankton composition in the English Channel", *Journal of Plankton Research*, vol. 22, pp. 2367–2371, 2000.

[IRI 02] IRIGOIEN X., HARRIS R.P., VERHEYE H.M. *et al.*, "Copepod hatching success in marine ecosystems with high diatom concentrations", *Nature*, vol. 419, pp. 387–389, 2002.

[IRW 12] IRWIN A.J., NELLES A.M., FINKEL Z.V., "Phytoplankton niches estimated from field data", *Limnology and Oceanography*, vol. 57, pp. 787–797, 2012.

[ITO 10] ITO S., ROSE K., MILLER A. *et al.*, "Ocean ecosystem responses to future global change scenarios: a way forward", in BARANGE M., WERNER R., FIELD J. *et al.* (eds), *Marine Ecosystems and Global Change*, Oxford University Press, New York, pp. 287–322, 2010.

[JET 12] JETZ W., MCPHERSON J.M., GURALNICK R.P., "Integrating biodiversity distribution knowledge: toward a global map of life", *Trends in Ecology and Evolution*, vol. 27, pp. 151–159, 2012.

[JON 04] JONES G.P., MCCORMICK M.I., SRINIVASAN M. *et al.*, "Coral decline threatens fish biodiversity in marine reserves", *Proceedings of the National Academy of Sciences of the United States of America*, vol. 101, pp. 8251–8253, 2004.

[KAH 11] KAHRU M., BROTAS V., MANZANO-SARABIA M. *et al.*, "Are phytoplankton blooms occurring earlier in the Arctic?", *Global Change Biology*, vol. 17, pp. 1733–1739, 2011.

[KEH 04] KEHAGIAS A., "A hidden Markov model segmentation procedure for hydrological and environmental time series", *Stochastic Environmental Research and Risk Assessment*, vol. 18, pp. 117–130, 2004.

[KIM 73] KIMURA T., MIZOKAMI A., HASHIMOTO T., "The red tide that caused severe damage to the fishery resources in Hiroshima Bay: outline of its occurrence and environmental conditions", *Bulletin of Plankton Society of Japan*, vol. 19, pp. 82–96, 1973.

[KLA 11] KLAIS R., TAMMINEN T., KREMP A. *et al.*, "Decadal-scale changes of dinoflagellates and diatoms in the anomalous Baltic Sea spring bloom", *PLoS ONE*, vol. 6, p. e21567, 2011.

[KLA 13] KLAIS R., TAMMINEN T., KREMP A. *et al.*, "Spring phytoplankton communities shaped by interannual weather variability and dispersal limitation: mechanisms of climate change effects on key coastal primary producer", *Limnology and Oceanography*, vol. 58, pp. 753–762, 2013.

[KRE 12] KREMP A., GODHE A., EGARDT J. *et al.*, "Intra-specific variability in the response of bloom-forming marine microalgae to changed climate conditions", *Ecology and Evolution*, vol. 2, no. 1 pp. 195–1207, 2012.

[KRU 10] KRUK C., HUSZAR V.L.M., PEETERS E.T.H.M. *et al.*, "A morphological classification capturing functional variation in phytoplankton", *Freshwater Biology*, vol. 55, pp. 614–627, 2010.

[LAS 91] LASSUS P., MAGGI P., PRONIEWSKI F. *et al.*, Distribution verticale de *Dinophysis spp* au cours d'un cycle de 48 heures dans le port d'Antifer (Normandie), Report IFREMER, 1991.

[LEE 06] LEES K., PITOIS S., SCOTT C. *et al.*, "Characterizing regime shifts in the marine environment", *Fish and Fisheries*, vol. 7, pp. 104–127, 2006.

[LEF 11] LEFEBVRE A., GUISELIN N., BARBET F. *et al.*, "Long-term hydrological and phytoplankton monitoring (1992–2007) of three potentially eutrophic systems in the eastern English Channel and the Southern Bight of the North Sea", *ICES Journal of Marine Science*, vol. 68, pp. 2029–2043, 2011.

[LEF 12] LEFEBVRE A., BELIN C., CAMUS P. *et al.*, La Directive Cadre Statégie pour le Milieu Marin (DCSMM: 2008/56/EU. Caractéristiques du Bon État Écologique (BEE) pour le descripteur 5 lié à l'eutrophisation, Report IFREMER, 2012.

[LEG 85] LEGENDRE P., DALLOT S., LEGENDRE L., "Succession of species within a community: chronological clustering, with applications to marine and freshwater zooplankton", *American Naturalist*, vol. 125, pp. 257–288, 1985.

[LEM 03] LEMOINE N., BÖHNING-GAESE K., "Potential impact of global climate change on species richness of long-distance migrants", *Conservation Biology*, vol. 17, pp. 577–586, 2003.

[LEO 06] LEONARD D., CLARKE K.R., SOMERFIELD P.J. *et al.*, "The application of an indicator based on taxonomic distinctness for UK marine biodiversity assessments", *Journal of Environmental Management*, vol. 78, pp. 52–62, 2006.

[LET 05] LETERME S.C., EDWARDS M., SEURONT L. *et al.*, "Decadal basin-scale changes in diatoms, dinoflagellates, and phytoplankton colour across the North Atlantic", *Limnology and Oceanography*, vol. 50, pp. 1244–1253, 2005.

[LIL 02] LILLY E.L., KULIS D.M., GENTIEN P. *et al.*, "Paralytic shell-fish poisoning toxins in France linked to a human-introduced strain of Alexandrium catenella from the western Pacific: evidence from DNA and toxin analysis", *Journal of Plankton Research*, vol. 24, pp. 443–452, 2002.

[LIT 08] LITCHMAN E., KLAUSMEIER C.A., "Trait-based community ecology of phytoplankton", *Annual Review of Ecology, Evolution, and Systematics*, vol. 39, pp. 615–639, 2008.

[LIT 12] LITCHMAN E., EDWARDS K.F., KLAUSMEIER C.A. *et al.*, "Phytoplankton niches, traits and eco-evolutionary responses to global environmental change", *Marine Ecology Progress Series*, vol. 470, pp. 235–248, 2012.

[LOH 12] LOHBECK K.T., RIEBESELL U., REUSCH T.B.H., "Adaptive evolution of a key phytoplankton species to ocean acidification", *Nature Geoscience*, vol. 5, pp. 346–351, 2012.

[MAG 88] MAGURRAN A.E., *Ecological Diversity and its Measurement*, Princeton University Press, Princeton, NJ, p. 179, 1988.

[MAG 10] MAGURRAN A.E., BAILLIE S.R., BUCKLAND S.T. *et al.*, "Long-term datasets in biodiversity research and monitoring: assessing change in ecological communities through time", *Trends in Ecology Evolution*, Special Issue: Long-term ecological research, vol. 25, pp. 574–582, 2010.

[MAR 78] MARGALEF R., "Life-forms of phytoplankton as survival alternatives in an unstable environment", *Oceanologica Acta*, vol. 1, pp. 493–509, 1978.

[MAR 97] MARGALEF R., "Our biosphere", in KINNE O.T. (ed.), *Excellence in Ecology Book 10*, Ecology Institute, Oldendorf and Luhe, 1997.

[MÉN 01] MÉNESGUEN A., AMINOT A., BELIN C. *et al.*, L'eutrophisation des eaux marines et saumâtres en Europe, en particulier en France, Report IFREMER, 2001.

[MEN 03] MENZEL A., "Plant phenological anomalies in Germany and their relation to air temperature and NAO", *Climatic Change*, vol. 57, pp. 243–263, 2003.

[MEN 14] MENESGUEN A., "Eutrophication of the Marine Environment", *in* MONACO A., PROUZET P. (eds), *The Land–Sea Interactions*, ISTE Ltd, London and John Wiley & Sons, New York, 2014.

[MET 96] METAXAS A., SCHEIBLING R.E., "Top-down and bottom-up regulation of phytoplankton assemblages in tide pools", *Marine Ecology Progress Series*, vol. 145, pp. 161–177, 1996.

[MIC 05] MICHELI F., HALPERN B.S., "Low functional redundancy in coastal marine assemblages", *Ecology Letters*, vol. 8, pp. 391–400, 2005.

[MIE 08] MIELEITNER J., BORSUK M., BÜRGI H.-R. *et al.*, "Identifying functional groups of phytoplankton using data from three lakes of different trophic state", *Aquatic Sciences*, vol. 70, pp. 30–46, 2008.

[MIE 13] MIETTINEN A., KOÇ N., HUSUM K., "Appearance of the Pacific diatom *Neodenticula seminae* in the northern Nordic Seas – An indication of changes in Arctic sea ice and ocean circulation", *Marine Micropaleontology*, vol. 99, pp. 2–7, 2013.

[MON 14a] MONACO A., PROUZET P., *Vulnerability of Coastal Ecosystems and Adaptation*, ISTE Ltd., London and John Wiley & Sons, 2014.

[MON 14b] MONACO A., PROUZET P., *Ocean in the Earth System*, ISTE Ltd., London and John Wiley & Sons, 2014.

[MOS 72] MOSSER J.L., FISHER N.S., WURSTER C.F., "Polychlorinated biphenyls and DDT alter species composition in mixed cultures of algae", *Science*, vol. 176, pp. 533–535, 1972.

[NEH 98] NEHRING S., "Non-indigenous phytoplankton species in the North Sea: supposed region of origin and possible transport vector", *Archive of Fishery and Marine Research*, vol. 46, pp. 181–194, 1998.

[NIC 11] NICHOLLS K.H., "Detection of regime shifts in multi-species communities: the Bay of Quinte phytoplankton example", *Methods in Ecology and Evolution*, vol. 2, pp. 416–426, 2011.

[NIX 09] NIXON S.W., FULWEILER R.W., BUCKLEY B.A. *et al.*, "The impact of changing climate on phenology, productivity, and benthic: pelagic coupling in Narragansett Bay", *Estuarine, Coastal and Shelf Science*, vol. 82, pp. 1–18, 2009.

[OSP 08] OSPAR, "Eutrophication status of the OSPAR maritime area", *Second OSPAR Integrated Report*, Technical Report, p. 107, 2008.

[PET 98] PETERSON G., ALLEN C.R., HOLLING C.S., "Ecological resilience, biodiversity, and scale", *Ecosystems*, vol. 1, pp. 6–18, 1998.

[PET 03] PETERSON A.T., "Predicting the geography of species' invasions via ecological niche modeling", *The Quarterly Review of Biology*, vol. 78, pp. 419–433, 2003.

[PET 06] PETCHEY O.L., GASTON K.J., "Functional diversity: back to basics and looking forward", *Ecology Letters*, vol. 9, pp. 741–758, 2006.

[PET 09] PETERSON A.T., "Shifting suitability for malaria vectors across Africa with warming climates", *BMC Infectious Diseases*, vol. 9, pp. 1–6, 2009.

[RAB 09] RABALAIS N.N., TURNER R.E., DÍAZ R.J. *et al.*, "Global change and eutrophication of coastal waters", *ICES Journal of Marine Science*, vol. 66, pp. 1–10, 2009.

[RAB 10] RABALAIS N.N., DÍAZ R.J., LEVIN L.A. *et al.*, "Dynamics and distribution of natural and human-caused hypoxia", *Biogeosciences*, vol. 7, pp. 585–619, 2010.

[RAC 12] RACAULT M.-F., QUÉRÉ C.L., BUITENHUIS E. *et al.*, "Phytoplankton phenology in the global ocean", *Ecological Indicators*, vol. 14, pp. 152–163, 2012.

[REI 07] REID P.C., JOHNS D.G., EDWARDS M. *et al.*, "A biological consequence of reducing Arctic ice cover: arrival of the Pacific diatom *Neodenticula seminae* in the North Atlantic for the first time in 800,000 years", *Global Change Biology*, vol. 13, pp. 1910–1921, 2007.

[REY 88] REYNOLDS C.S., "Functional morphology and the adaptive strategies of freshwater phytoplankton", in SANDGREN C.D. (ed.), *Growth and Reproductive Strategies of Freshwater Phytoplankton*, Cambridge University Press, 1988.

[REY 06] REYNOLDS C.S., *Ecology of Phytoplankton*, Cambridge University Press, London, 2006.

[RIC 04] RICHARDSON A.J., SCHOEMAN D.S., "Climate impact on plankton ecosystems in the Northeast Atlantic", *Science*, vol. 305, pp. 1609–1612, 2004.

[ROD 04] RODIONOV S.N., "A sequential algorithm for testing climate regime shifts", *Geophysical Research Letters*, vol. 31, pp. 1–4, 2004.

[ROM 13] ROMERO E., GARNIER J., LASSALETTA L. *et al.*, "Large-scale patterns of river inputs in south-western Europe: seasonal and interannual variations and potential eutrophication effects at the coastal zone", *Biogeochemistry*, vol. 113, pp. 481–505, 2013.

[ROS 07] ROSS O.N., SHARPLES J., "Phytoplankton motility and the competition for nutrients in the thermocline", *Marine Ecology Progress Series*, vol. 347, pp. 21–38, 2007.

[SCH 08] SCHLÜTER M.H., MERICO A., WILTSHIRE K.H. *et al.*, "A statistical assessment analysis of climate variability and ecosystem response in the German Bight", *Ocean Dynamics*, vol. 58, pp. 169–186, 2008.

[SCH 10] SCHLEUTER D., DAUFRESNE M., MASSOL F. *et al.*, "A user's guide to functional diversity indices", *Ecological Monographs*, vol. 80, pp. 469–484, 2010.

[SIA 07] SIANO R., The phytoplankton of the Campania coast: an ecological and taxonomic study, PhD Thesis, University of Messina, Italy, 2007.

[SIE 02] SIEGEL D.A., DONEY S.C., YODER J.A., "The North Atlantic spring phytoplankton bloom and Sverdrup's critical depth hypothesis", *Science*, vol. 296, pp. 730–733, 2002.

[SIM 09] SIMON N., CRAS A.-L., FOULON E. *et al.*, "Diversity and evolution of marine phytoplankton", *Comptes Rendus Biologies*, vol. 332, pp. 159–170, 2009.

[SOM 08] SOMMER U., LENGFELLNER K., "Climate change and the timing, magnitude, and composition of the phytoplankton spring bloom", *Global Change Biology*, vol. 14, pp. 1199–1208, 2008.

[SOM 11] SOMMER U., LEWANDOWSKA A., "Climate change and the phytoplankton spring bloom: warming and overwintering zooplankton have similar effects on phytoplankton", *Global Change Biology*, vol. 17, pp. 154–162, 2011.

[SOU 91] SOURNIA A., CHRDTIENNOT-DINET M.-J., RICARD M., "Marine phytoplankton: how many species in the world ocean?", *Journal of Plankton Research*, vol. 13, pp. 1093–1099, 1991.

[SOU 97] SOUDANT D., Application de modèles dynamiques bayésiens aux séries temporelles de *Dinophysis* à Antifer (Normandie, France), PhD Thesis, University of Paris 7, p. 123, 1997.

[STA 02] STACHOWICZ J.J., TERWIN J.R., WHITLATCH R.B. *et al.*, "Linking climate change and biological invasions: ocean warming facilitates nonindigenous species invasions", *Proceedings of the National Academy of Sciences*, vol. 99, pp. 15497–15500, 2002.

[SUI 07] SUIKKANEN S., LAAMANEN M., HUTTUNEN M., "Long-term changes in summer phytoplankton communities of the open northern Baltic Sea", *Estuarine, Coastal and Shelf Science*, vol. 71, pp. 580–592, 2007.

[SUT 05] SUTTON R.T., HODSON D.L.R., "Atlantic Ocean forcing of North American and European summer climate", *Science*, vol. 309, pp. 115–118, 2005.

[TET 95] TETT P., BARTON E., "Why are there about 5000 species of phytoplankton in the sea?", *Journal of Plankton Research*, vol. 17, pp. 1693–1704, 1995.

[THA 08] THACKERAY S.J., JONES I.D., MABERLY S.C., "Long-term change in the phenology of spring phytoplankton: species-specific responses to nutrient enrichment and climatic change", *Journal of Ecology*, vol. 96, pp. 523–535, 2008.

[THO 04] THOMAS C.D., CAMERON A., GREEN R.E. *et al.*, "Extinction risk from climate change", *Nature*, vol. 427, pp. 145–148, 2004.

[THO 12] THOMAS M.K., KREMER C.T., KLAUSMEIER C.A. *et al.*, "A global pattern of thermal adaptation in marine phytoplankton", *Science*, vol. 338, pp. 1085–1088, 2012.

[THU 05] THUILLER W., LAVOREL S., ARAUJO M.B. *et al.*, "Climate change threats to plant diversity in Europe", *Proceedings of the National Academy of Sciences of the United States of America*, vol. 102, pp. 8245–8250, 2005.

[TIL 94] TILMAN D., DOWNING J., "Biodiversity and stability in grasslands", *Nature*, vol. 367, pp. 363–365, 1994.

[TIL 01] TILMAN D., "Functional diversity", in LEVIN S.A. (ed.), *Encyclopedia of Biodiversity*, Academic Press, vol. 3, pp. 109–120, 2001.

[TRE 07] TRENBERTH K.E., JONES P.D., AMBENJE P.G. *et al.*, "Observations: surface and atmospheric climate change", in SOLOMON S., QIN D., MANNING M. *et al.* (eds), *Climate Change 2007: The Physical Science Basis. Contribution of Working Group I to the Fourth Assessment Report of the Intergovernmental Panel on Climate Change*, Cambridge University Press, Cambridge, pp. 235–336, 2007.

[VAN 11] VANDROMME P., STEMMANN L., BERLINE L., *et al.*, "Inter-annual fluctuations of zooplankton communities in the Bay of Villefranche-sur-mer from 1995 to 2005 (Northern Ligurian Sea, France)", *Biogeosciences*, vol. 8, pp. 3143–3158, 2011.

[VEH 11] VEHMAA A., LARSSON P., VIDOUDEZ C., *et al.*, "How will increased dinoflagellate:diatom ratios affect copepod egg production? A case study from the Baltic Sea", *Journal of Experimental Marine Biology and Ecology*, vol. 401, pp. 134–140, 2011.

[VIL 10] VILLÉGER S., MIRANDA J.R., HERNÁNDEZ D.F. *et al.*, "Contrasting changes in taxonomic vs. functional diversity of tropical fish communities after habitat degradation", *Ecological Applications*, vol. 20, pp. 1512–1522, 2010.

[WAL 13] WALTERS A.W., SAGRARIO M.D.L.A.G., SCHINDLER D.E., "Species and community-level responses combine to drive phenology of lake phytoplankton", *Ecology*, vol. 94, pp. 2188–2194, 2013.

[WAL 14] WALLACE R.B., BAUMANN H., GREAR J.S. *et al.*, "Coastal ocean acidification: the other eutrophication problem", *Estuarine, Coastal and Shelf Science*, vol. 148, pp. 1–13, 2014.

[WAR 88] WARWICK R., "The level of taxonomic discrimination required to detect pollution effects on marine benthic communities", *Marine Pollution Bulletin*, vol. 19, pp. 259–268, 1988.

[WAR 95] WARWICK R., CLARKE K., "New 'biodiversity' measures reveal a decrease in taxonomic distinctness with increasing stress", *Journal of Applied Ecology*, vol. 129, pp. 301–305, 1995.

[WAR 10] WARD E.J., CHIRAKKAL H., GONZÁLEZ-SUÁREZ M. *et al.*, "Inferring spatial structure from time-series data: using multivariate state-space models to detect metapopulation structure of California sea lions in the Gulf of California, Mexico", *Journal of Applied Ecology*, vol. 47, pp. 47–56, 2010.

[WAR 11] WARREN R., PRICE J., FISCHLIN A. *et al.*, "Increasing impacts of climate change upon ecosystems with increasing global mean temperature rise", *Climatic Change*, vol. 106, pp. 141–177, 2011.

[WAS 98] WASMUND N., NAUSCH G., MATTHÄUS W., "Phytoplankton spring blooms in the southern Baltic Sea, spatio-temporal development and long-term trends", *Journal of Plankton Research*, vol. 20, pp. 1099–1117, 1998.

[WAS 03] WASMUND N., UHLIG S., "Phytoplankton trends in the Baltic Sea", *ICES Journal of Marine Science*, vol. 60, pp. 177–186, 2003.

[WAS 08] WASMUND N., GÖBEL J., BODUNGEN B.V., "100-years-changes in the phytoplankton community of Kiel Bight (Baltic Sea)", *Journal of Marine Systems*, vol. 73 pp. 300–322, 2008.

[WAS 11] WASMUND N., TUIMALA J., SUIKKANEN S. *et al.*, "Long-term trends in phytoplankton composition in the western and central Baltic Sea", *Journal of Marine Systems*, vol. 87, pp. 145–159, 2011.

[WEI 03] WEITHOFF G., "The concepts of 'plant functional types' and 'functional diversity' in lake phytoplankton: a new understanding of phytoplankton ecology?", *Freshwater Biology*, vol. 48, pp. 1669–1675, 2003.

[WEI 13] WEIJERMAN M., LINDEBOOM H., ZUUR A.F., "Regime shifts in marine ecosystems of the North Sea and Wadden Sea", *Marine Ecology Progress Series*, vol. 298, pp. 21–39, 2013.

[WES 97] WEST M., HARRISON J., *Bayesian Forecasting and Dynamic Models*, 2nd ed., Springer Series in Statistics, New York, p. 680, 1997.

[WEY 01] WEYHENMEYER G.A., "Warmer winters: are planktonic algal populations in Sweden's largest lakes affected?", *AMBIO: A Journal of the Human Environment*, vol. 30, pp. 565–571, 2001.

[WHI 72] WHITTAKER R.H., "Evolution and measurement of species diversity", *Taxon*, vol. 21, pp. 213–251, 1972.

[WID 10] WIDDICOMBE C.E., ELOIRE D., HARBOUR D., *et al.*, "Long-term phytoplankton community dynamics in the Western Channel", *Journal of Plankton Research*, vol. 32, pp. 643–655, 2010.

[WIL 08] WILTSHIRE K.H., MALZAHN A.M., WIRTZ K. *et al.*, "Resilience of North Sea phytoplankton spring bloom dynamics: an analysis of long-term data at Helgoland Roads", *Limnology and Oceanography*, vol. 53, pp. 1294–1302, 2008.

[WIN 04a] WINDER M., SCHINDLER D.E., "Climate change uncouples trophic interactions in an aquatic ecosystem", *Ecology*, vol. 85, pp. 2100–2106, 2004.

[WIN 04b] WINDER M., SCHINDLER D.E., "Climatic effects on the phenology of lake processes", *Global Change Biology*, vol. 10, pp. 1844–1856, 2004.

[WIN 08] WINDER M., HUNTER D.A., "Temporal organization of phytoplankton communities linked to physical forcing", *Oecologia*, vol. 156, pp. 179–192, 2008.

[WIN 10] WINDER M., CLOERN J.E., "The annual cycles of phytoplankton biomass", *Philosophical Transactions of the Royal Society B: Biological Sciences*, vol. 365, pp. 3215–3226, 2010.

[WOR 14] WORMS, World Register of Marine Species, 2014. Available at http://www.marinespecies.org.

[ZHA 11] ZHAI L., PLATT T., TANG C. *et al.*, "Phytoplankton phenology on the Scotian Shelf", *ICES Journal of Marine Science*, vol. 68, pp. 781–791, 2011.

[ZHA 13] ZHAI L., PLATT T., TANG C. *et al.*, "The response of phytoplankton to climate variability associated with the North Atlantic Oscillation", *Deep Sea Research Part II: Topical Studies in Oceanography*, vol. 93, pp. 159–168, 2013.

[ZIV 14] ZIVERI P., PASSARO M., INCARBONA A. *et al.*, "Decline in coccolithophore diversity and impact on coccolith morphogenesis along a natural CO2 gradient", *The Biological Bulletin*, vol. 226, pp. 282–290, 2014.

3

Marine Seagrasses (Magnoliophyta) in the Intertropical Zone

3.1. From plant to habitat

3.1.1. *Taxonomy and morphology of marine magnoliophyta*

According to the new classification of living organisms, the marine seagrasses are formed by species of magnoliophytes situated in the group embryophytes from the green line of the chlorobionts. These species were recently placed in the phanerogam class of the angiosperms (also magnoliophytes), a subclass of themonocotyledons. Most of the bibliographic documents refer to this classification and describe the seagrasses as "marine phanerogams" or "marine angiosperms". The term seagrass has been used for a long time by coastal users and inhabitants. English-speaking countries use the term "seagrass beds" or "meadows". The term marine prairies is also used in the Spanish language: "praderas marinas".

Marine magnoliophytes have the same organs and tissues as other flowering plants: they have a nervous system composed of epigeanous parts (branches and leaves) and endogenous parts and roots (Figure 3.1). Therefore, these are plants that grow almost exclusively in zones of soft marine sediment. In most species, a basal body protects the apical meristem where the leaves emerge, which are organs of photosynthesis (acquisition of energy) and evapotranspiration (respiration). Although marine phanerogams are not as adapted as algae to ensure photosynthesis at depth, they are however more

---

Chapter written by Christian HILY.

efficient than freshwater species that are rarely found below 10 m. In fact, some seagrasses *Thalassodendron* or *Halophila decipiens* can be found at a depth of sixty meters in the clearest waters of tropical zones.

Figure 3.1. *Morphology of marine magnoliophytes – example of the species Thalassodendron ciliatum (photo C. Hily). For a color version of the figure, see color section*

Although marine seagrasses form a single ecological group, they do not form a single taxonomic group. In fact, marine phanerogams are distributed across four plant families in the super order *alismatiflorae* [DAH 85], which also includes many other species. Three of these families, *zosteraceae*, *cymodoceaceae* and *posidoniaceae* are only composed of marine magnoliophytes. The fourth family, *hydrocharitacae*, includes three genera of marine species (*Thalassia*, *Enhalus* and *Halophila*), but 14 other genera of freshwater species. Finally, other species, *Ruppia sp.* (*Ruppiaceae*) and *Lepilaena marina* (*Zannichelliaceae*) are not exclusively marine and are not systematically known in the strict sense as marine magnoliophytes.

Currently, 68 species of marine magnoliophytes have been described [LAR 06]. However, some are still undergoing genetic and morphometric research and many are considered to be conspecific. This classification can therefore still be rearranged, both in terms of taxonomy and nomenclature [GRE 03]. Among these species, 20 are present in French overseas territories, with nine genera and four families, whereas in mainland France and elsewhere in Europe, two species (*Zostera marina* and *Zostera noltii*) are present on the coast of the English channel and Atlantic and four on the Mediterranean coastal zone (*Posidonia oceanica, Zostera marina, Zostera noltii* and *Cymodocea nodosa).*

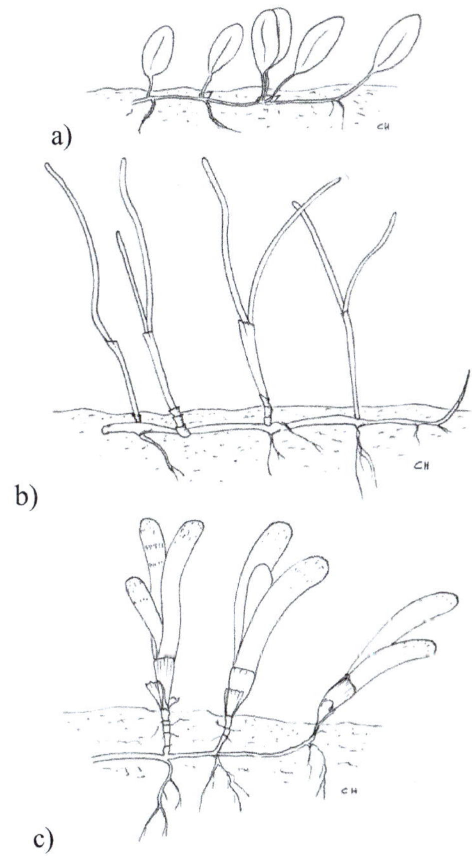

Figure 3.2. *Variability in the morphology of three genera of tropical magnoliophytes: a) Halophila, b) Syringodium, c) Cymodocea*

The general structure of a vegetative shoot (leaf–rhizome–root) is homogeneous but the morphological characteristics of the leaves, rhizomes and roots vary according to species (Figures 3.1 and 3.2). The majority have long banded leaves that are directly fixed to rhizomes via a node (e.g.

Enhalus, *Syringodium*, *Zostera*); others have small non-banded leaves without an erected stem (exclusively species of the genus *Halophila*); finally some such as *Thalassodendron* have leaf clusters at the extremity of erect, rigid and fibrous stem, which are an extension of the rhizomes. The roots emerge from the rhizomes often at a node, the morphology, rhizomes and roots vary from one species to another and are adapted to the type of substrates on which the species develop. *Phyllospadix*is is even capable of firmly attaching to hard substrates in high-energy environments hit by strong swell. The foliage leaf is a set of leaf blades enclosed by a leaf sheath and fixed at a rhizome node. A lacunar system (aerenchyme) containing gases (nitrogen, oxygen and carbon dioxide) reaches all organs and exchanges gases while contributing to the upright position of leaves in the water mass. All species are perennial except for some that locally have annual populations when environmental conditions are unfavorable.

Reproduction may be sexual or vegetative through the natural cutting of rhizome fragments or as clones through the extension of rhizomes. Vegetative reproduction is usually the main process by which seagrasses maintain themselves and regenerate.

Sexual reproduction occurs through flowering via pollination and fertilization. The flower parts and the method of fertilization are varied; depending on the species, the male and female flowers may be found on the same shoot or on separate shoots, supporting the hypothesis of several different origins for the ancestors of current phanerogams [LAR 89].

Reproductive shoots are often longer than vegetative shoots to facilitate pollination. Some species even undergo pollination at the sea surface, which facilitates fertilization by decreasing the risks of pollen dispersion into water mass (*Enhalus*, *Syringodium*). The germination of seeds, dispersed by currents, completes the lifecycle.

The seeds produced are classed in three large categories according to the type of stored energy, structure of the seed coat and method of germination [KUO 96]. Depending on the method, the seeds are dispersed by different physical or biological vectors that often concentrate and accumulate them over small areas (concept of seed bank); germination may occur rapidly after release or after a dormancy period of several years. If the conditions of temperature, salinity, hydrodynamism are favorable, a new plant may rapidly grow.

3.1.2. Biogeography

The geographic distribution of marine phanerogams is becoming clearer and clearer, without taking into account their dynamics and evolution forced by climate change; all the distribution zones have still not been completely described, due to the presence of some significant limitations. Marine magnoliophytes have colonized all the world seas, except the Antarctic. Currently, they cover a maximum of 10% of the coastal zones, that is to say approximately 500,000 km^2 [GRE 03]. Some species are present in all regions of the world, whereas some are found exclusively in temperate zones or pantropical zones: with regard to genera, four are found exclusively in tropical environments or similar (warm currents) (*Thalassia*, *Syringodium*, *Halodule* and *Enhalus*) and three of them are found in these tropical environments except for some species (*Cymodocea*, *Halophila* and *Thalassodendron*). Five genera are primarily found in temperate coastal environments (*Zostera*, *Phyllospadix*, *Heterozostera*, *Posidonia* and *Amphibolis*). Only *Zostera marina* is found in the Arctic Circle, Europe and the North Pacific. It is therefore the most northern species, whereas another species of the same genus, *Z. capensis*, has the most southern distribution.

Unlike many other marine plant groups, very few marine magnoliophytes accidentally grow in geographic zones other than their original distribution. Two introduced species, *Z. japonica* and *Halophila stipulacea*, are however known to occur perennially, on the Pacific coast of North-America and the Caribbean, respectively.

Recent advances in genetic analyses (DNA markers) have allowed us to better understand evolution and adaptive traits, reproduction strategies or responses to stress [REU 01]. Similarly, these techniques have recently allowed the status of species to be specified within genus. Genetic and phylogeography analyses also provide evidence for the connectivity between seagrasses via the dispersion of seeds and plant fragments and provide explanations for the geographic distributions, with consequences in terms of conservation (in particular identifying populations that fulfill the role of source or of a sink) and in terms of restoration of seagrasses through transplantation.

Some species are endemic: thirteen species in Australia are not found anywhere else in the world; six of the seven Caribbean species are endemic to this region. There are significant conservation challenges associated with these marine plants, since in order to ensure their perennity, the entire associated ecosystem is to be maintained.

At a regional level, the number of species of marine magnoliophytes that can grow at a given site is therefore highly variable. Mono- or plurispecific magnoliophytes may occur, both of which occur on different spatial and temporal scales. The simplest case is when a single species is represented over a wide geographic zone: seagrasses are obligatorily monospecific. This is the case on the island of La Reunion in the Indian Ocean where only one species *Syringodium isoetifolium* is present. When biogeographic ranges overlap, species potentially present at one site are not systematically classed as plurispecific seagrasses. Many factors can explain the species distribution in a given area and the method used to organize seagrasses. Firstly, each species has its own ecological requirements, and therefore the specific diversity of magnoliophytes will be highly variable depending on the local conditions of the environment, such as the granulometry of the substrate, the depth, hydrodynamism, sestonic load, water transparency, etc. So ideal conditions, i.e. moderate depth, fine sand, moderate hydrodynamism and sestonic load, are the conditions that facilitate the growth of plurispecific seagrasses. Nevertheless, as different traits vary according to species, competition for space often leads to natural successions in turn favoring the monopolization of space by only one species.

These are one-off natural and anthropogenic disturbances that will maintain apertures in vegetation cover by allowing the installation of less competitive species over time. Finally, the small-scale environmental variability may explain the small-scale monospecific spatial distribution of patches (= small areas of seagrasses isolated by bare sediment). The spectrum of tolerance to the substrate is wide for species such as *Cymodocea serrulata*, *Halophila decipiens*, *Thalassia testudinum* or *Syringodium filiforme*. These last two species often form mixed beds. *H. ovalis* is an opportunistic pioneer species capable of rapidly growing on bare or altered sediments [HIL 98], but is not adapted to inter specific competition. This is why, once the substrate is colonized, it disappears in favor of species with slower growth rates such as *S. isoetifolium*.

3.1.3. *The seagrass habitat*

The development of seeds, often concentrated in areas limited by marine currents, and in particular, the vegetative development of large clones, explain why these plants form dense populations [BEC 10]. The terms "meadows" and "prairies" describe the formation of a very characteristic marine landscape, distributed in patches that cover from several dozen square meters to several dozen hectares. In addition to this visual aspect, the morphology and high density of shoots also have consequences on their environment. Marine magnoliophytes are also considered as engineer species since their presence modifies the biotope and biocenosis in which they grow. These modifications are perennial and are stable and reproducible enough to change the characteristics of the sedimentary bed. From a vegetation-free sedimentary habitat such as "fine sand" and "coarse sand" or "muddy sand", it becomes a "meadow" habitat. A habitat is a spatially located, functional and temporally stable entity grouping together a biotope (physical-chemical and sedimentological environment) and the particular biocenosis (set of living species) associated with it. The "meadow" habitat has completely original physical, biogeochemical and biological characteristics that give it a set of particular interests both functional, heritage, landscape, scientific, economic and educational. Consequently, seagrass beds have a particular and prominent place in other marine habitats.

3.2. Role of seagrass beds in the coastal environment

Marine phanerogam beds are particularly important in tropical coastal environments, due to their functional and structural role. This role, essential for the coastal ecosystem dynamics, is also true in littoral zones where the primary production of macrophyte algae is low or modest, e.g. in the Mediterranean as well as along the sheltered coasts dominated by soft seabeds. The installation of a seagrass in a sedimentary zone greatly modifies the physical, sedimentological, physical-chemical and biological processes of this zone. The overall metabolism of seagrasses affects the carbon dynamics and nutrients in the environment, as well as the concentrations of oxygen in the water column and sediments. Seagrasses are important carbon sinks.

The structure of the seagrass leaf canopy mitigates the impact of waves and currents, thereby increasing sedimentation of suspended particles, and decreases their resuspension. Thus, the amount of fine silty sediments is usually higher in the meadow than in the plant-free sediment nearby. The dense network of rhizomes and roots (rhizosphere) considerably minimizes the erosion of sediments by hydrodynamism. With regard to the surrounding sediments, these two additional processes, less erosion and more sedimentation, coupled with the growth of rhizomes and leaves to remain on the surface of the sediment explain why the zone covered in vegetation is often raised compared with surrounding sedimentary zones (Figure 3.3), (which is described as seagrass "mat").Within the seagrass meadow, the non-vegetated sediments often appear as depressions that reach several dozen centimeters. During phases of erosion the limits of the seagrass/bare sediment often appear as micro bluffs, revealing the rhizomes and roots. In this rhizosphere, a particular microbial fauna controls the mineralization of organic matter and the regeneration of nutrients. Seagrass beds therefore act as a sink of organic matter, by acting as a filter. However, at other times they can act as a source for surrounding ecosystems through the export of plant debris from the meadow during episodes of big swell or strong currents.

Figure 3.3. *Mitigation of the effect of coastal surges: seagrass mat (Thalassia testudinum) eroded by surges (photo C. Hily)*

In tropical environments, seagrass beds are known to limit the devastating effect of cyclonic swells that pass over reef barriers or that enter directly into bays without corals. By absorbing part of the wave energy, they decrease the impact and protect the coast from erosion. However, they withstand the worst of these surges, which are probably the most determining natural factor explaining their dynamics. Entire grass beds are torn and scattered or accumulate on the coast forming protective foreshores on the beaches. Following this, individual shoots of fragments of the mat can survive and begin to develop, explaining the high inter-annual variability often observed and changes in the surface covered by seagrass in a given area.

3.3. Functioning of seagrass beds

3.3.1. *Production and metabolism*

Marine seagrass beds form part of the most productive ecosystems in the world. Primary production occurs through the production of seagrass shoots, but also from the production of associated algae (macrophytes and microphytes), attached as epibionts on leaves, and develop on the sediment between the shoots. Plant biomass is often very high and can reach record levels, particularly in seagrass beds composed of mats of *Posidonia oceanica* or *Thalassodendron ciliatum*. For the seagrasses *Posidonia oceanica* in the Mediterranean, the biomass can reach 900 g of dry matter (DM) per m^2 for leaves, up to 470 g DM/m^2 for epiphytes on the leaves, and 50 g DM for epiphytes on the rhizomes, and up to 5,500 g DM for rhizomes and roots [BOU 06]. If the biomass stored is high, the production is also high and ensures an abundant supply of organic matter, which is the basal energy source for a very complex trophic network.

Metabolic activities and seagrass structure modify the physical and chemical conditions of the water column and sediments in colonized zones [LAR 06]. In the same way, respiration and photosynthesis modify the concentrations of oxygen in the two environments. This production can also exceed 10 g O_2 m^{-2} j^{-1} [HEM 00]. The production of oxygen by photosynthesis exceeds by one order of magnitude, under normal conditions, the requirements for respiration. This excess is partially stored in internal tissues and air-lacunae of leaves.

Seagrasses also play a role in the calcareous cycle, on the one hand by increasing the dissolution of carbonates within sediments and on the other hand by increasing the precipitation of carbonates through the development of algae and other calcifying epibiontic organisms. Rhizomes and roots release oxygen, which facilitates the mineralization process through the aerobic pathway, and therefore produced carbon dioxide, accelerating the dissolution of carbonates [MAR 02]. Seagrasses can play different roles in the biogeochemistry of the water and sediments, which depend on the species themselves, since the development of root and rhizome systems is highly variable. Therefore, *Halophila* have a poorly developed root system compared with that of larger species forming mats such as posidonia.

3.3.2. *Trophic network*

The functional biodiversity of seagrasses is very high, favored by the three-dimensional (3D) structure of the vegetation developed in an inextricable network of leaves, stems, rhizomes and roots. This structure creates many microhabitats allowing the presence of a wide diversity of trophic groups and species. Leaf composition is often rich in lignin and cellulose, which limits the number of herbivores that directly consume them. These herbivores are large organisms such as tortoises, sirenian mammals (dugongs and manatees), some fish and macro-invertebrates (sea urchins, sea hares (aplysia)). Locally, sea urchins may be very abundant. Through grazing, they are capable of considerably modifying the ecological functioning of seagrass beds, creating zones of reduced biodiversity and thereby facilitating wave erosion.

Other herbivores, also called micrograzers, primarily gastropod mollusks, mainly consume epiphytic algae or, in the sediment, small macrophytes and microphytes.

Figure 3.4. *Seagrass beds, biodiversity hotspots: epibiosis under a canopy of Talassodendron ciliatum (photo C. Hily). For a color version of the figure, see color section*

The role of epibionts is important for the functioning of seagrass beds. The leaves of marine magnoliophytes are in fact excellent supports for many living organisms forming the epiphyton (Figure 3.4). These are of a varied nature, bacteria, fungi and protozoa, microphytic algae, meiofauna, macrophytic algae and macro invertebrates. Micro organisms congregate organic and inorganic particulate elements (grains of sand or mud) and form what is commonly known as a biofilm. This biofilm progressively colonizes leaves in the form of ecological successions, punctuated by the seasons, throughout the life of each leaf. The lifespan of each leaf varies from 11 days for *Halophila*, to approximately 130 days for *Posidonia*, with the exception of *Amphipolis Antarctica* whose leaves can last for three years. Epiphytic algae are an important part of the primary production of seagrass beds, equivalent in some cases to the primary production of phanerogams themselves. These macroalgae often have a high species richness; for example, *Amphibolis griffithii* seagrass beds can support up to 150 different species of epiphytes.

Invertebrates that are attached to leaves have been studied much less than algae, especially in tropical zones. These invertebrates include many groups of bryozoa, tunicates, sponges, foraminifera, hydra. Logically, the longer the lifespan of the leaves; the larger the colonizable area (long and wide leaves), the greater the diversity and average size of epiphytes. Mobile species that reach the leaves, such as many gastropods, with some species being almost exclusive, feed on the biofilm and young macroalgae; they form part of the trophic group called micrograzers. Many detrivorous amphipod and isopod crustaceans are also directly associated with the epiphyton.

Interactions between seagrass/epiphytic algae/micrograzers is therefore a key process in the functioning of seagrass beds, since they control the primary production of epiphytes through micrograzers, depending on good leaf growth. Some herbivorous fish such as Mugilidae can also control the biomass of epiphytic algae in seagrass beds. The decline in these grazers due to pollution or fishing can mean dramatic consequences for seagrass beds. In fact, the development of too many epiphytes is known to create a "shadowing" effect limiting photosynthesis at the leaves. This process, which is becoming more frequent in coastal waters that are becoming increasingly rich in nutrients, can lead to the complete disappearance of seagrasses. In fact, these nutrient-rich waters favor the rapid growth of brown and green filamentous algae that have high biomass. The abundance and diversity of epiphyte grazers are therefore essential for the resilience of seagrass beds to degradation in the quality of coastal waters, particularly visible in lagoon environments of many zones.

However, in most seagrasses, the primary production is not directly consumed but enters the trophic web via remains. This is accumulated at the base of leaves and forms litter or is exported by currents or swell often far from their site of origin. Dead leaves may become stranded on beaches, accumulated on the high-water mark thus forming benches along the shore or in submarine zones of other ecosystems, corals, mangroves, mudflats, etc. They will slowly be degraded by bacteria and fungi, favored by fragmentation under the action of hydrodynamism and through consumption by detritivores (primarily amphipod and isopod crustaceans, but also sea urchins). Progressively, the remaining fragments become increasingly finer and appear similar to particles of organic matter accumulated at the surface or buried in sediment. These particles are consumed by depositivores (primarily annelids but also holothurians). All these primary consumers are prey for a wide range of secondary consumers that provide diversity to the trophic web.

There are several levels of predators, dominated by fish on the one hand which hunt in seagrasses, mainly at night and come from neighboring ecosystems, coral reefs, rocky seabeds, or the deep sea, and on the other hand, by invertebrates that come out at night from their burrows or hiding places (crustaceans, crabs, lobsters, gastropods and cephalopods).

3.4. Challenges in the conservation of seagrass beds

Seagrass beds are ecosystems with a particularly high biodiversity, which is fragile since their existence depends on one or several living organisms, and house rare and endangered species; they therefore have a high heritage value.

3.4.1. *Rare and endangered species*

Marine magnoliophytes have an intrinsic heritage value due to the low diversity of species that exist globally. The disappearance of a single species would correspond to a 0.5% decrease in the species diversity. The unique biological functioning, mode of reproduction, adaptation to the marine environment make them rare botanical curiosities that, as a species, deserve systematic protection. Their global sporadic distribution, and low number of ubiquitous species, compared to endemic species (in regional basins), reinforce this value. At local and regional scales, the heritage value of species is reinforced since the number of species is often very low (two for the north-east Atlantic) and does not exceed a dozen (Caribbean basin: seven species), the maximum probably occurs in Australia (29 species).

Seagrasses as engineer species allow rather specific animal and plant populations to develop and a large species diversity and associated functional diversity. Among the species recorded in seagrass beds, some are completely subservient, others are present in a greater abundance than in other habitats where they naturally occur, and come across different habitats. Hundreds are rare and endangered and therefore have a high heritage value.

As previously mentioned in this chapter, the direct consumption of seagrass leaves by macroherbivores is a process limited to several species and only some types of seagrasses. In tropical zones, sirenian mammals (dugong, manatees) are almost the only consumers of marine phanerogams. Symbolic and mythical creatures of tropical seagrasses, they have become very rare and endangered, and have a high heritage value. The remaining populations are dependent on seagrass beds for their survival. Preserving seagrasses in areas where turtles and sirenians are present has become a priority in the management and conservation of coastal ecosystems. They comprise umbrella species whose protection, by preserving their food resources, also allows the preservation of biodiversity associated with seagrasses.

However, as with terrestrial ecosystems, the conservation of ecosystems controlled by large herbivores, actions in favor of turtles, manatees and dugongs must be carried out in a highly monitored framework to prevent overgrazing and loss of biodiversity associated with the fragmentation or loss of biomass of seagrasses. The consumption of seagrasses by turtles causes defoliation whereas dugongs may also consume part of the rhizomes and roots. As these herbivores prefer to consume certain species, selectivity can sustainably modify the species composition of seagrasses, by favoring rapidly developing species, which are primarily small pioneer species such as *Halophila*, and therefore unlikely to establish an associated rich and biodiverse biocenosis.

3.4.2. *Specific role of seagrasses: ichthyofauna of coastal environments*

The abundance and diversity of fish species are greater than in zones with vegetation-free sediments. Depending on the species of marine phanerogam that form the seagrass bed, their role with regard to fish is also highly variable. Generally speaking, the diversity and abundance of ichthyofauna increase with complexity of the seagrass bed and size of the canopy. The growth rate of fish and invertebrates are greater overall in seagrass beds than in vegetation-free sediments. Seagrasses also act as a source for vegetation-free sediments, the abundance and diversity of fish on the sediments decrease as the distance from the seagrass increases [FER 91].

Many studies have focused on the role of seagrass beds on fish populations, particularly with regard to species of commercial value that use them in one or more of their life stages. In fact, the role of seagrass bed on fish varies greatly depending on the species. Although some small species are permanent residents (Syngnatidae, Atherinidae, etc.), most of them are only temporarily present.

3.4.2.1. *Spawning area and nursery*

For some species, seagrass beds are favored reproduction sites or even nurseries. The development of juveniles and their survival are favored by, however, the abundance of small easily accessible prey (amphipods and shrimp) and on the other hand due to the dense canopy that protects them relatively well from predators [JAC 01]. However, the proportion of juvenile stages in the seagrass ichthyofauna varies according to the types of sites and the other species present in the seagrass bed, and also, according to the presence of coral reefs and mangroves nearby.

3.4.2.2. *Feeding zone*

Some species of fish are herbivores and even if they usually graze on epiphytes such as Mugilidae, some consume seagrass leaves. They are therefore dependent on this habitat even if they can find refuge at night in hard or rocky coral substrates.

On the other hand, seagrass beds form hunting grounds for many carnivorous fishes, primarily nocturnal species (Labridae). Many coral reef fishes occasionally frequent seagrass beds. Seagrass beds are therefore feeding grounds for many commercial species of fish, such as goliaths, tarpons, snooks (*Centropomus sp.*), parrot fish, barracudas, as well as invertebrates such as lobsters, crabs, crayfish shrimp, sea urchins, spider conchs, etc.

3.4.2.3. *Refuge zone*

Seagrass beds are favored hunting grounds, since they house many prey. In fact, many fish find refuge and shelter here: small species and juveniles hide in the leaves, in the interlacing rhizomes and benefit from the movement of leaves by hydrodynamism to escape predators.

3.4.2.4. *Temporal variation in the frequentation of seagrass beds by ichthyofauna*

The tidal cycle is essential in the frequentation of ichthyofauna: if water levels become low at low tide, most of the motile fauna will move; then, only small species will seek refuge in the pools or in inundated galleries in the sediment. The seasons also modify the presence of fish and their periods of reproduction, recruitment and movements of migratory species. For the ichthyofauna, all these factors cause a high variability at different spatiotemporal scales within and between seagrass beds. This high variability also explains the contrasting results from studies carried out to find correlations between the loss of seagrass area and the decrease in fish captured: the presence of substitute habitats in proximity to seagrass beds in regression can, in particular, moderate the decrease of fish stocks.

3.4.3. *Services provided by seagrass beds to humans*

Their high biodiversity but also the particular structure of seagrasses themselves increase the capacity of seagrass beds to perform different ecological functions, which provide ecological services from which humans can benefit directly or indirectly, from the goods produced, used and consumed by humans with a high economic and/or social value. The services provided to humans by seagrass beds are plentiful and used for different types of services (support, regulation, sampling and cultural). They are one of the natural ecosystems that provide the greatest number of services to humans, making their need for conservation much greater:

– support services: bioturbation, primary productivity, secondary productivity, nutrient cycle and mineralization, carbon and oxygen cycle;

– sampling services: fish for food, crustaceans for food, building materials (drywall), genetic resources;

– regulation services: spawning area for species, refuge zone for species, control of waves and energy from currents, regulation of erosion and siltation, sedimentation dynamics required by micro-organisms and benthic fauna, carbon storage;

– cultural services: recreational fishing, preservation of marine biodiversity for ethical reasons, school excursions, source of knowledge, view (landscape), underwater diving.

3.5. Pressures on and threats to seagrasses

Marine magnoliophytes developed in coastal zones close to the shore are subject to many natural and anthropogenic pressures, direct or indirect, local or global (such as climate change). According to the species, the tolerance levels and resilience to diverse pressures are highly variable.

3.5.1. *Cyclones*

All tropical regions experience cyclone seasons, whose force depends on the zone considered. They affect seagrass beds in two ways: coming from land through the rainwater run off and from the ocean through the action of swells. The flow of freshwater from high rainfall considerably decreases the salinity of lagoons. Main phanerogam species are euryhaline and do not all equally tolerate variations in salinity. *Halodule uninervis* and *Halophila stipulacea* can adapt rather well to these rapid changes in salinity, whereas *Cymodocea serrulata* for example tolerates this less well [PHI 88]. Moreover, these runoff waters are very muddy and loaded with agricultural pollutants and diverse macrowaste.

Sedimentary erosion caused by cyclonic surges damage seagrass beds. Even if they show better resistance than bare sediments due to their root and rhizome systems, erosion varies according to the system developed by each species. For example, *Thalassia testudinum* has a very solid root network that is highly resistant to stormy events [CRU 05]. However, species with short leaves, such as *Halophila ovalis*, which also have shallow rhizomes and fine roots, are not very resistant to swells and surges.

3.5.2. *Emersion*

Species of marine phanerogams do not all equally tolerate desiccation and UV radiation. Emersion at low tide is a determining factor for the distribution of seagrasses. Particular climate events (barometric low tides during cyclonic periods, large coefficients associated with periods of high temperature and wind, etc.), which tends to increase the emersion and desiccation of leaves, damage seagrass beds. Some species cope less well than others in extended periods of shore exposure, for example *Thalassia testudinum*.

3.5.3. *Global climate change*

Accelerated by anthropogenic activities, climate warming will be a key factor in the future seagrass dynamics in tropical zones. The increase in sea level and temperatures of coastal waters are two factors that may modify the distribution of marine phanerogams. These modifications will be complex and vary according to site, involving processes of inhibition, tolerance and facilitation depending on the species. We must therefore expect variations (extensions or regressions) in the area of seagrasses but also modifications in their species composition, some species are disadvantaged whereas others will benefit. The secondary effects caused by warming of water bodies will also be determinant: climate models predict many cyclonic events associating wells and changes to tide regimes.

3.5.4. *Fishing*

The existence of commercial populations (fish, lambis and sea urchins) lead to strong fishing pressure with significant impacts not only on the biodiversity in its entirety but also on the seagrass biotope, especially under the action of dredges, trawls and other gears (Figure 3.5a). The marine phanerogams' shoots are deteriorated by the use of towed gears, creel and other equipment used during fishing on foot and trampling [KEN 02, LER 06].

a) b)

Figure 3.5. *Seagrasses are widely used by local populations and, under the impact of trampling, sampling and uprooting, weaken the biocenosis of seagrasses and the erosion of plant biomass. a) Subsistence fishing in Madagascar; b) recreational water activities in New Caledonia (photos C. Hily)*

3.5.5. *Boating and water sports*

In tourist areas where boating is highly developed, the impact of moorings and boat propellers on seagrass beds is one of the most influential factors. Anchors tear rhizomes and roots, and chains tear the leaves. Organized moorings also greatly affect this zone through the movements of boat chains. Moreover, for the well-being of tourists, seagrass beds are often removed from swimming areas in front of hotels: creating "white seabeds" and channels to access docks. Visitors to lagoons who undertake swimming and leisure activities (wind surfing and kite surfing) trample the seagrass beds (Figure 3.5b).

3.5.6. *The extraction of sandy or coral materials*

The extraction of coral or sandy materials, for use in construction, directly affects the seagrass beds, or indirectly by causing the erosion in nearby areas, extraction and increase in turbidity and hypersedimentation.

3.5.7. *Amenities*

In the most popular tourist areas or areas with high population growth, the development of infrastructure such as ports, defenses against the sea, artificial beaches in front of hotels, navigation channels or even the infill of land for building purposes, are usually developed to the detriment of seagrasses. Indirectly, these amenities modify the currents and therefore the shoreline and cause more intense sedimentation.

3.5.8. *Terrigenous effluents*

Like most coastal habitats in the world, for several decades, seagrasses have faced an increase in particle input into water bodies and the sedimentation of fine particles on the seafloor. Deforestation to create areas for accommodation or agriculture, as well, the creation of amenities or poor agricultural practices, favor water runoff and soil erosion. Consequently, there is a decrease in the level of light available for plant species and therefore a decrease in the depth at which seagrasses can grow. Hypersedimentation on leaves also decreases photosynthesis and decreases the growth and survival of seedlings. As a consequence of uncontrolled economic development, the greater input of particles decreases the capacity of wetlands and mangroves to filter and purify. Therefore, lagoons receive the majority of terrigenous effluents and suffer the effects of hyper sedimentation. Effluents of fish farms located along sea shores also play a role in this phenomena, inducing mangrove regression and being a source of pollutants. Fertilizers, used on land crops, leached from the soil after heavy rain, end up in the sea, causing high nutrient inputs, particularly of phosphate and nitrate, which stimulate the growth of marine plants and therefore seagrasses. In the oligotrophic zone, the growth of seagrasses may be a sign of moderate nutrient input of anthropogenic origin. However, in excess, nutrients also favor the development of phytoplankton and macroalgae, including epiphytes on the leaves of phanerogams, and have a negative effect on seagrasses. Via the same process, pesticides are transported into lagoons. These chemical contaminants can cause leaf loss and a decrease in photosynthesis in phanerogams [HAY 00] in regions of intensive cultivation of sugar cane, bananas, pineapples, etc.

3.5.9. *Harvesting*

In many parts of the world, marine phanerogams have long been used by local populations for many different uses. Today, if these exploitations are in sharp decline, they may promptly affect many poor and remote areas of the world, where there is still a cultural and economic dependence on seagrasses. The rhizomes of *E. acoroïdes* provide food to Kenya and Thailand; the Seri people in Mexico transform them into flour; the high nutrient content of *Zostera sp.* makes it a good food source for sheep and pigs around the Black Sea, etc. Thai people use the biological properties of some species in medicine. Leaf fibers may be used to form clothes, ropes, carpets, upholstery, mattresses, thermal and sound insulation in houses, to feed livestock in winter and to replace thatch in some parts of the world (Europe, Asia and Africa). They are used as fertilizer in Tunisia, as mulch in agriculture in Japan, as a packaging material in Australia or even as embankments to prevent erosion.

Today, under these pressures, marine phanerogam beds are vulnerable, or even declining [WAL 06]: situated at the sea–land interface, they withstand the worst of the effects caused by increasing human activities. In highly populated areas of the world, scientists have seen a decrease in the area occupied by seagrasses for the first time in a century. Although many areas were destroyed despite the existence of seagrass beds being unknown; the authors [SHO 00] extrapolated a loss of approximately 12,000 km^2 when only 2,900 km^2 was referenced between the mid-1980s and the mid-1990s. Waycott *et al.* [WAY 09] reported a loss of 110 km^2 per year in the area occupied by seagrasses since the beginning of the 1980s and a disappearance of 30% in this area since they were first recorded in 1879. Moreover, they highlighted a decrease of more than 0.9% per year prior to 1940 and since 1990 it has reached 7% per year.

Land use to accommodate human populations is at the expense of some misunderstood ecosystems, such as mangroves and seagrass beds. The lack of public awareness with regard to the loss of seagrasses contributes to their decline. It has therefore become essential to increase communication about seagrasses to ensure their effective management [DUA 08]. We must back up this information and raise public awareness about the importance of seagrasses and the threats they are faced with, so that measures can be put in place for their protection, as well as for coral reefs and mangroves [ORT 06]. Moreover, all these marine biocenoses are associated and are interdependent. Share research results with the public should be a priority, since if one of these biocenoses becomes threatened, the consequences on one will have repercussions on the others and the whole ecosystem will be endangered.

3.6. Restoration of seagrass beds

This involves improving their growth conditions, for example by reducing the input of harmful effluents into their environment, by transplanting then or by sowing their seeds in a given location. This type of action is often proposed as a compensatory measure against the impact of a coastal development. Transplantation trials usually have a low success rate, are quite costly and are labor intensive, since the location must be appropriate. Current research into different species for the improvement of conditions that confer success and into the restoration of seagrass habitats is underway; however, in many situations, poor water quality is common and is often a limiting factor. Therefore, it is obviously preferable to preserve existing seagrass beds instead of trying to restore them once destroyed. The results also vary greatly depending on the species: those with high growth rates will have better recovery capabilities than slow-growing species.

3.7. The functional role of seagrasses in the lagoon ecosystem

Marine ecosystems are open ecosystems and have many interactions between one another. Tropical seagrass beds interact with other lagoon ecosystems and coastal waters: mangroves, soft sediment communities and coral reefs. The term interaction implies that seagrass fulfills a functional role in neighboring ecosystems but also that these ecosystems influence the ecological functioning of seagrass beds. This therefore also implies that any degradation to one of these ecosystems will significantly disrupt the associated ecosystems and also improving the ecological quality of one ecosystem will favor the others.

With regard to seagrass beds, their interactions with coral reefs and mangroves can therefore be considered in both directions: the role of seagrasses on associated ecosystems and the role of associated ecosystems on seagrasses.

a)

b)

Figure 3.6. *Functional continuity between a) seagrass and mangrove and b) seagrass and coral reef (photos C. Hily)*

As we go from the coast to offshore, the theoretical transect of a lagoon can be simplified in the form of a mangrove-seagrass-reef succession (Figure 3.6). The complexity of this diagram can, of course, be increased through the addition of a central channel, non-vegetated sediment and the presence of fringing reefs on the coast in the absence of mangroves. However, seagrasses are often centrally located between mangroves and reefs, which gives them a special role as both a transit area, sink and source for flow of matter and energy into the lagoons. In fact, seagrass beds are open systems that directly export rhizomes, root and leaves as debris, and indirectly, through the transport of particulate and dissolved organic matter, elements important in the functioning of adjacent ecosystems, lagoon sediments, reefs and mangroves.

During periods of heavy swells, particularly during cyclones, reefs protect seagrass beds by preventing them from being ripped from the sediment. In addition, seagrasses stabilize sediments and therefore mitigate the impacts of abrasion, the turbidity and the sedimentation of particles on corals. Coral reefs and well-developed seagrass beds therefore play a synergistic role in the protection of beaches and shores against the actions of erosion caused by waves.

Seagrass beds act as natural filters between terrigenous waters (that have passed or not through mangroves) and coral reefs. These processes that clarify the water and decrease the sestonic load over a coast-reef gradient are highly beneficial to corals. If seagrass leaves are transported to the barrier reef as an organic matter supply, this will be slowly degraded with no notable effect on the water clarity. Nutrients transported by storm water run-off and sewage pass by seagrass beds before reaching the barrier reefs. They are partially absorbed by seagrass leaves, epiphytic algae and microalgae and root systems. This absorption of nitrogen and phosphorus by seagrasses reduces the opportunities for algae to grow on coral reefs and improves environmental conditions for the survival and development of coral. Obviously, fine particles trapped by seagrasses can be resuspended in cyclonic swells and re-distributed in coastal zones during these periods. It is important to note that mangroves also play an equivalent role in the sestonic load decrease and in nutrient absorption. The most efficient filtration system resulting in favorable waters for the development of corals is therefore a continuation of the mangrove and seagrass prior to the barrier reef. Mangroves act as a primary filter and facilitate the action of seagrasses by preventing clogging with fine particles and the proliferation of epiphytic algae when terrigenous waters contain high amounts of particles and nutrients.

Highly oligotrophic waters hinder seagrass production. When sites are little affected by human activities and terrigenous inputs, mangroves grow in proximity to seagrass beds, organic debris is produced abundantly by the mangrove, and they store particles and nutrients, favoring the growth of marine phanerogams. It is under these conditions for example that *Thalassia testudinum* forms extremely dense beds.

Seagrass beds are areas of tidal and day/light migrations of many species that are dependent on the reefs and mangroves: part of the mobile mangrove macrofauna seeks refuge in seagrass beds when mangroves are uncovered at low tide, whereas many reef species, particularly predators, hunt at night in seagrass beds where many species of small fishes and juveniles are found [NAG 00].These juveniles are also young species that will spend their adult life in coral reefs or other ecosystems. Seagrasses may play a trophic role for some mangrove organisms: it has been found that the larvae of organisms that depend on seagrasses might form part of the food of filter feeders attached in large numbers to the roots of the mangroves trees.

In addition, studies have shown that in each of these three ecosystems, the movements of their characteristic species often overlap into adjacent ecosystems, forming a "halo" around their preferred habitat. This process also explains the high biodiversity in seagrass beds that temporarily house species typical of both reefs and mangroves. Finally, other species such as turtles use several ecosystems in their biological cycle, frequently visiting coral reefs outside of feeding periods spent in seagrass beds.

Much still remains to be understood about the interactions involved in these three major ecosystems in tropical coastal environments. This is a rapidly developing research topic since there is a true awareness; it is only by considering this in a wider context that the conservation of the functioning of this ecosystem, its biodiversity and associated services will be assured.

3.8. Conclusion

The population is still unaware of the fact that seagrass beds play an important role in the functioning of coastal ecosystems, particularly those in intertropical zones. Even if efforts are supported by local authorities and scientists, on the one hand, to minimize pressures, and on the other hand, to better understand the functional processes that are leading their dynamics toward their degradation or restoration, it is still not enough. We must do more to reverse the recurring losses on a global scale before it is too late. It is for this reason that, in the few countries where initiatives are already underway, such as Australia, information about their value, originality, richness and threats must be conveyed to local populations on a large scale through educational activities and actions to raise awareness.

3.9. Bibliography

[BEC 10] BECHELER R., DIEKMANN O., HILY C. *et al.*, "The concept of population in clonal organisms: mosaics of temporally colonized patches are forming highly diverse meadows of Zostera marina in Brittany", *Molecular E cology*, vol. 19, pp. 2394–2407, 2010.

[BOU 06] BOUDOURESQUE C.F., BERNARD G., BONHOMME P. *et al.*, *Préservation et conservation des herbiers à Posidonia oceanica*, Ramonge pub, Monaco, pp. 1–102, 2006.

[CRU 05] CRUZ-PALACIOS V., VAN TUSSENBROEK B.I., "Simulation of hurricane-like disturbances on caribean seagrass bed", *Journal of experimental Marine Biology and Ecology*, vol. 324, pp. 44–60, 2005.

[DAH 85] DAHLGREN R.M.T., CLIFFORD H.T., YEO P.F., *The Families of the Monocotyledons: Structure, Evolution and Axonomy*, Springer Verlag, Berlin, 1985.

[DUA 08] DUARTE C., DENNISON W., ORTH R., CARRUTHERS T., "The Charisma of Coastal Ecosystems: Addressing the Imbalance", *Estuaries and Coasts: Journal of the Coastal* and *Estuarine*, Research Federation, vol. 31, pp. 233–238, 2008.

[FER 91] FERREL D.J., BELL J.D., "Differences among assemblages of fish associated with Zostera capricorni and bare sand over a large spatial scale", *Marine Ecology Progress Series*, vol. 72,pp. 15–24, 1991.

[GRE 03] GREEN E.P., SHORT F.T., *World Atlas of Seagrasses*, UNEP World Conservation Monitoring Centre, University of California Press, 2003.

[HAY 00] HAYNES D., RALPH P., PRANGE J. *et al.*, "The impact of the Herbicide Diuron on Photosynthesis in three species of tropical seagrass", *Marine Pollution Bulletin*, vol. 41, pp. 288–293, 2000.

[HEM 00] HEMMINGA M., DUARTEC.M., *Seagrass Ecology*, Cambridge University Press, Cambridge,2000.

[HIL 88] HILLMAN K., MCCOMB A.J., "The primary production of the Seagrass Halophila ovalis in the Swan/Canning estuary, Western Australia", *Biology of Seagrasses*, pp. 635–685, 1988.

[JAC 01] JACKSON J.B.C., KIRBY M.X., BERGER W.H. *et al.*, "Historical overfishing and the recent collopse of coastal ecosystems", *Science*, vol. 293,pp. 629–638, 2001.

[KEN 02] KENDRICK G.A., AYLWARD M.J., HEGGE B.J. *et al.*, "Changes in seagrass coverage in Cockburn Sound, Western Australia between 1967 and 1999", *Aquatic Botany*, 73, pp. 75–87, 2002.

[KUO 96] KUO J., KIRKMAN H., "Seedling development of selected Posidonia species from southwest Australia", in KUO J., PHILLIPS R.C., WALKER D.I. *et al.* (eds), *Seagrass Biology: Proceedings of an International Workshop*, University Western Australia, Perth, pp. 57–64, 1996.

[LAR 89] LARKUM A.W.D., DEN HARTOG C., "Evolutionand biogeography of seagrasses", in LARKUM A.W.D., MC COMB A.J., SHEPERD S.A. (eds.), *Biology of Seagrasses*, Elsevier, Amsterdam, pp. 143–165, 1989.

[LAR 06] LARKUM A.W.D., ORTH R.J., DUARTE C.M. (eds), *Seagrasses: Biology, Ecology and Conservation*, Springer, Dordrecht, 2006.

[LER 06] LERICHE A., PASQUALINI V., BOUDOURESQUE C.F., *et al.*, Spatial, temporal and tructural variations of Posidonia oceanica seagrass meadow faing human activities", *Aquatic Botany*, vol. 84, pp. 287–293, 2006.

[MAR 94] MARBÀ N., CEBRIÀN J., ENRIQUEZ S., DUARTE C.M., "Migration of large-scale subaqueous bedforms measured using seagrass (Cymodocea nodosa) as tracers", *Limnology And Oceanography*, vol. 39, pp. 126–133, 1994.

[MAR 02] MARBÀ N., HEMMINGA M.A., MATEO M.A., *et al.*, Carbon and nutrient translocation between seagrass ramet", *Marine Ecology Progress Series*, vol. 226, pp. 287300,2002.

[NAG 00] NAGELKERKEN I., VAN DER VELDE G., GORISSEN M.W. *et al.*, "Importance of mangroves, seagrasbeds and the shallow coral reef as a nursery for important coral reef fishes, using a visual census technique", *Estuarine, Coastal and Shelf Science*, vol. 51, pp. 31–44, 2000.

[ORT 06] ORTH R.J. CARRUTHERS T.J.B., DENNISON W.C., *et al.*, "A Global Crisis for Seagrass Ecosystems", *Bio Science*,vol. 56, no. 12, pp. 987–996, 2006.

[PHI 88] PHILLIPS R.C., MEÑEZ E., "Seagrasses", *Smithsonian Contributions to the Marine Science*, vol. 34, pp. 27–84, 1988.

[REU 01] REUSCH T.B.H., "New markers – old questions: population genetics of seagrasses", *Marine Ecology Progress Series*, vol. 211, pp. 261–274, 2001.

[SHO 00] SHORT F.T., WYLLIE-ESCHEVERRIA S., "Global seagrass declines and effects of climate change", in SHEPPARD C.R.C. (ed.), *Seas at the Millennium: An Environmental Evaluation*, Pergamon, Elsevier, Amsterdam, vol. 3, pp. 10–11, 2000.

[WAL 06] WALKER D.I., KENDRICK G.A., MCCOMB A.J., *Decline and Recovery of seagrass ecosystems: The Dynamics of Change, Seagrasses: Biology, Ecology, and conservation*, Springer, Dordrecht, pp. 551–565, 2006.

[WAY 09] WAYCOTT M., DUARTE C., CARRUTHERS T. *et al.*, "Accelerating loss of seagrasses across the globe threatens coastal ecosystems", *PNAS*, vol. 106, no. 30, pp. 12377–12381, 2009.

4

Biocomplexity of Coral Ecosystems: Diversity in All its States

4.1. Introduction

Biocomplexity is a concept that has underpinned ecology for a long time but has only recently been defined [COL 99]. The basic idea assumes that everything is different from the sum of its parts and that it depends on a set of rules that unite these parties and have them interact within a rather complex environment. This concept has conciliated several key areas in ecology (Figure 4.1), namely:

– the structure of life, from genes to the ecosystem;

– interactions between the various elements in this structure and with external and environmental factors;

– the integration of time in the construction of these elements and their interactions.

We direct the reader to [MIC 01] for a very general approach to biocomplexity, and to [CAD 06] for a more complete overview.

In the first of the key areas of biocomplexity, life is presented as being composed of multiple organization levels, the main levels being genes, the individual, the population, the species, the community and the meta-community. The boundary between these levels is often difficult to

---

Chapter written by Michel KULBICKI, Mehdi ADJEROUD, Laure CARASSOU, Pascale CHABANET, Valeriano PARRAVICINI, Dominique PONTON, Fanny HOULBREQUE and Laurent VIGLIOLA.

distinguish, and even more so as you go up the organization levels. Therefore, it is usually easy to define an individual, but for colonial species such as corals, this already poses many questions. Defining a population can be complex since it is necessary to define at what level the flow of individuals or genes with neighboring populations can be used to identify this population. The concept of species can appear to be simpler, at least for the vertebrates, but even for these there are many challenges as we will see in this chapter. Defining a community or meta-community will be based on the boundaries allocated to habitats, which is often complex and subjective since many species occupy several habitats throughout their life and at each moment in life, each species has a range that does not necessarily coincide with the boundaries of a single habitat. In addition to these organization levels, another characteristic of living organisms is that of ecological functions, which are much more difficult to define. It is reasonable to think that each species plays a number of roles in the functioning of a system. Each of these roles or functions may be accomplished by several species as a whole or in parts. This brings us to the concept of functional group, which is a set of species (or individuals) that perform a similar function within a community or meta-community.

This concept of a functional group represents a transition between the second key area of biocomplexity, the interactions between the elements of this structure. The individuals of a community interact with one another in many ways according to highly complex interactions. This involves intraspecific (e.g. reproduction, social behavior) and interspecific interactions (e.g. predation, parasitism, symbiosis, commensalism), all influenced by environmental factors on many scales. Human interaction is a separate case, since man is in some sense part of the community as a super-predator, but man is also a single actor through direct and indirect impacts on the environment at all scales. The action of man is a great example for the third key area of biocomplexity, the importance of time. Many studies (e.g. [JAC 01a, JAC 01b]) have shown that the current structure of biodiversity as well as the current levels of factors even the habitat boundaries are the result of human pressures, whose effects began even when man was present in low population densities. The history of human colonization therefore plays a central role in the current structure and function of natural communities. Other historical influences have equally important roles, such as, for example, the geological history of biogeographic regions as we will see in this chapter.

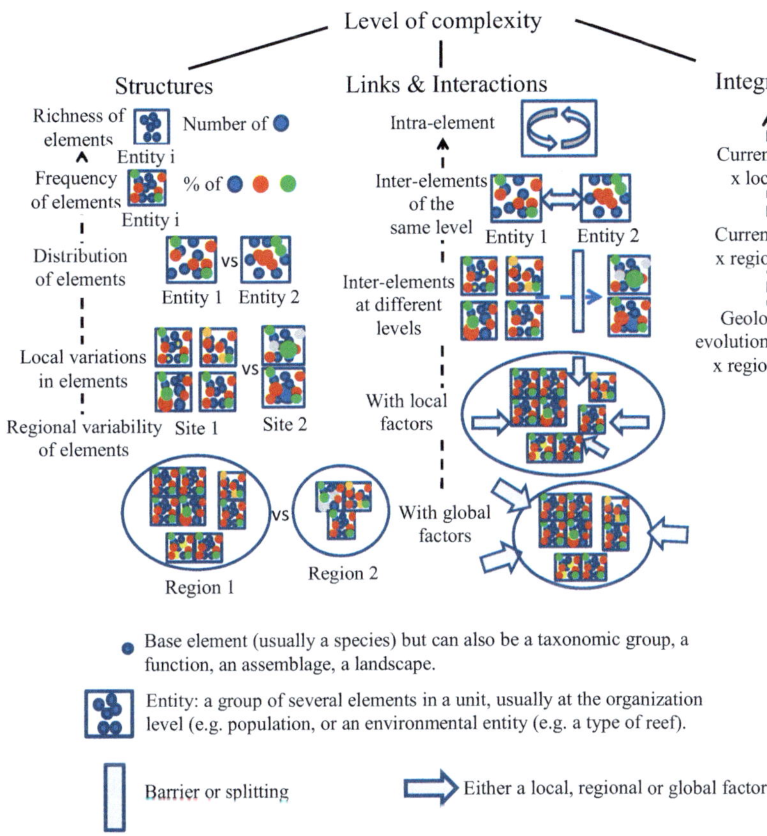

Figure 4.1. *Conceptual diagram of biocomplexity, adapted from [CAD 06]. vs: versus. The vertical dotted arrows indicate interaction between levels. Gray arrows indicate the actions of the factors (either global, therefore external to the system, or local, therefore internal to the system)*

Why should we look at biocomplexity in the context of coral reefs? Our knowledge of these systems is relatively recent compared to that of temperate or cold marine systems. The first scientific observations on fish populations in a coral environment date back to the end of the 1950s, whereas observations in a temperate environment[1] date back to the end of the

1 Reductionist approach: here this involves deducing "laws" from an experimental approach, for example, observations in tanks or cages or in an open but controlled environment.

19th Century. Therefore, coral reefs have not been studied as much as other systems via reductionist approaches even if this type of approach is becoming increasingly used [MIC 01] and involves analyzing the phenomena in an isolated manner, trying as much as possible to manage factors interacting with these phenomena. Moreover, coral reefs, as we will see in this chapter, have a remarkable spatial homogeneity[2] at many levels, which favors comparative approaches, one way to analyze biocomplexity. Coral reefs are also one of the most diversified systems on the planet and this high diversity does not lent itself to reductionist approaches and therefore requires analyses that can integrate this complexity of structures.

What are the main characteristics of coral reefs? These complex ecosystems (600,000 km^2 in the world) are found throughout the entire tropical belt (Figure 4.2), the latitudinal limits being 35°N (in Japan, [YAM 12]) and 32°S (Lord Howe Island). The main characteristic of these reefs is the presence of reef-building corals. The geographic limit of the distribution range of coral reefs is linked to the ecological requirements of corals, (see details in section 4.2.2.2). Corals create a complex habitat whose flora and fauna are usually highly diverse. Coral reefs are present in more than a hundred countries and provide a range of services for nearby populations (e.g. fishing, coastal protection and tourism) but also for the planet (e.g. carbon sequestration, biodiversity). These reefs are currently threatened by many stresses, both local (e.g. fishing, pollution, coastal developments, invasive species), and global (e.g. increase in sea surface temperature and ocean acidity, changes in the magnitude and frequency of cyclones). With polar regions and tropical forests, they are probably one of the most sensitive ecosystems and therefore their decline is rapid. At present, an estimated 20% of reefs have been permanently destroyed, and 35% are in a critical state and are seriously endangered [WIL 08].

In what follows, we will therefore try to show how a biocomplex approach can be useful to understand a system such as coral reefs. Besides this example, we will try to show how useful these concepts are and how they complement reductionist analyses.

---

2 Spatial homogeneity: possibility of finding the same species, the same structures or the same functions from one site to another.

Figure 4.2. *Map showing the main coral reefs in the world. Source UNEP, WCMC-008. Black lines represent the areas with high freshwater input forming a barrier for corals and associated ecosystems. In black are other natural barriers, essentially linked to cold currents or upwellings. The dots represent the western limit of the oceanic barrier between the Indo-Pacific and East Pacific*

4.2. Diversity in the coral world

In what follows, we are mainly going to focus on two groups of organisms, corals and fish. These two groups are by far the best-known in terms of diversity, biology and ecology in the coral world, even though our knowledge is still far from being complete. Furthermore, the matrix of coral reefs is generated mainly by corals that offer a complexity and diversity of habitats at different spatial scales, which form the basis of all coral reef diversity. Fish are the main exploited resource in these reefs and they are characteristic of this environment. This does not of course mean that other groups of organisms such as algae, mollusks, crustaceans, echinoderms, worms etc. are not equally diverse, if not more so, and do not play key roles in their environment. Our current knowledge of these other groups is generally less complete, and it is likely that the outline developed for corals and fish may also apply to the other groups, although each group has its own specific features.

4.2.1. *Overview*

4.2.1.1. *Different diversities*

Diversity is generally perceived and measured at the species level, but the situation is in fact much more complex. The genetic variability within a species can be seen as the first level of diversity. It is essential as it conditions the adaptability of species and consequently the stability, robustness and resilience of populations to environmental changes. The genetic variability is on the one hand intrinsic to the species (some species have a naturally low genetic diversity, for example due to their reproductive behavior), and on the other hand related to the environment, with isolated populations or with few individuals displaying a lower genetic diversity.

The next level of diversity is usually the population. The definition of a population is most often linked to genetic exchange. Within a community, that is to say the set of species found in a given habitat, usually there is only one population per species. When we consider several communities, a single species may have one or several populations based on the degree of exchange between these communities. Many organisms living in a coral environment have larvae that are frequently found in the pelagic environment for periods ranging from several days to several months. Usually species with a short larval stage will tend to form distinct populations more easily than species with a longer larval stage. The smaller and more isolated an island is, the greater the number of species with longer larval stages [LUI 13]. As a result, on these islands, local populations will be undoubtedly more related to distant populations than in continental systems where species have shorter larval life stages and a high self-recruitment rate [JON 99], i.e. the return of juveniles to their parents' areas of origin. Another term often used in coral reef ecology is associated with exploitation, individuals are considered to belong to the same population if they are targeted by the same fishing unit. This corresponds to the concept of "stock" in fisheries terms. Being very important in resource management, this definition is mainly used for its practicality, as it is relatively easy to define the action limits of a fishing unit in a coral environment. However, this concept has a conceptual flaw, as the local dynamics of a population is partially linked to its recruitment sources; these can be very remote or found in a non-exchange context (source populations and sink populations).

The subsequent levels, diversity, community and meta-community, are a matter of the definition of habitats. In the coral environment, most species are found in several habitats throughout their lives, as larvae, juveniles and then adults. A single species, during a given life stage, for example in the adult stage, may also be found in several habitats. For example, many coral fish visit adjacent areas, either regularly (for food or refuge), or occasionally (e.g. for reproduction). The same goes for non-reef species (e.g. some sardines) which occasionally visit reefs and may thus also participate in the dynamics of these reefs (e.g. food for reef predators). Zones adjacent to the reefs, whether they are seagrass beds, algal zones, mangroves or soft bottoms, therefore significantly participate in a variety of ways to the presence of species in reef communities. Meta-communities are no longer defined by their habitat boundaries but by their biogeographical limits, which are still not fully understood in the coral environment despite recent advances [KUL 13, VER 09]. Moreover, these biogeographical limits, which integrate their geological history and the evolutionary history of species, are not the same for different group of organisms within the same meta-community. Therefore, coral and fish species, even if they have many common biogeographical limits, do not all share the same boundaries. Currently, it is therefore difficult to define the biogeographical limits for meta-communities based solely on the species distribution. Consequently, biogeographical regions used in reef meta-communities are in general currently based on the limits defined by the physical characteristics of the environment, and expert opinions [SPA 07].

4.2.1.2. *The orders of magnitude of diversity in the coral world versus other biomes*

Coral environments are considered to be one of the most diverse ecosystems in the world, the marine equivalent of tropical forests on land. It is, however, difficult to quantify the difference between coral environments and other environments, because on the one hand, outside of the vertebrates and a few other groups, the biodiversity data is poorly known [APP 12, PLA 11], and on the other hand, there are distinct ideas of biodiversity:

– the overall diversity, i.e. all species living in a given ecosystem;

– the diversity density i.e. the number of species per unit area (or unit volume).

These two concepts are used to assess diversity. Conceptually, it is possible to have a high overall diversity but a relatively low density diversity, which would imply that, from one place to another, the species differ (high beta diversity[3] [KOL 03]). In the attempt to locate the diversity of coral reefs compared with other ecosystems, we are going to analyze two examples: fish and mollusks.

There are approximately 32,800 species of fish currently described in the world [FIS 10], 17,200 of which are marine species. There are 7,300 species of reef fish [KUL 13]; given that reefs represent an area of 0.02% of the marine environment, these figures show that the number of species per unit area is high. At present, we do not know of any other marine environment with such densities of fish. Some geographical comparisons, however, provide an overview of this diversity. Around New Caledonia, there are approximately 1,800 species of reef fish [FRI 11] with a coastline of approximately 1,500 km. In the Europe-Mediterranean zone, from the north of Norway to the mouth of the Nile, there are approximately 1,100 species within the same depth range for a linear coastline of over 30,000 km. However, a comparison with continental waters offers a very different perspective. In the Indo-Pacific, there are approximately 4,800 species of reef fish. The Amazon basin has less species (approximately 3,500 [JUN 07]) but in a biogeogrqaphical area that is approximately twenty times smaller.

Mollusks are also very diverse in the coral environment. Most species are small (<8 mm) and currently only a limited number of these species have been described. [BOU 02] point out that for New Caledonia alone (20,000 km^2 of lagoons) 2,738 species have been described, a similar number for that found in Europe, [BOU 06] indicating that 3,600 species of mollusks are known in the whole North-East Atlantic (approximately 1.5 million km^2). One of the most remarkable features of the mollusk diversity in the coral environment is, firstly, the possibility of incredibly high concentrations; for example, [BOU 02] indicate that 32% of 2,738 species of New Caledonia were recorded at a single site of only a few m^2 (which is more than the total number of marine mollusks recorded in Britain); however, as in most highly diverse fauna [MOU 13], the proportion of rare species (represented by a single example) is very high.

It is possible to extend this type of comparison to many other groups (e.g. brittle stars [STO 12], algae [KER 06, KEI 14], stomatopods [BAR 06],

3 Beta diversity: this is a variation rate in the diversity that measures the differences in the number of species or in the species composition between different periods or places.

hermit crabs [WIL 04]). The diversity of the coral world, however, still remains to be determined in most cases due to the challenges faced when sampling and identifying species (see section 4.2.1.4).

4.2.1.3. *The great patterns of diversity in the coral world*

The diversity of species is not evenly distributed in the coral world. The species distribution is a function of biogeographical regions, which can be seen as the result of the interaction between the biological and ecological characteristics of species and their environment over a geological time scale. There have been many attempts to split the world into such regions, essentially using one of three approaches:

– boundaries fixed by significant changes in the parameters of the environment, such as for example temperature, salinity, currents or the existence of natural barriers (e.g. [BEL 05]);

– expert opinions based on the level of endemism, [BRI 74] with a minimum level of endemism arbitrarily set at 10% for the marine environment;

– use of species composition (lists of flora and fauna) (e.g. [FLO 08, ROB 09].

More than one of these approaches can be combined [SPA 07]. On a larger scale, that is to say > 5,000 km, different approaches often give similar results (e.g. [KUL 13]). However, the results are usually "unstable" for zones less than 5,000 km, the boundaries or even the existence of biogeographical regions vary from one author to another. A good understanding of the limits of these regions is important since meta-communities are defined by these regions. As we will analyze later, there are some very important links between these meta-communities, i.e. regionally, and what will be observed in a given reef, i.e. locally. For example, we will establish that the levels of diversity observed on a reef are a function of the regional pool of species, that the robustness and resilience of local communities to change are also associated with the species available at the regional level, or even that the fish production on a reef also partly depends on the species present in the region.

We still do not know enough to be able to establish biogeographical limits for all phyla in coral environments. At present, there are few comprehensive analyses of these limits; the best documented, as we have already mentioned, are for corals [KEI 13, VER 09] and reef fish [KUL 13]. For these two phyla, the limits obtained have many similarities, despite the different methods used. The distribution of other groups shows either a strong correlation (e.g. gastropods and lobsters [ROB 02], shrimp [REA 08], seagrasses [GRE 03]) or divergences (e.g. sharks, marine phanerogams [TIT 10]).

The diversity of most organisms, however, peaks in the "Coral Triangle" which refers to an area in Indonesia, the Philippines and the China Sea (Figures 4.8 and 4.13 on corals and fish; also see [ROB 02]). The diversity decreases as we move away from this diversity center, with a greater decrease toward the center of the Pacific Ocean than the Indian Ocean, in which the diversity is relatively stable from India to East Africa. In the Indo-Pacific, there are few boundaries that are difficult to cross, even if there are several major obstacles (Figure 4.2). The main boundaries are:

– the boundary between the Indian and the Pacific Ocean in Indonesia (a limit corresponding to the Wallace line for terrestrial organisms);

– the boundary between the Australian and Pacific tectonic plate;

– the boundaries of freshwater from Himalayan rivers (Ganges, Brahmaputra and Indus) and the China Sea (Mekong, Yangtse-Kiang and Huang-Ho);

– the straits of the Red Sea and Persian Gulf.

An even greater barrier is the oceanic space between the islands of the central Pacific (Hawaii, Line Islands, Marquesas Islands and Easter Island) and the continents of America and the Galapagos Islands (5,300 km between the Marquesas Islands and the Galapagos Islands, the two closest archipelagos). Even if many species manage to cross this space (20% of reef fish in the East Pacific come from the central Pacific), the flora and fauna of the East Pacific are much less diverse than the rest of the Indo-Pacific. However, this flora and fauna, in many cases, show closer phylogenetic affinities with those of the Atlantic than with those of the Indo-Pacific. The flora and fauna of the East Pacific reach their maximum diversity around the Isthmus of Panama (e.g. [ROB 09]) with a decrease as the latitude increases. The limits of this region are associated with cold currents (California and South of Peru). In the Atlantic, the maximum diversity is observed in the Caribbean where the flora and fauna are relatively homogeneous (e.g. [FLO 08]). The mouths of the Amazon Rivers (Orinoco and Amazon) are barriers that are very difficult to cross for species associated with coral formations. Similarly, the diversity decreases significantly from west to east in the Atlantic, since there are no longer any true coral reefs in the East Atlantic or south of the Amazon [LUI 12].

4.2.1.4. *Other limits to our knowledge*

Analyzing the diversity involves many stages, two of which are essential:

– identification of species;

– knowing the spatial distribution of species.

As for other extremely diverse ecosystems, these two stages pose many methodological problems in coral ecosystems. The identification and description of species are done by taxonomists and systematicians using specimens whose origin should be defined with maximum precision. Some groups, such as sponges, tunicates, polychaete worms or mollusks have a large number of species but not enough taxonomists to describe the species that have already been collected. The number of taxonomists and systematicians has declined, with little recruitment over the past few years, meaning that the gaps in our knowledge will remain for some time yet. However, recent advances in genetics offer hope. With relatively limited gene sequences, it is nowadays possible to distinguish one species from another.

Despite some limitations, several international initiatives based on these sequences, such as *Consortium for the Barcode of Life* (CBOLD[4]) or Genbank have been implemented. For each specimen collected, a tissue sample is analyzed for a predetermined sequence of one or several genes, standardized photos are taken and, whenever possible, the specimens are sent to a museum. This approach, although it does not allow species to be described, allows them to be distinguished from one another and therefore named. Of course, there are other more powerful genetic analyses, for example when reviewing a family or genus. Many species, known as cryptic species as they are impossible to distinguish based on anatomical criteria, have been identified over the past few years. To give some examples, in fish of the genus *Trimma*, genetic analysis has allowed the number of species to increase from 52 to 94 and there are still many samples to process [WIN 14]. In corals, there are also numerous examples (e.g. [BUD 10, FUK 04, SOU 10]).

4 www.barcodeoflife.org/content/about/what-cbol.

Our knowledge of species distribution is often limited by difficulties in the collection of specimens. Beyond the regulations implemented over the past few years to prevent the looting of flora and fauna (and that mitigate or even prevent the export of samples), many species are difficult to collect, and there are many reasons for this. It may be due to the habitat (e.g. depth, reef matrix, zones with strong currents) or to the behavior of organisms (e.g. nocturnal animals, burrowers, animals capable of avoiding fishing gear) or size (very small organisms); the sampling effort is also very different from one region to another. In the Pacific, the flora and fauna of Japan, Hawaii, Great Barrier Reef or New Caledonia are quite well known. However, large regions are still poorly described, such as for example the entire Asiatic coast between Sumatra and the Persian Gulf, and are undersampled, with the exception of the Maldives and part of Thailand. The same goes for most of the African coast (from Somalia to Mozambique) or to the south of the Red Sea. It is very difficult to assess the proportion of species sampled in a given place. For some groups, such as fish, corals or some echinoderms, it is possible to proceed through comparison between well-known and less-studied groups. Let us take reef fish as an example. The family *Chaetodontidae* (butterfly fish) is composed of species that are easy to identify, collect, and therefore represent a reference family. However, the family *Apogonidae* (cardinal fish) requires a greater sampling effort to obtain a good estimate of their local diversity. Modeling (taking into account biogeographic factors such as island size or isolation) of the diversity ratio between these two families for a number of correctly sampled regions allows us, through extrapolation, to roughly estimate the diversity of cardinal fish in the less well sampled regions. Another method involves extrapolating the distribution of a species according to its known distribution and ecological envelope, which is given by the limits of the environmental conditions (temperature, salinity, turbidity, habitat and connectivity) between which a species is known.

This section shows that the diversity of coral reefs is still poorly understood, with the exception of some groups of organisms. It is unlikely that our knowledge on this diversity will improve significantly beyond the already explored groups in the next 10 years, especially with the lack of specialists and large exploration programs. Despite these uncertainties, the overall distribution of this diversity can probably be trusted, with the existence of biogeographic regions common to many phyla.

4.2.2. *Diversity of the main constituents of coral reefs*

In this section, we will focus in more detail on the distribution and characteristics of several large groups characterizing coral reefs. Here the aim is not to be exhaustive but to show how this diversity underpins the biocomplexity. The diversity is a function of habitat; therefore, the first step in this analysis will be to assess how the coral reef habitat is structured in general.

4.2.2.1. *Habitats*

Even if no "standard" coral reef exists since the geomorphology of a reef depends on many factors, it is possible to define the typical elements of a coral reef [BAT 75] for the Indo-Pacific, given that the organization of reefs in the Atlantic and East Pacific has many similarities. Most coral reefs are organized along two gradients, a gradient of depth and gradient of exposure to swell (Figure 4.3). The top of coral reefs is usually composed of a relatively flat zone, or plateau, often bounded by a reef crest, also called the algal crest, beyond which it slopes considerably; in some cases, this can become vertical. At the base of this slope is the accumulation zone, or reef embankment, then a transition zone toward adjacent habitats, often dominated by soft bottoms. The gradient of exposure usually includes a "windward" side and a "leeward" side. The windward zone withstands the effects of swell with little relief on the plateau, encrusting corals and crustose algae. The windward slope is often characterized by "spurs and grooves", the weathering of ridges parallel to swell in which the back and forth of water surges over the reef. Coral is sparse in the initial few meters of this zone and increases in diversity and structural complexity with depth. On the leeward side, the plateau is enriched with formations of branching, finger or tabular corals as the distance from the swell increases. The reef slope is usually quite gentle and dotted with diverse coral formations. This zone is also a sedimentation zone with the accumulation of sand and debris. On a larger scale (Figure 4.4), different types of reefs can exist along a coastal-offshore gradient, with or without the existence of a lagoon zone bounded by a more or less continuous line of reefs that form a barrier to the ocean.

The diversity of flora and fauna in coral ecosystems is probably due to their evolutionary history, habitat diversity and cycles of disturbance. The diversity of habitats in the coral environment is largely due to the diversity of coral forms (see section 4.2.2.2).

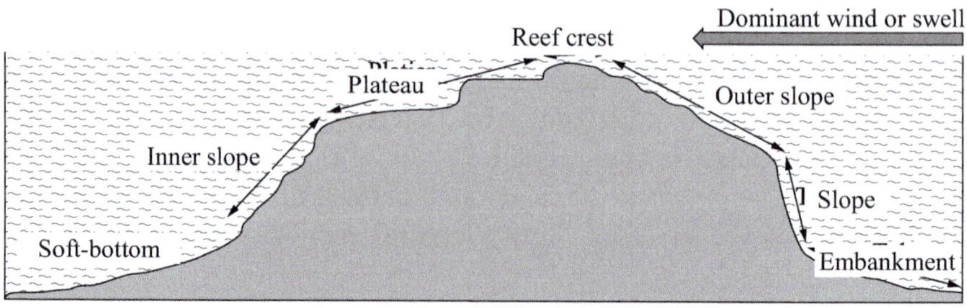

Figure 4.3. *Diagram of a coral reef showing both the gradients of depth and the gradients of exposure in the spatial organization of a reef*

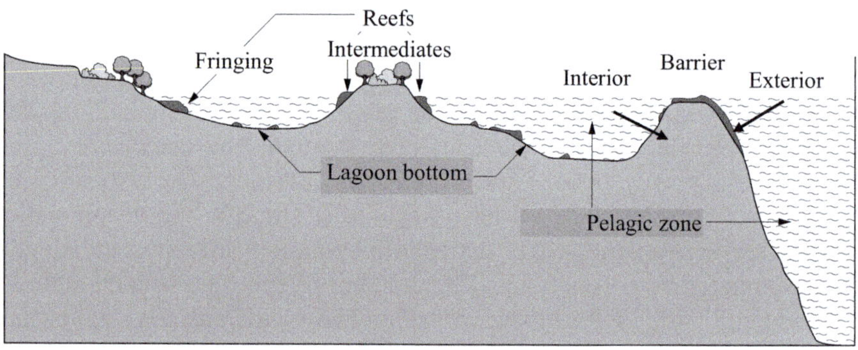

Figure 4.4. *Cross-section of a reef system. Note that the lagoon system may be absent or not include an intermediate reef*

Corals may either be branched, digitate, tabulate foliose, massive, encrusting, etc. these different forms can be found in complex combinations and therefore form a reef matrix with shelters of all sizes. Many other sessile organisms grow on this matrix, including encrusting algae, which play an essential role in the construction and resilience of coral reefs (see section 4.2.2.2); however, sponges, tunicates, soft corals, sea fans, etc. also play an important role in the reef structure dynamics and increase the complexity of the habitat. The evolutionary history probably plays a central role in the level of diversity of a reef. In fact, the diversity that we currently observe is the result of many interactions over extremely long periods. Therefore, it is highly likely that, previously, the tropical Atlantic would have had a greater diversity than what is currently observed (e.g. [BUD 96]). This decrease in diversity may be either due to the progressive isolation of the center of biodiversity in the Caribbean, particularly following the closure of the Isthmus of Panama, or due to unfavorable conditions in terms of refuge for reef species during glaciations in the Quaternary period [COW 13, PEL 14]. During these periods of global cooling, regions supporting the presence of coral reefs would become considerably reduced, especially in the Caribbean. As a result, with all other conditions being equal, Caribbean reefs were less diverse than those in the Indo-Pacific, which saw a progressive increase in zones favorable to corals and more refuges during ice ages, with a better geographic distribution than in the Atlantic [COW 13, PEL 14].

Correlations between coral diversity and those of other large coral reef phyla are very high, showing the high dependence of this system on the coral diversity. On a more local scale, one hypothesis shows that environmental changes – if their frequency and intensity are not too high or too low ("intermediate") – may also increase the diversity (intermediate disturbance hypothesis [CON 78, ROG 93, ROX 04]). According to this hypothesis, events such as cyclones, invasions by coral predators (*Acanthaster planci*), episodes of high sedimentation, etc. by destroying the coral matrix, can locally create new niches for pioneer species, which bit by bit are replaced by species highly subservient to corals. This hypothesis, although challenged [FOX 13], can partly explain the high spatial and temporal variability of diversity in the coral environment. There are, however, different patterns of diversity in the community organization between the Indo-Pacific and the Atlantic or the East Pacific. The Indo-Pacific is characterized by a high beta diversity [BEL 08]. However, the number of species found over small areas in the Indo-Pacific is not necessarily greater than in the Atlantic or East Pacific, where the number of species in a small area can also be very high, but unlike the Indo-Pacific, the species composition of communities remains very stable from one place to another ([ARI 08, Barneche *et al.* in press]).

The concept of habitat is a function of size and behavior of the species but also the organization level being addressed, the habitat of a population is not that of a community. With regard to the biocomplexity, this poses the problem of interlocking scales, as it would be naive to think that organisms, or higher organization levels, can be seen as living in complete autarky, in a fixed environmental envelope and limited in time and space. Moreover, as there are many levels of organization from the gene to the meta-community, there are habitat levels ranging from several microns to entire biogeographic regions. The succession of these habitats has a sort of fractal nature, the degree of complexity increases as the scales reduce. It is difficult to propose a universal nomenclature as the fractal dimensions are not homogeneous from one site to another. To highly simplify this, successive orders of magnitude can be considered, beginning with the inframetric scale (microhabitat), followed by metric (the corals measuring between 0.5 and 20 m^2), hectometric (the reef element such as for example the plateau or slope), then kilometric (the isolated reef, for example a portion of the barrier reef), decakilometric (a reef complex grouping barrier reefs and fringing reefs as well as adjacent environments such as soft bottoms, seagrass beds or algal beds). To give an idea of the complexity involved, a geomorphological classification of the world reefs has been carried out by the program *Millennium Coral Reef Mapping Project* [AND 06, AND 08]. This classification involves five levels of increasing complexity, the simplest involves two classes and the most complex has 795 classes (note that for the highest level, more classes may be added). Not only does the number of organisms concerned increase with increasing scales, but the number and complexity of links and interactions between them or with their environment also rapidly increase with each change of scale.

4.2.2.2. *Corals*

A reservoir of biodiversity, complexity and diversity of trophic networks and of biotic interactions, source of products and richness, coastal protection, etc. are some characteristics of tropical coral ecosystems. This high diversity is partly associated with the particular biology of the main bioconstructors of the reef, hermatypic corals [SCH 85] unlike slowly growing ahermatypic asymbiotic corals, which do not truly build reefs. The main founding species, reef-building corals, also represent food source and a habitat for many reef species, which gives them a key role in the maintenance and biodiversity of the reef [BEL 06, LES 04]. Moreover, most corals are sessile and have a particularly high longevity (several hundred years for some species), and, therefore, integrate environmental conditions over long periods. Sensitive to environmental variations, corals are therefore good indicators to detect the effects of both anthropogenic and natural disturbances [HOE 04, RIC 93, SAL 92].

Over the centuries, the term "coral" has been used in different ways, which has resulted in a slight ambiguity in its definition. Currently, most authors limit themselves to only cnidarians that secrete a rigid, well-developed calcareous exoskeleton (of aragonite, one form of calcium carbonate); this way "soft corals" and other little or uncalcified octocorals such as Alcyonacea are excluded. The "hard corals" include hydrozoa such as the fire coral *Millepora spp.*, entirely calcified octocorals such as the Mediterranean red coral *Corallium rubrum* (Linnaeus, 1758), the blue coral *Heliopora coerulea* (Pallas, 1766), the organ pipe coral *Tubipora musica* Linnaeus, 1758, and scleractinians (also known as stony corals). These, along with sea anemones, zoanthids, tube-dwelling anemones, black corals and coral limorpharians, form the subclass Hexacorallia. Reef-building corals, which are mainly scleractinians (Figure 4.5), live in symbiosis with unicellular algae (dinoflagellates of the genus *Symbiodinium*) and zooxanthellae [HOE 04]. These microalgae live in the living tissues of corals, more specifically in the endoderm of polyps, in incredibly high densities (several millions per cm^2). Note that corals also live in direct association with bacterial communities, present in the external mucus of colonies; so, the term "holobiont" is used to refer to the unit formed by the coral, its zooxanthellae and its bacteria [ROS 07]. By facilitating the growth of coral colonies, symbiosis between corals and algae largely explains their extraordinary capacity to bioconstruct reefs in waters that are often highly oligotrophic. Following the death of a colony, the accumulation and chemical transformation of calcareous skeletons form the framework of the reef.

Most reef-building corals are colonial (a key exception being mushroom corals, of the family Fungiidae); an adult colony produces several hundred or thousand polyps through the process of budding and is therefore genetically identical. Coral colonies are characterized by their highly diverse growth forms, of which there are eight main categories (Figure 4.6): branched/arborescent, tabulate, digitate, corymbose, foliose/laminar, encrusting, massive, free (colony not fixed to substrate). Note that some colonies have intermediate forms. These growth forms vary from one species to another, but intraspecific variations in the skeleton in response to environmental conditions are also observed and are called ecomorphs [VER 76]. This partly explains the difficulties encountered when identifying species in some coral genera [VER 13].

Corals are characterized by their high diversity, complexity and plasticity in their life history traits, and in particular their modes of reproduction: sexual reproduction through the release of gametes (external fecundation; the majority of species) or planula larvae (internal fecundation), and asexual reproduction through fragmentation, fission or different types of budding [HAR 90, HAR 11]. Some species can also release asexual larvae (parthenogenesis), which will therefore have an identical genotype to that of the parents. Asexual reproduction in corals leads to the formation of clones (colonies physically separated but genetically identical) within a population, which poses problems when defining the individual and the longevity of these organisms. Some species of corals are hermaphroditic (a colony with both male and female gonads) and gonochoric species (colonies with separate sexes), but all are iteroparic. Each coral species can reproduce in different ways as mentioned. The relative importance of sexual and asexual reproduction can vary, within a species, with the geographic location of a population [AYR 00, BAI 09].

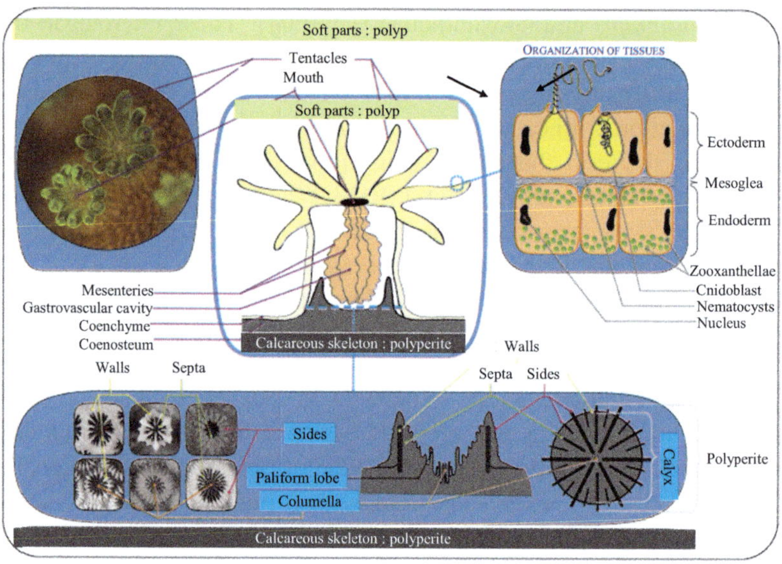

Figure 4.5. *Diagram showing the structure and main elements of a polyp (soft part) and the polypierite (skeleton) of scleractinian reef-building corals (illustration: P. Bosserelle)*

Figure 4.6. *Main growth forms of coral colonies (illustration: P. Bosserelle)*

Coral larvae, whether produced by means of internal or external fecundation, or even parthenogenetic reproduction, will for the most part live a pelagic life. The release of larvae is usually an episodic phenomenon (many coral species reproduce only once a year and sometimes in a synchronous manner) and larvae are difficult to observe *in situ* due to their small size (< 2 mm).

At present, many questions have been raised about their lifespan and their spatial dispersion which ensures connectivity between populations. The few existing studies show the high variation in larval lifespan within and between species (from several hours to several weeks [HAR 90]). However larval lifespan, often estimated in laboratory experiments, is not always correlated with the spatial scale of dispersion in the natural environment [FAU 12, SEL 11]. Larvae with short lifespans are also capable of attaching to floating debris and can therefore travel thousands of kilometers in the open ocean. However, there are sometimes local currents that keep larvae in proximity to the source reef, which consequently limits the dispersion of larvae, even if they have large lifespans. Studying the genetic structure of populations is an approach that often avoids a certain number of difficulties linked to the study of the larval phase, and also allows the relative contribution of asexual versus sexual reproduction to be assessed [ADJ 14, AYR 00].

After the pelagic larval phase, the *planula* will attach themselves to the reef substrate (installation phase) and will rapidly begin to develop their calcareous skeleton. During this stage, the recruited juveniles are only composed of a polyp and are less than a millimeter in size. Intracolonial budding rapidly multiplies the number of polyps. These young colonies (i.e. juveniles), composed of several dozen/hundreds of polyps, are still not sexually mature. It is only after three or four years, when they reach a diameter of approximately 5 cm, that these colonies reach the adult stage and can sexually reproduce. This critical period between the installation of larvae on the substrate and the beginning of the adult phase corresponds to the recruitment phase [PEN 13]. During this phase, young corals are particularly sensitive to the physico-chemical conditions of the environment, predation by corallivorous species that feed on polyps or degrade their skeleton, and spatial competition with algae and other invertebrates such as sponges or soft corals [PEN 10]. The mortality over the first few years of life is particularly high, and ultimately, only a handful of larvae out of thousands produced by each colony during spawning will survive and reach the adult stage.

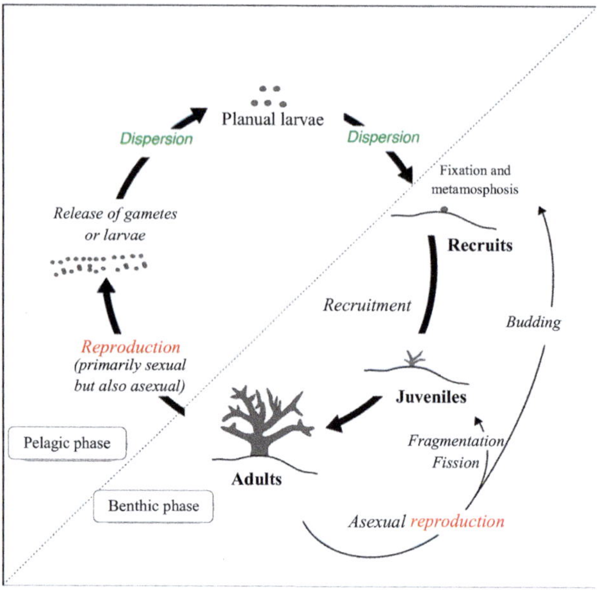

Figure 4.7. *Diagram showing the coral life cycle and their main modes of reproduction (illustration: M. Adjeroud)*

COMMENTS ON FIGURE 4.7.– Except for rare exceptions, adult corals are fixed to a substrate. Sexual reproduction leads to the release of larvae (internal fecundation) or gametes (external fecundation). Corals also reproduce asexually (fragmentation, fission, budding, and even the production of parthenogenetic larvae). Planula larvae have a pelagic phase, whose lifespan varies according to each species. Larvae, which are often less than a millimeter in size, attach to the substrate and metamorphose rapidly. Sexual maturity is reached after approximately four years.

The development and distribution of reefs throughout the globe depend on the ecological requirements of their main builders, the corals [COU 12]. On a global scale, coral growth is dependent on temperature and salinity. Corals are stenothermal organisms whose lower temperature limit is approximately 18°C. However, they do have an upper temperature limit (approximately 30°C) beyond which the physiology of corals is affected, which causes the mass expulsion of zooxanthellae: this phenomenon is known as coral bleaching, which is currently one of the most serious effects of global change on coral reefs [DOU 03, HUG 03a]. There are some rare cases of coral species that are well adapted and capable of withstanding temperatures higher than 36°C or less than 18°C, such as in the Persian Gulf. Most corals tolerate a salinity range between 28 and 40. Large decreases in salinity cause significant stress and can cause mass death. This is what, on a global scale, explains the lack of coral reefs close to large rivers such as the Amazon or Ganges. With regard to light, hermatypic corals, due to their symbiotic association with zooxanthellae, can only live in the euphotic zone. Reefs are therefore only located in the first 50 m, and beyond this depth, coral colonies are too rare and dispersed to form a true reef structure.

It is difficult to give an exact figure, but it is estimated that the number of hermatypic coral species is approximately 800, with an estimated 110 genera and 18 families. Despite recent progress due to the advances in genetic techniques and molecular biology, the identification of corals is still a particularly difficult task [VER 13]. Morphoanatomical criteria of the skeleton, on which the identification of most species is based, are still difficult to assess and are often quite subjective. The presence of ecomorphs in some genera also poses a serious problem when establishing species boundaries [VER 76]. The systematic and taxonomy of corals is currently experiencing a certain turmoil. Recent studies combining morphometric and genetic analyses (integrative taxonomy) led to a thorough review of the systematics of scleractinians, with changes being made to many genera and families [VER 13], M. Pichon – personal communication). In this context, some species, yet widespread and which have been subject to many analyses over the past decades, such as *Pocillopora damicornis* and *Stylophora pistillata*, have proved to be cryptic species complexes [KES 13, SCH 14, SOU 10].

On a biogeographic scale, corals are grouped into two provinces, in which several regions are distinguished [ROS 88, HUG 03b]. The Atlantic province includes reefs in the Caribbean, Mesoamerica, the coast of Brazil and those of West Africa. The Indo-Pacific province includes reefs of the Indian Ocean, Red Sea and Pacific, which is split into reefs of the East Pacific, Central Pacific and West Pacific. Note that the coral diversity is significantly higher in the Indo-Pacific province (780 species versus 70 reef species in the Atlantic), and there is no common species shared between the Indo-Pacific and Atlantic province. Within each of these provinces, the coral diversity is not homogeneous, but instead varies greatly by geographic location [HUG 03, ROS 88]. In the Indo-Pacific province, a diversity gradient is observed: the center of maximal diversity (the coral triangle (section 4.2.1.3) has approximately 600 species of coral (80 genera)). Coral diversity decreases as we move away from this region, toward the east or west. Recorded are approximately 380 species on the Great Barrier Reef, 350 in New Caledonia, 170 in French Polynesia, 50 in Hawaii, and only about 20 on Clipperton Island, at the eastern extremity of the Pacific Ocean.

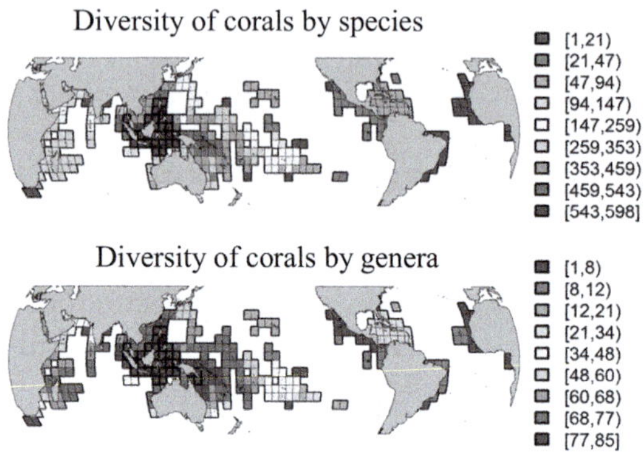

Figure 4.8. *Diversity gradient of corals by number of species and genera. Data taken from the IUCN database on corals*

4.2.2.3. *Fish*

A key issue is to define what a coral fish is [BEL 98]. For some sedentary species, whose territory is entirely within coral zones, such a topic is redundant; this is not so for many species that only frequent coral zones part of the time or whose territory includes habitats other than coral reefs. This question becomes complicated when we consider undeveloped reefs, such as in many areas of the tropical Atlantic or East Pacific, since even species living in proximity to corals are not necessarily coral species. To simplify this, we will consider a coral fish, any species that spend part of their time on a coral reef or a reef where there are corals. Therefore, some coastal pelagic species, such as sprats or trevally, will be considered as coral species as they migrate seasonally or daily to reefs, acting both as prey for many reef predators and predators of many small reef species.

Approximately 7,300 species (42% of the marine species) are associated with coral reefs according to our definition. This number of species is an under-estimate as it is difficult to give an accurate number; however, Mora *et al.* [MOR 11] estimate that 92% of existing species have been described.

These estimates are of course not homogeneous between the different regions. In the Atlantic and East Pacific, most species in the 0–50 m zone have been described whereas in the Indo-Pacific many regions have been under-explored. Among the families or genera that require further sampling, identification or description include *Gobiidae*, *Trypterygiidae*, *Blenniidae*, *Apogonidae* represented by small species (<15 cm) or most anguilliforms or *Bythitidae* which hide in the reef matrix.

Coral fish, like most living organisms, are split into families; few families are very diverse and there are many families with very few species, or even just one (Figure 4.9). Therefore, the 10 most diverse families represent 51% of the total diversity, whereas 43 families have only one species (21% of families) only representing 0.6% of the diversity. These figures change slightly according to region (Figure 4.9). Families with only one species often have life history traits far from average [MOU 13], and consequently, these rare species often play specific roles in the functioning of reef fish communities.

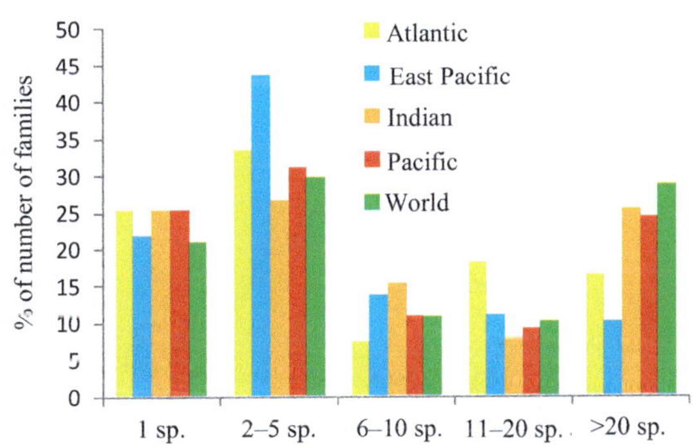

Figure 4.9. *Proportion of families according to number of species, in the four oceans and globally*

As with corals, many species of coral fish have a vast geographical range. [LET 97] have shown that fish communities of La Reunion and French Polynesia have more than 50% species in common, even though these two regions are separated by a distance of 15,000 km. This means that fish communities from highly separated regions can show many taxonomic similarities and therefore functional ones. This brings us onto endemism and speciation zones, an important specificity of coral reefs is that there is no overlap, for corals and fish, between regions of high diversity and regions of high endemism [HUG 02]. Endemism in reef fish is distributed differently according to the ocean (Figure 4.10), with much greater endemism in the Atlantic and East Pacific than in the Indo-Pacific. In the Atlantic, the endemism is due to the significant isolation between different basins [FLO 08], with non-endemic species showing the ability to disperse [LUI 12]. In the Indo-Pacific, the endemism is concentrated around the peripheral regions, such as the Red Sea, Hawaii, the Marquesas Islands and Easter Island. The abundance of endemic species in the Indo-Pacific is distributed differently according to whether we consider central or peripheral regions (Figure 4.11). In central regions, endemic species are not abundant, whereas in peripheral regions these species form a significant part of the communities, and in some cases become the majority. Endemic species do not differ greatly from other species in most life history traits, except for size and to a lesser extent diet. In central regions of the Indo-Pacific, endemic species are smaller than their counterparts in peripheral regions (Figure 4.12); also note that in these peripheral regions, endemic species include macro-herbivorous or corallivorous species whereas these trophic groups are absent from endemic species in central regions.

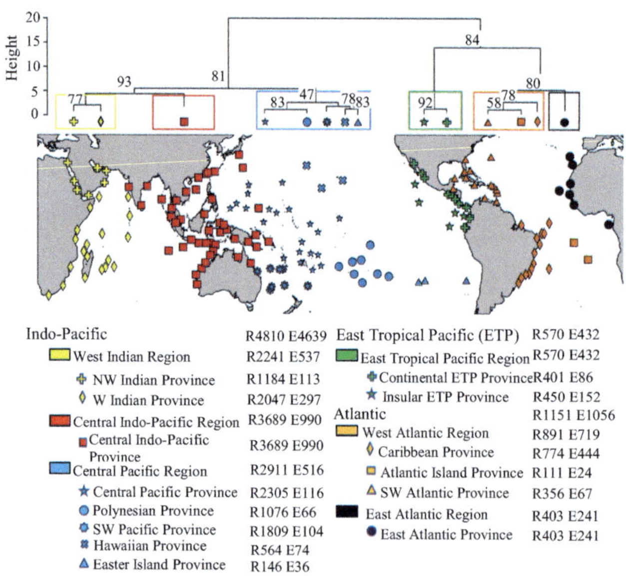

Figure 4.10. *Biogeographic regions defined by the species composition of reef fish [KUL 13]. The R numbers indicate the number of known species for the geographic area and the E numbers indicate the number of endemic species. For a color version of the figure, see color section*

Endemic species are more vulnerable in the long-term than non-endemic species due to the low connectivity in their populations. The abundance of these endemic species in peripheral regions of the Indo-Pacific indicates, however, that these species are better adapted to local conditions than equivalent species with a wider geographic distribution. These, however, contribute most to the abundance of communities in most regions. This may be associated with their ability to recruit and colonize which are probably above average. Given that the larger the species are, the greater the distribution range [LUI 13], usually large species in a community have wide distribution ranges, which protects them from local or general extinction. As fishing primarily targets larger species, the wide geographical distribution of these species therefore gives them some protection against extinction due to fishing.

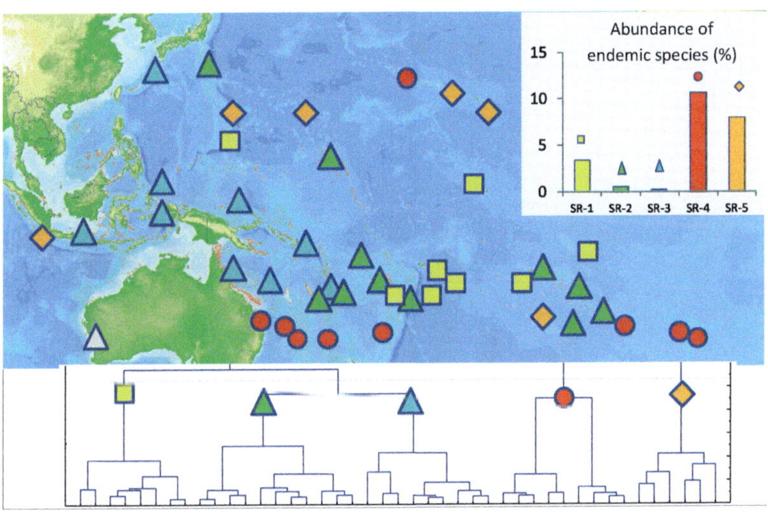

Figure 4.11. *Grouping of sites according to the relationship between abundance and size of geographical distribution range of reef fish species in the Pacific (unpublished analysis on 118 islands and 8,000 transects – GASPAR program of FRB)*

COMMENTS ON FIGURE 4.12.– SR-1 to SR-5 refer to site groups defined in Figure 4.11. Note that endemic species greater than 50 cm are only found in groups SR-4 and SR-5. On the graph to the right, the categories represent trophic groups dominated by the following food items: HM: macro-algae; SI: sessile invertebrates; FC: fish; HD: turf and microalgae; OM: omnivores; PK: plankton; MI: mobile invertebrates. Note that HM and SI only include endemic species in groups SR-4 and SR-5.

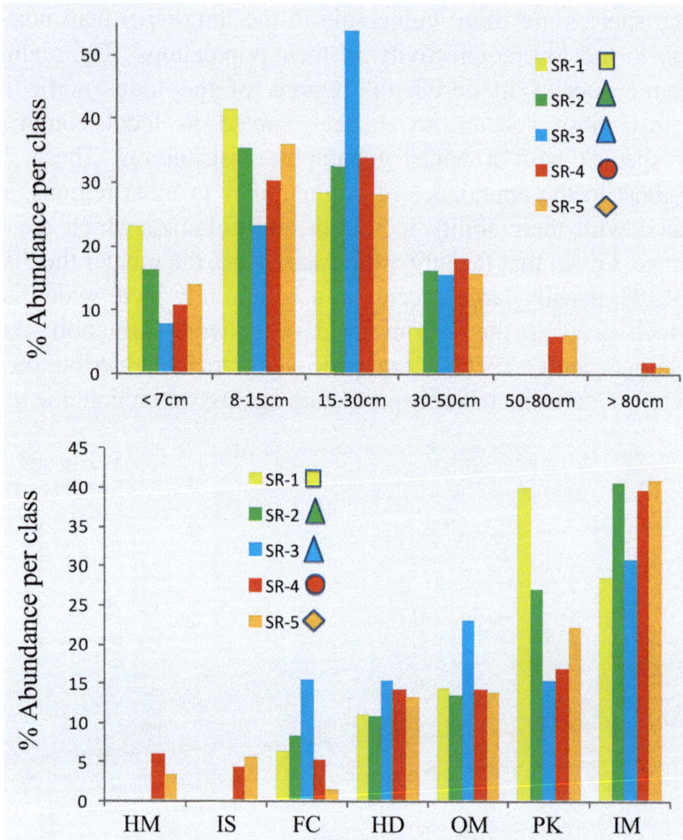

Figure 4.12. *Species distribution of endemic reef fish according to their size and diet in the Pacific*

The number of reef fish species is a function of several large-scale factors. [PAR 13] proposed a model that classifies these factors into four main groups of hypotheses:

– hypotheses associated with energy, particularly sea surface temperature (SST);

– the biogeography, including size of the islands and size of reef area as well as the connectivity (distance to the closest island or continental shelf);

– the geological history, essentially represented by biogeographical region;

– the mid-domain effect which is the accumulation of species at the center of an area of distribution following the overlap of ecological envelopes of species present.

The model by [PAR 13] accurately described the patterns of diversity observed (80% of variance explained) (Figure 4.13). The coral area and the biogeographic region are factors that have the greatest impact on the diversity of fish, followed by the coast line length, the connectivity and mid-domain effect; the surface temperature only plays a minor role once the other factors have been taken into account. Diversity increases with reef area, which is consistent with the biogeography theory of islands [MAC 67].

The geological history plays a key role, with, as mentioned above, fewer species in the Atlantic and East Pacific. These two regions have experienced large extinctions in the past, caused by the closure of the Isthmus of Panama but also the small area and low geomorphological diversity of refuge zones [PEL 14]. These repeated extinctions not only caused the number of species to plummet but also selected larger species in these two regions with regard to the rest of the coral world. This has many repercussions in terms of conservation and management (see Chapter 5). The coast line length and connectivity play an essential role in term of recruitment; the longer the coast line length the larger the connectivity and the easier it will be for species to recruit in adjacent zones or even auto recruit (juveniles colonize zones where their parents inhabit) [BER 12]. The mid-domain effect on the diversity of reef fish in the Indo-Pacific is contested [CON 03], but the measurement used (distance to the closest center of biodiversity) has a significant effect; the diversity decreases as we move away from the center of biodiversity (once the other factors are taken into account).

[BEL 01] found that the proportion of the diversity represented by a given family, whether corals or fish, remains more or less constant from one end of the Indo-Pacific to the other as long as there is a sufficient number of species. For example, the *Labridae* represent 7.7–8.2% of the total diversity both on the coasts of East Africa, as well as in Indonesia and French Polynesia. As there is a relatively strong link between taxonomy and life history traits, i.e. the latter are preserved within a taxonomic group, we can expect the proportion of different traits to also remain relatively constant from one region to another.

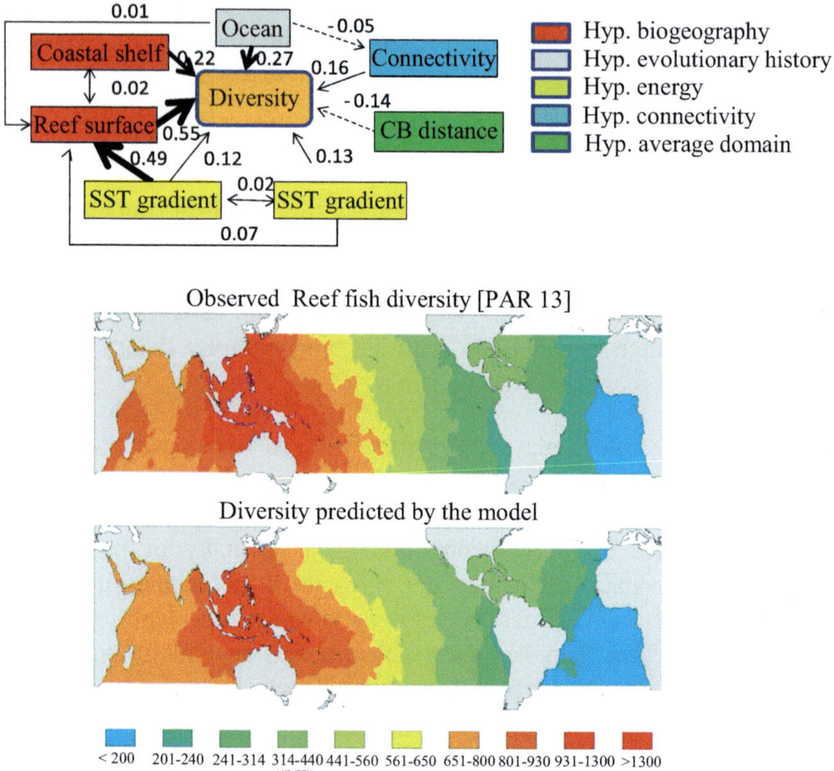

Figure 4.13. *Model (SEM: structural equation model) representing the relative role of different factors to explain the level of reef fish diversity throughout the world [PAR 13]. The maps show, on the one hand, the distribution observed in the diversity of these fish, and on the other hand the diversity predicted by the model. The adjustment is optimal in the zones where the diversity is at its maximum; the most significant differences are observed around the periphery (south-west Indian Ocean; oceanic islands in the East Pacific). For a color version of the figure, see color section*

Recent studies (Parravicini *et al.* in preparation) effectively confirm that the contribution to the total diversity of most complies with a null model; in other words, the proportions remain constant within a given ocean. However, trophic structures and size (Figure 4.14) do not comply with a null model. In particular, the smallest sizes are preferentially found in regions with high diversity and large sizes are preferentially found on zones of low connectivity. With regard to diet, the proportion of invertebrate feeders decreases as diversity increases, with the proportions of piscivores, planktonophages and omnivores increasing simultaneously. The biogeographic region plays a very important role in the proportions observed. Therefore, as already mentioned, the average overall size is higher in the East Pacific and Atlantic than in the remaining coral reefs.

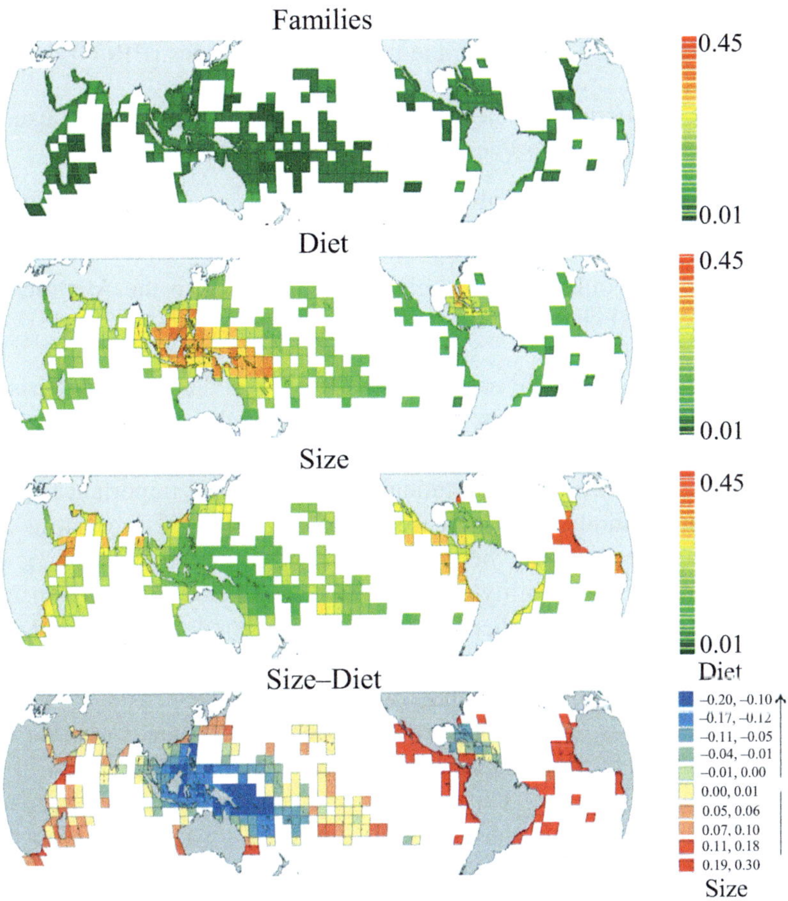

Figure 4.14. *Deviation from a neutral model for the spatial distribution of reef fish for taxonomic composition (at the family level), trophic composition (seven feeding classes) and size (six size groups). For a color version of the figure, see color section*

COMMENTS ON FIGURE 4.14.– The closer the values are to zero the more the proportions observed indicate a distribution similar to being selected at random. The last map represents the differences between the maps showing distribution by size and by diet, blue indicates that the communities are structured by diet, and red, by size.

Coral fish communities are mainly composed of small fish (<15 cm) in most regions [KUL 15]. The high abundance of small sizes is a common phenomenon in ecology, primarily due to metabolic reasons [BRO 04].

As already mentioned, species in the Atlantic or East Pacific are usually bigger than those in the Indo-Pacific. This has two main consequences:

– the abundance will be higher in the Indo-Pacific any other additional factor being equal;

– communities should be more vulnerable to fishing in the Atlantic and East Pacific (see next chapter).

Coral fish have a much more varied diet than their counterparts in temperate or cold environments do, in particular with herbivorous species and species that consume sessile invertebrates. Herbivorous fish represent between 3.5% and 7.3% of the regional diversity. Their importance varies from one ocean to another and increases with temperature. Herbivorous fish partly control the balance between algae and corals [CAR 13], an equilibrium vital for the stability of reefs in the long term. Fish that consume sessile invertebrates are another characteristic of coral fish communities. There are relatively few of these species, which can be split into species that consume coral and species that consume sponges and tunicates. Corralivorous species, primarily *Chaetodontidae* and some *Gobiidae* and *Labridae*, are small in size (usually <15–20 cm) and are usually found in the Indo-Pacific, whereas those that consume sponges and tunicates, primarily *Pomacanthidae*, *Monacanthidae* and *Tetraodontidae*, are found in all oceans and have a much greater average size (>25–30 cm). However, these fish are not abundant enough to regulate corals, sponges or tunicates. Nevertheless, reef fish have the capacity to adapt their diets and they will eat what is available, however respecting the physiological or morphological constraints of their species. For this reason, there is a large number of omnivorous species, i.e. species that consume both plants and animal matter, and the proportion varies according to the level of availability. This flexibility in diet is probably a consequence of the high level of competition associated with the high diversity of these communities. We therefore find, on the one hand, specialized species with a narrow dietary niche and on the other hand, opportunistic species with a wide dietary niche and a high capacity to adapt, showing that diversity does not necessarily lead to competition toward ultra-specialization.

We do not have space in this chapter to discuss other life history traits of reef fish in detail, but there are two traits that are important in order to understand the characteristics of these communities: pelagic larval duration (PLD) and the home range of adults. In the coral environment, most species of fish have a pelagic larval stage after which the larvae transform into juveniles and select a reef habitat. The pelagic phase of larvae varies according to the species and to a lesser extent environmental conditions, and is usually between one and ten weeks (in some cases it is longer). This pelagic stage allows the dispersion of species often over large distances. Generally speaking, there is a correlation between size of species, their PLD and size of their geographical range; small species tend to have short PLD and small geographical ranges.

Once settled on the reef, the species may have different strategies to occupy space. Many species are sedentary or territorial, living all of their adult life in only a few m^2, at most several hundred m^2. This is the case for example with gobies, blennies, butterfly fish and damselfish. Most of these species are small, for metabolic reasons (a small territory does not provide enough energy to a large species, except if the metabolism is very low).

A second group includes species that remain on the same reef for long periods. These are primarily medium to large species as well as some schooling species. This group includes parrotfish, surgeonfishes, some grouper and triggerfish. These species will, however, leave the reef if conditions become unfavorable or also to reproduce, unlike sedentary or territorial species.

The last group includes fish that frequently change reefs and cover large distances. Most of these species are, in fact, pelagic, such as trevally, which use reefs as a food source, and less frequently, as a refuge.

The last two groups are mainly composed of species with large geographic range and long PLD. They are much less vulnerable to global changes in the long-term than sedentary or territorial species due to their high capacity to disperse and colonize. However, most species targeted by fishing are situated in these two groups, with a large home range, which can cause local fragility (see next chapter).

4.3. Links between diversities

4.3.1. *From local to regional*

The spatial organization of diversity can be thought of as a funnel model with a reservoir of species linked to the biogeographic region at the top and a series of environmental filters, which restrict species bit by bit as this funnel descends, to ultimately give the species composition of a given location. This local species composition is not, however, constant because environmental conditions are constantly changing and the filtering effects are not always the same. There is a correlation between the number of species of a region (often called gamma diversity) and the number of species found in a given location (usually called alpha diversity); the relationship is generally asymptotic (Figure 4.15) according to [LOR 00].

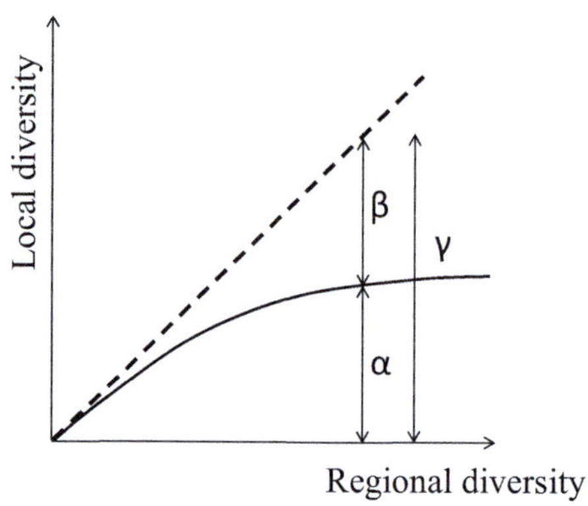

Figure 4.15. *Relationship between local (alpha) and regional (gamma) diversity. The dotted like indicates equality between local and regional diversity (all known species in the region are locally present). The beta diversity represents the difference between a mean local diversity and the available regional diversity (according to [LOR 00])*

As we have already seen, the regional diversity is a function of large-scale factors such as biogeographic region, connectivity, surface temperature, reef area or coast line length. Consequently, the local diversity will therefore do the same. Therefore, the local diversity (alpha), on a small island will be lower than on a larger island (Figure 4.16) since the gamma diversity of a small island is lower than that on a large island. The next question we will cover is how species that live locally are "filtered" in the regional species reservoir. Many studies on fish show that local factors, such as depth, reef type and live coral cover determine the species composition of a given reef. This species composition, for a given environment, is not however unchangeable in time or space. Some species are replaced by other species that occupy similar functions, or even some species may temporarily disappear. These variations in species composition as well as relative abundance of species are usually attributed to variations in recruitment [CAL 96, DOH 91]. However, recruitment alone cannot fully explain this [CAL 96]. Another possibility is associated with the reef load capacity. At the base, a given reef receives a certain amount of energy and nutrients that condition the abundance and diversity of herbivores (or primary consumers) which in turn determine the secondary consumers and then tertiary. The relationships between these different trophic levels are controlled by a set of rules resulting in three key functioning diagrams: *top-down*, *bottom-up* and *wasp-waist* [CUR 03]. The *top-down* model is a system where predation controls the structure; *bottom-up* is a system controlled by primary production or consumers of primary production; *wasp-waist* is a double action of both *top-down* and *bottom-up* leading to a bottleneck in the middle of the trophic pyramid.

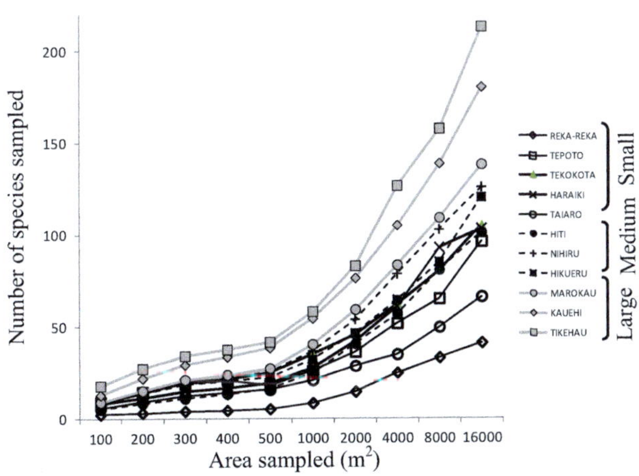

Figure 4.16. *Relationship between alpha diversity and sampled area for twelve islands on French Polynesia [KUL 00]*

COMMENTS ON FIGURE 4.16.– All the islands are atolls of Tuamotu with the exception of Mooréa, which is a high island part of the Society Islands. Areas less than 500 m^2 correspond to "intra-transect" values, those over 500 m^2 to "inter-transect" values.

The variability in diversity, whether spatial or temporal, is called the beta diversity. This variability can be measured in many ways [KOL 03]. The diversities are *a priori* linked to one another; the greater the gamma diversity the greater the alpha and beta diversity will be. There are, however, exceptions, which must be understood, as they allow a better understanding of how reef communities function. As previously mentioned, the alpha diversity is associated with scale. Current studies on reef fish show that the gamma diversity is lower in the Atlantic and East Pacific than in the West Pacific (by a factor of 3 to 10). However, on a smaller scale (40 m^2), the diversity observed on reefs in the Atlantic or East Pacific is greater or equal to that of the West Pacific. Then, as we consider large scales, the difference decreases to finally reverse in favor of the western Pacific.

The explanation of this phenomenon is found within the biology of the species. In the West Pacific, a large proportion of species are small and territorial or sedentary whereas in the Atlantic and East Pacific, there is a greater proportion of larger species with large home ranges. Consequently, in these regions, fish occupy the whole area, but the same species are always found from one site to another (high alpha diversity compared with gamma diversity and low beta diversity), whereas in the West Pacific, each species occupies a reduced space but the species constantly change from one place to another (low alpha diversity compared with gamma diversity and high beta diversity). This results in the resources being used in completely different ways depending on the region, but also a very different species vulnerability (see next chapter).

Diversity is not randomly distributed in space or time; they follow rules that are still poorly understood in the coral environment. Generally speaking, the local biological diversity increases with the complexity and diversity of landscapes and energy input. The complexity and diversity of landscapes depend on the geology and geomorphology on the one hand, but also reef-building organisms such as corals, calcifying algae and sponges. These reef-building organisms depend on the input of energy, nutrients and environmental conditions. There is therefore a dual effect of energy input, i.e. the water temperature. This brings up two "rules", that of Bergmann and that of Rapoport, which have some similarities. Bergmann's law states that within the same group of organisms (for example fish or corals), the mean size of individuals and/or species will be greater in cold environments than in warm environments. This law was originally used for warm-blooded organisms, but seems to apply regularly to other organisms [BLA 99], fish presenting however, numerous exceptions [MIL 06]. In fact, this law is often confused with a second phenomenon, the mean size of species decreases when the diversity increases [KUL 15]. In the coral environment, diversity increases with temperature and for many organisms, the best example being fish, the proportion of small species increases with diversity, and therefore with temperature. However, if we carefully separate the effect of diversity and temperature, there is no strong correlation between size and temperature in coral fish, particularly in the Atlantic and East Pacific [KUL 15].

Rapoport's rule states that the geographic range of species decreases with latitude, with tropical species having on average smaller distribution ranges than their temperate or subtropical counterparts. This law is also associated with the relationship between diversity and size of species. In fact, smaller species, usually when diversity is high, have smaller geographical ranges than larger species [LUI 13], as their capacity to disperse and colonize are more limited than larger species, associated with their more specialized habitat (and also sometimes diet). Both Bergmann and Rapoport's rules therefore refer to the relationship between diversity and size of individuals or species. From this point of view, coral reefs are quite specific, due to the reef-building nature of many organisms, such as corals, and the increase in diversity of organisms with temperature; the benthic landscapes become more complex as waters become warmer. A more complex habitat provides refuges and intermediate trophic levels, and therefore increases diversity, for example, of fish. There is therefore a double trigger effect of water temperature on the diversity and size of species and individuals (at least for fish).

We have already mentioned other effects on diversity such as the geological history, which determines the available species reservoir, in quantity and quality. In particular, the regional diversity (gamma) will be somehow "blocked" by the species reservoir capacities in the ocean considered. So, for coral reef fish, there are approximately 1,100 species available in the Atlantic, with up to 630 species in a given region (northern Caribbean), whereas in the Indo-Pacific, there are more than 5,000 species available with a regional peak of 2,300 species in the Coral Triangle. It is, however, important to note that our knowledge on the spatial distribution of diversity in coral environments is still very limited; only some groups (fish, corals, algae and echinoderms) are understood enough to be analyzed. It is likely that all organisms do not have the same distribution patterns and some factors have more weight than others in the distribution of their diversity.

One frequently asked question when analyzing the relationships between local and regional diversity, is the conservation of structures. An example will better explain this. If in a species reservoir of a region there are 10% herbivores, will there be 10% herbivores in all areas of this region? This has been analyzed essentially in coral fish, but it is likely that in broader terms the results are reproducible with other groups of organisms. The scale of analysis is large in this case. For a large scale, are structures conserved when we move from the ocean entity to the region? Current research (Paraviccini *et al.* [PAR 13]) shows that for the size of the species and for the diet, the responses are opposite (Figure 4.14). In particular, smaller sizes are found preferentially in regions of high diversity, once the positive effect of diversity on the proportion of small sizes is taken into account [JON 02]; large sizes are preferentially found in areas with poor connectivity. With regard to diet, the proportion of invertebrate eaters decreases as the diversity increases with a simultaneous increase in piscivores, planktonophages and omnivores.

Within a region, it is possible to test multiple factors; the following example shows the effect of island size on the conservation of proportions of size and feeding regime among 12 atolls in French Polynesia. On these atolls, 354 species have been recorded during diving expeditions (according to a standardized method). On the histograms in Figure 4.17, the first bar to the left indicates the values for the species reservoir (354), both graphs above show the values at the local level (500 m^2 transect); the two following graphs show the values per atoll according to size. These figures indicate that as we move from 500 m^2 to several thousand km^2, the trophic or size structures hardly change. Therefore, at these scales, there is a high conservation of the diversity structure, even if the species that insures these structures are not the same, since in this case, no atoll had more than 150 species and no transect had more than 60 species. This conservation of life history traits is also found in functional groups, and therefore processes that insure the functioning of these assemblies, as we will see in the following sections.

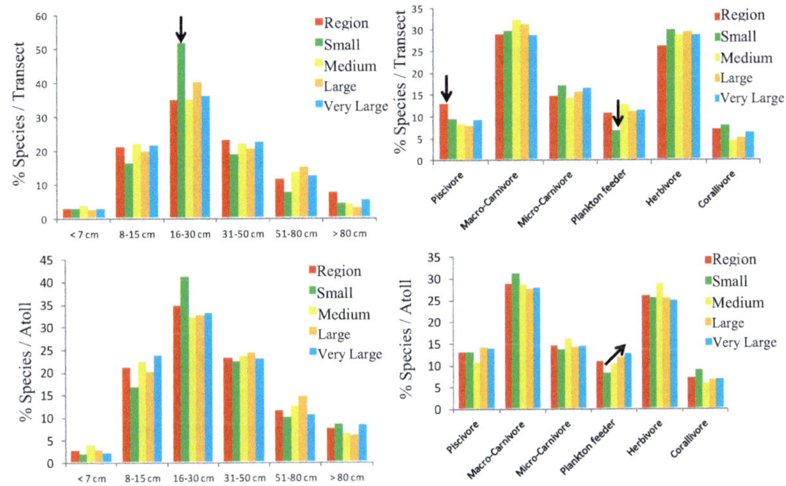

Figure 4.17. *Size and trophic structure in diversity according to atoll size (small, medium, large and very large), the two top histograms are for a local scale (500 m²) and the two bottom histograms are for the integrality of atolls. The notable values or trends are indicated by black arrows. The bar for the region is constant on both series and represents the proportions of all species in the region*

4.3.2. *From species to function*

In Ecology, an ecosystem is defined as a group of living organisms (biocenosis) in their habitat (biotope) and their interactions. These two interdependent components are intimately linked. The biotope provides the conditions required for living organisms, and the living organisms modify the conditions of the biotope. An ecosystem is therefore a non-fixed living entity allowing an ecological succession from an initial state, with a biocenosis composed of pioneer organisms, toward a final complex state where the maximum energy available is transferred to as many species as possible. When this capacity is optimal, ecological succession ends and results in a biocenosis in equilibrium with the environment. This state known as the "climax" is assumed to be relatively stable, at least without radical variations in climate, environmental conditions or human interventions. Currently, we know that the concept of climax can only be relative, since the ecosystem continues to dynamically adjust to natural variations in its populations.

The functional links between the biotope and the biocenosis are insured by the actions of the species. Therefore, each species will have its own role, that is to say, to insure a function within the ecosystem. The equilibrium between these functions allows the ecosystem to be maintained in a certain climax state whereas a functional imbalance may result in a change in state, sometimes reversible or irreversible. If irreversible, the ecosystem changes trajectory and may either tend toward another climax state, or decline. Following a set of disturbances, regressive evolutionary series may appear resulting in a biocenosis with fewer species, each time deviating lightly more from the climax.

Therefore, the role species play is vital for the functioning and existence of ecosystems. The more the functions within an ecosystem, i.e. the greater the functional diversity, the more complex, productive, robust and richer the ecosystem will be [CAD 12, CAD 13]. This is associated with the concept of ecological niches [HUT 57, LEI 95], which states that each species has a place (habitat) and a specific role within the ecosystem, aiming to best exploit the environment according to its dietary needs, place in the trophic network and relationships with other species, pace of activity and conditions required for reproduction. Therefore, the ecological niche determines a functional relationship that unites an organism with its ecosystem. The greater the number of functions in a biocenosis, the better it will be able to exploit its biotope, which will result in high species richness and productivity [HOO 05, LOR 01]. Moreover, it will be able to better resist disturbances by means of the functions required to give this robustness. Several species can also play the same role, that is to say, ensure the same function. This is known as functional redundancy. Therefore, some species may disappear from an ecosystem and some may appear due to chance and disturbances. However, the function that they provide usually persists. As the set of functions is maintained, whatever the exact identity of species forming the unit, the ecosystem will be maintained. Functional redundancy therefore increases the resilience of an ecosystem since it will affect several species that provide the same function and are hence "functionally interchangeable". However, functions that are not very redundant are vulnerable, the most vulnerable being those provided by a single species. When this is the case and the function is essential for the functioning of the ecosystem, this is known as the "keystone species" whose presence or absence will influence the whole ecosystem via the function it provides.

After many centuries studying species diversity, ecologists have come to understand the importance of the functions provided by the species in an ecosystem. Functional ecology has therefore become a blooming sector. In the following sections, we will present some concrete examples from our research on coral reef fish to provide the reader with an initial appreciation of functional ecology applied to coral reefs.

4.3.2.1. *Defining and estimating functional diversity*

4.3.2.1.1. Definition and functional traits

Functional diversity is defined as the diversity of roles or functions provided by species in an ecosystem. This simple definition is actually quite complex to integrate as a whole. What is the exact function of a species in an ecosystem? This is a very broad question that is difficult to answer. If we take an example of a species that acts as the prey, it is beneficial in an ecosystem because it provides food for predators. Being the prey is therefore one of the functions of this species. This same species also has to feed, and by feeding, they provide several other functions. For example, herbivores control algal growth and help in regulating competition between algae and reef-building corals. There are, however, many variations at the functional level in herbivores, with essentially three complementary groups: scrapers (or excavators), grazers and browsers [GRE 09]. The first group (scrapers) scrape and erode the coral substrate by feeding. They therefore create clear spaces on which sessile organisms can grow, for example coral larvae. The second (grazers) consume algal turf (or microalgae) and associated sediments, but without disturbing the underlying reef surface. They can also consume young macroalgal shoots, thereby preventing the development of large algal beds competing for space with reef-building corals. The final group, composed of browsers, are the only herbivores capable of consuming large macroalgae. Whereas scrapers feed on the epilithic microalgal matrix, browsers of large algae effectively reduce the biomass of fleshy macroalgae, therefore playing an essential role in maintaining equilibrium of the coral ecosystem [BEL 06a, BEL 06b, JOH 13]. On a coral reef, erosion of the herbivore function can lead to a phase shift, coral areas are progressively replaced by microalgae as seen in the Caribbean; this is known as coral–algal phase shift [HUG 94]). The existence of browsing herbivores helps mitigate these phase shifts. By feeding on macroalgae, this type of herbivore makes the algae retreat and therefore allows the action of other herbivores (scrapers and grazers) that feed on the algal turf on the coral substrate, leaving space for coral larvae and thereby favoring the replacement of algae by coral (algal–coral phase shift). Using this example, we can establish a link between the characteristics – or life history traits – here feeding traits, of species and their functions within the ecosystem. Moreover, according to the hypothesis of functional conservation [DAR 59, POW 09], evolutionary close species often have similar life history traits. Therefore, the functional diversity is associated with the diversity of genetic lineages or phylogenetic diversity. These two facets of biodiversity are at the heart of current concerns as they are intimately linked to the functioning of ecosystems, and are also highly degraded by human activities [CAD 13, DAG 14].

4.3.2.1.2. Functional classification

Species have many life history traits, not just related to feeding. At present, classifying reef fish into functional groups is based on the different combinations of these traits [MOU 13, PAR 14]. As shown in the previous example, the feeding regime is an important functional trait. For fish, we typically have piscivores, carnivores, herbivores-detritivores, omnivores and plankton feeders, with each of these categories being composed of even more specific groups (e.g. scrapers, browsers or grazers for herbivores). Other important traits of fish are species size, home range, position in the water column, nycthemeral activity and schooling. Size has a primordial role since it largely defines the predator–prey relationships and determines the energy required for metabolism of individuals and therefore the quantity of food required for survival [COS 09, FIS 10]. Size also plays a role in reproductive strategies (e.g. size at maturity, reproductive capacity). Home range (mobility) is also a trait associated with energetic aspects, as mobile species require a larger amount of energy than sedentary species. Moreover, by moving, they transfer energy between different habitats in an ecosystem [NOR 84]. The nycthemeral activity also has functional implications since nocturnal and diurnal species are not accessible to the same predators and do not feed on the same prey [FOX 11]. The position in the water column is also a functional trait that is involved in energy flows between different horizontal strata from the bottom to the surface and vertical strata from the coral ecosystem, pelagic area to the reef area [SCH 00]. Finally, schooling is associated with the functional fingerprint of species, individuals living in large schools having a major impact on the transfers of nutrients within an ecosystem [ROB 76, MEY 85].

Each of these life history traits can be quantified (e.g. a species can reach a maximum size of 48 cm as an adult) or split into categories (e.g. species of an average size). The combination of these traits helps define increasingly complex functional classification diagrams. For example, a simple diagram is obtained by combining size and diet, with functions such as large piscivore, average sized herbivore or small carnivore, and a complex diagram by combining all the available traits (e.g. large solitary nocturnal highly mobile piscivore that hunts in open water). The main advantage of these functional classifications is their instinct. We can easily identify functions and fairly simply measure the diversity and functional redundancy, by counting the number of functional entities in a classification diagram and the number of species in each function. The downfall of this method is its subjectivity since defining these functions depends on the number and identity of life history traits chosen in the classification [GUI 11].

One way of overcoming this subjectivity, linked to the number of life history traits chosen in a functional classification diagram, is to group species according to life history trait similarity. For this, we form functional dendrograms, i.e. a diagram that is composed of functional branches based on species life history traits, with species on closer branches having similar traits, and therefore performing similar functions in an ecosystem. The functional diversity of a group of species is therefore given by the sum of the lengths of branches on a functional tree in the assembly. One advantage of this technique is that it eliminates the subjectivity from the number of life history traits per classification. Moreover, it is applicable to all types of trees, in particular phylogenetic trees which rely on evolutionary links, which help measure the phylogenetic diversity [ALL 09]. The main disadvantage of this method is that it is difficult to concretely determine the identity of these functions.

4.3.2.1.3. Diversity, redundancy, vulnerability and functional insurance

Diversity, redundancy, vulnerability and functional insurance are directly associated but are not synonyms. The functional diversity represents the diversity of functions performed by species in an ecosystem. It can easily be measured by counting the functions present in a given assembly according to a pre-established functional classification diagram. It is natural to assume that, for a given number of species, the greater the functional diversity, the lower the functional redundancy. In fact, for a fixed number of species, there will be fewer species per function (redundancy) if there are many functions. If the redundancy is low, this means that some functions are performed by very few species, or even by a single species, and consequently these functions may be vulnerable to the disappearance of one or few species. Nevertheless, if the functional diversity is high, then the ecosystem may possess fauna whose functions allow it to adapt to disturbances. From this point of view, functional diversity provides insurance to the ecosystem. Finally, to simplify this, the presence of many species usually leads to a high functional diversity (the more species there are, the more functions there will be), and a high functional redundancy (the more species there are, the more species there are per function), and a lower vulnerability and a higher robustness/resilience. Today, although we know that there is a relationship between species diversity and functional diversity, there are many uncertainties with regard to its form and effects in terms of redundancy, vulnerability and functional insurance.

4.3.2.2. *Relationship between species and functional diversity*

4.3.2.2.1. Theoretical forms of the relationship between species and functional diversity

The relationship between species and functional diversity increases as a straight line or curve passing through the origin. In fact, adding new species to an assembly can only add new functions or increase the redundancy of existing functions. A steep slope in this relationship indicates that new functions rapidly emerge when the number of species increases, whereas a gentle slope indicates an increase in the redundancy of functions already present in the assembly. On this basis, [MIC 05] proposed four general forms for this relationship (Figure 4.18).

The first theoretical form (A1) is a straight line with a slope of 1 passing through the origin. It implies that each species provides a unique function in the assembly. The second (A2) is a straight line with a slope of <1 according to the hypothesis that new functions appear at a constant rate in the assembly of species. The third form (B) is based on the principle that new functions rapidly appear in assemblies with few species, but it is the functional redundancy that tends to increase in diverse assemblies. The final theoretical form (C) assumes that the relationship between the functional and species diversity varies according to the environment. Whereas only some species that share functional traits can coexist in a simple environment, the transition toward a new more complex environment results in the rapid increase in the number of species and functions. This S-shaped scenario is assumed to occur when an ecosystem comes back to its initial status following a disturbance.

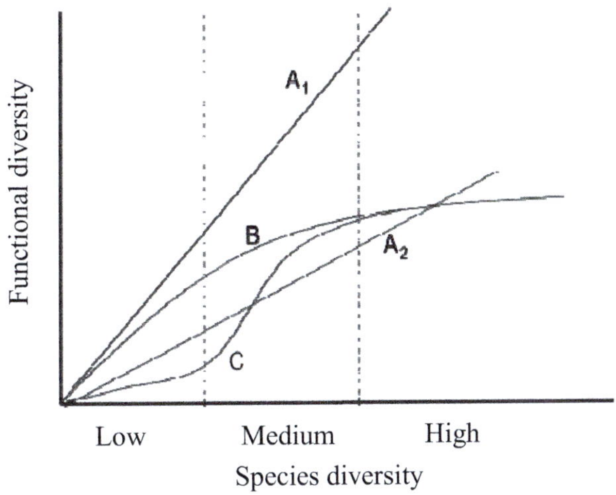

Figure 4.18. *Theoretical forms of the relationship between species diversity (number of species) and functional diversity (number of functions performed by the species) (adapted from [MIC 05])*

4.3.2.2.2. Case study on lagoon-reef fish in New Caledonia

Analyzing the shape of the relationship between species and functional diversity for a very diverse ecosystem such as a coral reef has been carried out by [GUI 11]. Results from measuring the high biodiversity of coral fish in New Caledonia have shown that the functional diversity increased rapidly from low levels of species diversity, and then slowed down when the number of species was large (Figure 4.19). Even though scenario (B) from [MIC 05] was proven, it was not completely realized as no plateau was reached in this relationship. This result is important as it means that new functions continue to appear even when the species diversity is very high, and that some functions, very rare and vulnerable, can only appear in exceptional levels of diversity.

4.3.2.2.3. Case study on coral reef fish on a global scale

The relationship between species and functional diversity observed at the local level of an assembly of coral reef fish (Figure 4.19) is maintained at the global scale (Figure 4.20). Regions in the world which have, for historical reasons, few species (Atlantic Ocean and East Pacific province) also have few functions compared with regions with many species (Indian Ocean and Pacific Ocean). Given the relationship between functional diversity and the vulnerability of ecosystems, we can therefore say that the high species diversity observed on coral reefs in the Indo-Pacific provides this region with a functional insurance that allows it to better withstand disturbances.

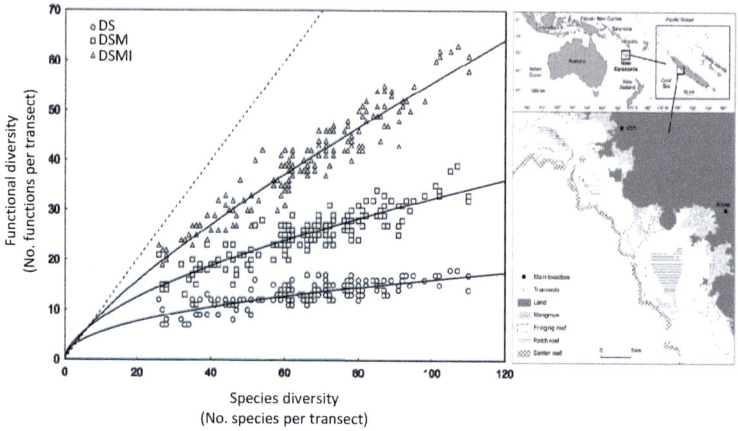

Figure 4.19. *Relationship between species and functional diversity for coral reef fish in New Caledonia (Voh-Koné-Pouembout region, Northern Province) for different functional classification diagrams, DS, DSH and DSHG(D: diet; S: size; H: Home range: G; schooling). The dotted line represents a straight line with a slope of 1 (adapted from [GUI 11])*

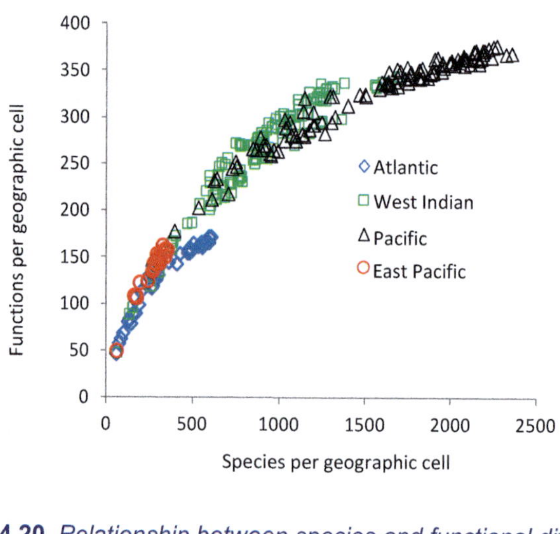

Figure 4.20. *Relationship between species and functional diversity on a global scale. Each point corresponds to a geographic cell of 5° x 5° (data from GASPAR database (FRB))*

4.3.2.3. *Diversity, redundancy, vulnerability and functional insurance*

4.3.2.3.1. Case study of lagoon-reef fish in New Caledonia

As we have seen previously, the more functions an ecosystem has, the more likely it will possess one or several functions that allow it to withstand a disturbance. In other words, the functional diversity provides ecosystems with insurance. For example, assemblages with several types of herbivores will have a greater robustness/resilience capacity against the algae-coral phase shift. Similarly, many species per function, i.e. a high functional redundancy, will insure the ecosystem against the local disappearance of a function.

There is therefore a certain antagonism between functional redundancy and functional diversity. In fact, for each new species in a system, whether there is an increase in redundancy and therefore a greater insurance per function but a conservatism of functions, or a new function is created, and therefore a better use of the environment and its resources, but to the detriment of redundancy. A recent study shows that these two forces are in action [GUI 11].

On reefs in New Caledonia, the number of functions observed in fish assemblages is lower than predicted by a neutral model (random sampling), given the number of species present per station below a threshold of about 90 (Figure 4.21). Below this threshold, the species diversity tends to strengthen the functional redundancy and therefore helps the ecosystem withstand the loss of local species. Beyond this threshold of 90 species per station, the number of functions observed in the assemblies does not differ from what was expected by chance given the species diversity. Consequently, beyond a threshold of 90, the functional diversity tends to increase with the arrival of new rare functions, potentially useful in the case of disturbance of an unknown nature.

This same study also suggests the existence of functions essential for the functioning of the coral ecosystem. These functions would be present even at a low level of diversity so that a coral reef can function, then they will become redundant making the reefs more resilient with higher diversity. This would imply that reefs, in areas of low diversity, have many functions in common with high-diversity areas.

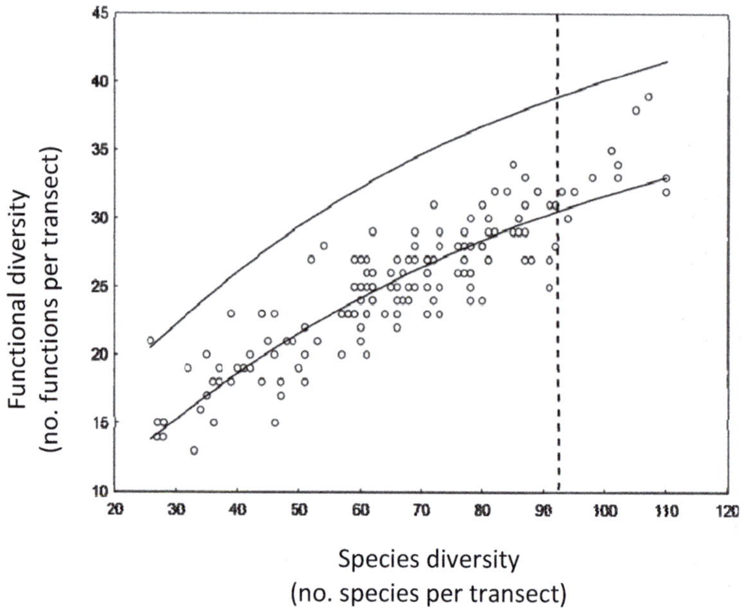

Figure 4.21. *Relationship between species and function diversity of coral reef fish in New Caledonia [GUI 11]*

COMMENTS ON FIGURE 4.21.– The curves represent the interval between which observations (points) should be found if functions were selected by chance among species. For points below the lower curve, the species entering the system tend to belong to functional groups already present rather than expected by chance. The dotted line indicates the threshold beyond which the number of functions observed in assemblies corresponds to random sampling.

4.3.2.3.2. Case study on reef fish on a global scale

The results from this study on a local scale in New Caledonia [GUI 11] were confirmed by other studies on a global scale [MOU 14] which used the concept of functional volume, which corresponds to the dispersion of species in a multidimensional space of life history traits. It is possible to calculate the overall functional volume for a group of coral reef fish (i.e. more than 7,000 species). Once this overall volume is defined, it is possible to determine the proportion of this overall space which is occupied by coral reef fish from different biogeographical provinces of the world (as defined by [KUL 13] (Figure 4.22).

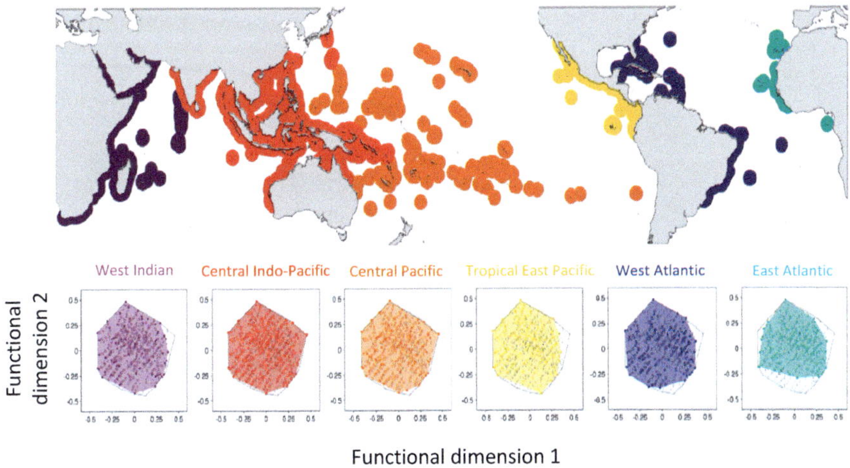

Figure 4.22. *Space of functional volume occupied by coral fish species from six biogeographical provinces of the world [KUL 13] (adapted from [MOU 14]). For a color version of the figure, see color section*

Regions with low species diversity (Atlantic and East Pacific) occupy a nearly similar functional volume to that of regions with high species diversity (Indian Ocean and Pacific Ocean). In other words, the extent of functions insured by the species present in each ocean is similar despite the large differences in species number and differences in species compositions. Another aspect of this study is the importance of functions with only one species (Figure 4.23). The proportion of these functions in regional species pools is higher than 40%, which implies that many functions are potentially vulnerable, in all oceans, regardless of the overall species richness. However, some functions have a large redundancy that can often exceed a hundred species. This redundancy is largely a function of species size; functions grouping together smaller species are generally more redundant than those insured by larger species, with a clear split at 30 cm. This relationship between functional redundancy and species size is likely to be linked with the relationship between species richness and species size [BRO 04].

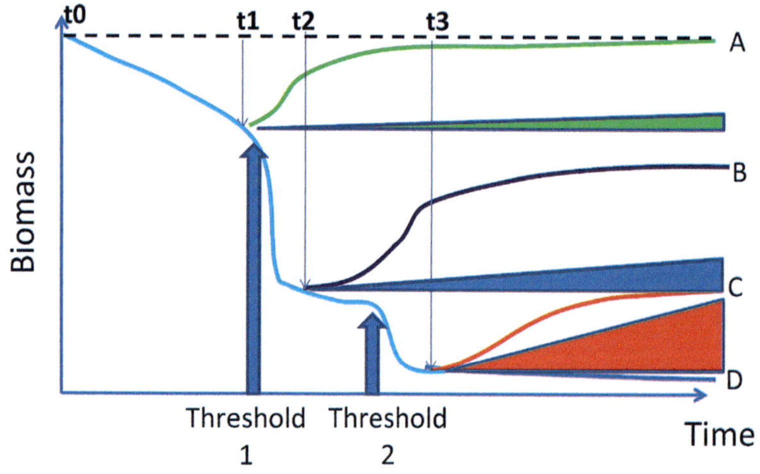

Figure 4.23. *Redundancy of functional groups by biogeographic region (see Figure 4.6 for the regions), organized by increasing diversity. EA: East Atlantic ; EP: East Pacific; WA: West Atlantic; WI: West Indian; CP: Central Pacific; CIP: Central Indo-Pacific. The figures on the abscissa represent the number of species per region*

4.4. Conclusion

This chapter on biocomplexity has shown that the organization of communities in extreme diverse environments are still highly structured and largely predictable if the analysis takes place at a high enough level of organization. Fish, which were the main model species in this chapter, do not, however, represent all aspects of this organization and the organization of other main groups, such as corals, must be better understood to validate these organization diagrams. The fundamental mechanisms of functioning, however, show a remarkable consistency from one biogeographic region to another if we consider that species composition can change dramatically between these regions. This does not mean that the functioning of these communities is identical from one region to another. In fact, the evolutionary history has caused very significant differences between regions, both in the number of species present and their characteristics. It is therefore a kind of antagonism between large interregional convergences on the one hand, and on the other hand adaptations in the species communities in each region. This contrast has not been formally verified as far as we know on other highly complex ecosystems, such as tropical forests, but it is likely that similarities with coral reef systems exist.

This organization has important implications for managers and users of these systems as we will try to show in the following chapter. In fact, it is likely that rules of management or usage may be globally applied, but it is essential to adapt them to the local context. The first lesson of this chapter is therefore that despite the low level of our current understanding of the taxonomy of species, their biology and ecology, not to mention the interactions of all genera involved in these ecosystems, it is already possible to understand some essential concepts regarding the functioning of these systems. The second lesson is that we cannot afford to conduct a more detailed analysis because each region has specific characteristics, especially with regard to the species present and the set of interactions with the local environment.

4.5. Bibliography

[ADJ 14] ADJEROUD M., GUÉRÉCHEAU A., VIDAL-DUPIOL J. *et al.*, "Genetic diversity, clonality and connectivity in the scleractinian coral *Pocillopora damicornis*: a multi-scale analysis in an insular, fragmented reef system", *Marine Biology*, no. 161, pp. 531–541, 2014.

[ALL 09] ALLEN B., KON M., BAR-YAM Y., "A new phylogenetic diversity measure generalizing the shannon index and its application to phyllostomid bats", *American Naturalist*, no. 174, pp. 236–243, 2009.

[AND 06] ANDRÉFOUËT S., MULLER-KARGER F.E., ROBINSON J.A. *et al.*, "Global assessment of modern coral reef extent and diversity for regional science and management applications: a view from space", in SUZUKI Y., NAKAMORI T., HIDAKA M. *et al.* (eds), *10th ICRS*, Japanese Coral Reef Society, Okinawa, Japan, pp. 1732–1745, 2006.

[AND 08] ANDRÉFOUËT S., CHAGNAUD N., CHAUVIN C. *et al.*, *Atlas of French Overseas Coral Reefs*, Centre IRD de Nouméa, 2008.

[APP 12] APPELTANS W., AHYONG S.T., ANDERSON G. *et al.*, "The magnitude of global marine species diversity", *Current Biology*, vol. 22, no. 23, pp. 2189–2202, 2012.

[ARI 08] ARIAS-GONZÁLEZ J.E., LEGENDRE P., RODRÍGUEZ-ZARAGOZA F.A., "Scaling up beta diversity on Caribbean coral reefs", *Journal of Experimental Marine Biology and Ecology*, vol. 366, no. 1, pp. 28–36, 2008.

[AYR 00] AYRE D.J., HUGHES T.P., "Genotypic diversity and gene flow in brooding and spawning corals along the Great Barrier Reef, Australia", *Evolution*, vol. 54, pp. 1590–1605, 2000.

[BAI 09] BAIRD A.H., GUEST J.R., WILLIS B.L., "Systematic and biogeographical patterns in the reproductive biology of scleractinian corals", *Annual Review of Ecology, Evolution and Systematic*, vol. 40, pp. 551–571, 2009.

[BAR 06] BARBER P., BOYCE S.L., "Estimating diversity of Indo-Pacific coral reef stomatopods through DNA barcoding of stomatopod larvae", *Proceedings of the Royal Society B: Biological Sciences*, vol. 273, no. 1597, pp. 2053–2061, 2006.

[BEL 98] BELLWOOD D.R., "What are reef fishes? – Comment on the report by DR Robertson: do coral-reef fish faunas have a distinctive taxonomic structure?", *Coral Reefs*, vol. 17, no. 2, pp. 187–189, 1998.

[BEL 01] BELLWOOD D.R., HUGHES T.P., "Regional-scale assembly rules and biodiversity of coral reefs", *Science*, vol. 292, no. 5521, pp. 1532–1535, 2001.

[BEL 05] BELLWOOD D.R., HUGHES T.P., CONNOLLY S.R. *et al.*, "Environmental and geometric constraints on Indo-Pacific coral reef biodiversity", *Ecology Letters*, vol. 8, no. 6, pp. 643–651, 2005.

[BEL 06a] BELLWOOD D.R., HOEY A.S., ACKERMAN J.L. *et al.*, "Coral bleaching, reef fish community phase shifts and the resilience of coral reefs", *Global Change Biology*, vol. 12, pp. 1587–1594, 2006.

[BEL 06b] BELLWOOD D.R., HUGHES T.P., HOEY A.S., "Sleeping functional group drives coral reef recovery", *Current Biology*, vol. 16, pp. 2434–2439, 2006.

[BEL 08] BELMAKER J., ZIV Y., SHASHAR N. *et al.*, "Regional variation in the hierarchical partitioning of diversity in coral-dwelling fishes", *Ecology*, vol. 89, no. 10, pp. 2829–2840, 2008.

[BER 12] BERUMEN M.L., ALMANY G.R., PLANES S. *et al.*, "Persistence of self-recruitment and patterns of larval connectivity in a marine protected area network", *Ecology and Evolution*, vol. 2, no. 2, pp. 444–452, 2012.

[BOU 02] BOUCHET P., LOZOUET P., MAESTRATI P. *et al.*, "Assessing the magnitude of species richness in tropical marine environments: exceptionally high numbers of molluscs at a New Caledonia site", *Biological Journal of the Linnean Society*, vol. 75, no. 4, pp. 421–436, 2002.

[BOU 06] BOUCHET P., DUARTE C.M., *The Exploration of Marine Biodiversity Scientific and Technological Challenges,* Fundación BBVA., vol. 33, 2006.

[BRO 04] BROWN J.H., GILLOOLY J.F., ALLEN A.P. *et al.*, "Toward a metabolic theory of ecology", *Ecology*, no. 85, pp. 1771–1789, 2004.

[BUD 96] BUDD A.F., JOHNSON K.G., STEMANN T.A., "Plio-Pleistocene turnover and extinctions in the Caribbean reef coral fauna", *Evolution and Environment in Tropical America*, pp. 168–204, 1996.

[BUD 10] BUDD A.F., ROMANO S.L., SMITH N.D. *et al.*, "Rethinking the phylogeny of scleractinian corals: a review of morphological and molecular data", *Integrative and Comparative Biology*, vol. 50, no. 3, pp. 411–427, 2010.

[CAL 96] CALEY M.J., CARR M.H., HIXON M.A. *et al.*, "Recruitment and the local dynamics of open marine populations", *Annual Review of Ecology and Systematics*, vol. 27, no. 1, pp. 477–500, 1996.

[CAD 06] CADENASSO M.L., PICKETT S.T.A., GROVE J.M., "Dimensions of ecosystem complexity: heterogeneity, connectivity, and history", *Ecological Complexity*, vol. 3, no. 1, pp. 1–12, 2006.

[CAD 12] CADOTTE M.W., DINNAGE R., TILMAN D., "Phylogenetic diversity promotes ecosystem stability", *Ecology*, no. 93, pp. 223–233, 2012.

[CAD 13] CADOTTE M.W., "Experimental evidence that evolutionarily diverse assemblages result in higher productivity", *Proceedings of the National Academy of Sciences*, vol. 110, pp. 8996–9000, 2013.

[CAR 13] CARASSOU L., LÉOPOLD M., GUILLEMOT N. *et al.*, "Does herbivorous fish protection really improve coral reef resilience? A case study from New Caledonia (South Pacific)", *PLoS ONE*, vol. 8, no. 4, p. e60564, 2013.

[COL 99] COLWELL R.R., "Complexity and connectivity: a new cartography for science and engineering", *Address to AGU Fall Meeting, San Francisco, CA, World Wide Web,* 3 December 2003, 1999.

[CON 78] CONNELL J.H., "Diversity in tropical rain forests and coral reefs", *Science*, vol. 199, no. 4335, pp. 1302–1310, 1978.

[CON 03] CONNOLLY S.R., BELLWOOD D.R., HUGHES T.P., "Indo-Pacific biodiversity of coral reefs: deviations from a mid-domain model", *Ecology*, vol. 84, no. 8, pp. 2178–2190, 2003.

[COS 09] COSTA G.C., "Predator size, prey size, and dietary niche breadth relationships in marine predators", *Ecology*, vol. 90, pp. 2014–2019, 2009.

[COU 12] COUCE E., RIDGWELL A., HENDY E.J., "Environmental controls on the global distribution of shallow-water coral reefs", *Journal of Biogeography*, vol. 39, pp. 1508–1523, 2012.

[COW 13] COWMAN P.F., BELLWOOD D.R., "The historical biogeography of coral reef fishes: global patterns of origination and dispersal", *Journal of Biogeography*, vol. 40, no. 2, pp. 209–224, 2013.

[CUR 03] CURY P., SHANNON L., SINCLAIR M. *et al.*, "The functioning of marine ecosystems: a fisheries perspective", *Responsible Fisheries in the Marine Ecosystem*, pp. 103–123, 2003.

[DAG 14] D'AGATA S., MOUILLOT D., KULBICKI M. *et al.*, "Human-mediated loss of phylogenetic and functional diversity in coral reef fishes", *Current Biology*, available at http://dx.doi.org/10.1016/j.cub.2014.01.049, 2014.

[DAR 59] DARWIN C., *On the Origin of Species*, Harvard University Press, Cambridge, MA, 1859.

[DOH 91] DOHERTY P.J., "Spatial and temporal patterns in recruitment", *Sale PF Ecology of Coral Reef Fishes*, Academic Press, pp. 261–293, 1991.

[DOU 03] DOUGLAS A.E., "Coral bleaching – how and why?", *Marine Pollution Bulletin*, vol. 46, pp. 385–392, 2003.

[FAU 12] FAURBY S., BARBER P.H., "Theoretical limits to the correlation between pelagic larval duration and population genetic structure", *Molecular Ecology*, vol. 21, pp. 3419–3432, 2012.

[FIS 10] FISHER J., FRANK K., LEGGETT W., "Global variation in marine fish body size and its role in biodiversity–ecosystem functioning", *Marine Ecology Progress Series*, vol. 405, pp. 1–13, 2010.

[FLO 08] FLOETER S.R., ROCHA L.A., ROBERTSON D.R. *et al.*, "Atlantic reef fish biogeography and evolution", *Journal of Biogeography*, vol. 35, no. 1, pp. 22–47, 2008.

[FOX 11] FOX R.J., BELLWOOD D.R., "Unconstrained by the clock? Plasticity of diel activity rhythm in a tropical reef fish, *Siganus lineatus*", *Functional Ecology*, vol. 2, pp. 1096–1105, 2011.

[FOX 13] FOX J.W., "The intermediate disturbance hypothesis should be abandoned", *Trends in Ecology and Evolution*, vol. 28, pp. 86–92, 2013.

[FRI 11] FRICKE R. KULBICKI M., WANTIEZ L., "Checklist of the fishes of New Caledonia, and their distribution in the Southwest Pacific Ocean (Pisces)", *Stuttg Beitr Natkd Ser A (Biol)*, vol. 4, pp. 341–463, 2011.

[FUK 04] FUKAMI H., BUDD A.F., PAULAY G. *et al.*, "Conventional taxonomy obscures deep divergence between Pacific and Atlantic corals", *Nature*, vol. 427, no. 6977, pp. 832–835, 2004.

[GRE 03] GREEN E.P., SHORT F.T., *World Atlas of Seagrasses*, UNEP World Conservation Monitoring Centre, Berkeley (California, CA): University of California, 2003.

[GRE 09] GREEN A.L., BELLWOOD D.R., *Monitoring Functional Groups of Herbivorous Reef Fishes as Indicators of Coral Reef Resilience: A Practical Guide for Coral Reef Managers in the Asia Pacific Region*, IUCN, 2009.

[GUI 11] GUILLEMOT N., KULBICKI M., CHABANET P. *et al.*, "Functional redundancy patterns reveal non-random assembly rules in a species-rich marine assemblage", *PLoS ONE*, vol. 6, no. 10, p. e26735, available at doi:10.1371/journal.pone.0026735, 2011.

[HAR 90] HARRISON P.L., WALLACE C.C., "Reproduction, dispersal and recruitment of scleractinian corals", in DUBINSKY Z. (ed.), *Coral Reefs*, Elsevier, Amsterdam, pp. 133–207, 1990.

[HAR 11] HARRISON P.L., "Sexual reproduction of scleractinian corals", in DUBINSKY Z., STAMBLER N. (eds), *Coral Reefs: An Ecosystem in Transition*, Springer Science+Business Media, pp. 59–85, 2011.

[HOE 04] HOEGH-GULDBERG O., "Coral reefs in a century of rapid environmental change", *Symbiosis*, vol. 37, pp. 1–31, 2004.

[HOO 05] HOOPER., D.U., CHAPIN III F.S., EWEL J.J. *et al.*, "Effects of biodiversity on ecosystem functioning: a consensus of current knowledge", *Ecological Monographs*, vol. 75, no. 1, pp. 3–35, 2005.

[HUG 94] HUGHES T.P., "Catastrophes, phase shifts, and large scale degradation of a Caribbean coral reef", *Science*, vol. 265, pp. 1547–1551, 1994.

[HUG 02] HUGHES T.P., BELLWOOD D.R., CONNOLLY S.R., "Biodiversity hotspots, centres of endemicity, and the conservation of coral reefs", *Ecology Letters*, vol. 5, no. 6, pp. 775–784, 2002.

[HUG 03a] HUGHES T.P., BAIRD A.H., BELLWOOD D.R. *et al.*, "Climate change, human impacts, and the resilience of coral reefs", *Science*, vol. 301, pp. 929–933, 2003.

[HUG 03b] HUGHES T.P., "Biogeographic comparisons of coral faunas", *Oceanis*, vol. 29, pp. 291–301, 2003.

[HUT 57] HUTCHINSON G.E., "Concluding remarks", *Cold Spring Harbor Symposia on Quantitative Biology*, vol. 22, pp. 415–427, 1957.

[JAC 01a] JACKSON J.B., "What was natural in the coastal oceans?", *Proceedings of the National Academy of Sciences*, vol. 98, no. 10, pp. 5411–5418, 2001.

[JAC 01b] JACKSON J.B., KIRBY M.X., BERGER W.H. *et al.*, "Historical overfishing and the recent collapse of coastal ecosystems", *Science,* vol. 293, no. 5530, pp. 629–637, 2001.

[JOH 13] JOHANSSON C.L., BELWOOD D.R., DEPCZYNSKI M. *et al.*, "The distribution of the sea urchin Echinometra mathaei (de Blainville) and its predators on Ningaloo Reef, Western Australia: the implications for top-down control in an intact reef system", *Journal of Experimental Marine Biology and Ecology*, vol. 442, pp. 39–46, 2013.

[JON 99] JONES G.P., MILICICH M.J., EMSLIE M.J. *et al.*, "Self-recruitment in a coral reef fish population", *Nature*, vol. 402, no. 6763, pp. 802–804, 1999.

[JON 02] JONES G.P., CALEY J., MUNDAY P., "Rarity in coral reef fish communities", in SALE P. (ed.), *Coral Reef Fishes: Dynamics and Diversity in a Complex Ecosystem*, Academic Press, pp. 81–101, 2002.

[JUN 07] JUNK W.J., SOARES M.G.M., BAYLEY P.B., "Freshwater fishes of the Amazon River basin: their biodiversity, fisheries, and habitats", *Aquatic Ecosystem Health & Management*, vol. 10, no. 2, pp. 153–173, 2007.

[KEI 13] KEITH S.A., BAIRD A.H., HUGHES T.P. *et al.*, "Faunal breaks and species composition of Indo-Pacific corals: the role of plate tectonics, environment and habitat distribution", *Proceedings of the Royal Society B: Biological Sciences*, vol. 280, no. 1763, p. 20130818, 2013.

[KEI 14] KEITH S.A., KERSWELL A.P., CONNOLLY S.R., "Global diversity of marine macroalgae: environmental conditions explain less variation in the tropics", *Global Ecology and Biogeography*, vol. 23, no. 5, pp. 517–529, 2014.

[KES 13] KESHAVMURTHY S., YANG S.Y., ALAMARU A. *et al.*, "DNA barcoding reveals the coral 'laboratory rat":, *Stylophora pistillata* encompasses multiple identities", *Scientific Reports*, vol. 3, p. 1520, 2013.

[KOL 03] KOLEFF P., GASTON K.J., LENNON J.J., "Measuring beta diversity for presence–absence data", *Journal of Animal Ecology*, vol. 72, no. 3, pp. 367–382, 2003.

[KUL 13] KULBICKI M., PARRAVICINI V., BELLWOOD D.R. *et al.*, "Global biogeography of reef fishes: a hierarchical quantitative delineation of regions", PLoS ONE, 2013.

[KUL 15] KULBICKI M., PARRAVICINI V., MOUILLOT D., "Patterns and processes linked to body size", in MORA C. (ed.), *Ecology of Fishes on Coral Reefs*, Cambridge University Press, 2015.

[LEI 95] LEIBOLD M.A., "The niche concept revisited: mechanistic models and community context", *Ecology*, vol. 76, pp. 1371–1382, 1995.

[LES 04] LESSER M.P., "Experimental biology of coral reef ecosystems", *Journal of Experimental Marine Biology and Ecology*, vol. 300, pp. 217–252, 2004.

[LET 97] LETOURNEUR Y., KULBICKI M., GALZIN R. *et al.*, "Comparaison des peuplements de poissons marins des récifs frangeants de trois îles océaniques de l'Indo-Pacifique (La Réunion, Moorea et la Nouvelle-Calédonie)", *Cybium*, vol. 21, no. 1, pp. 129–145, 1997.

[LOR 00] LOREAU M., "Are communities saturated, on the relationship between α, β and γ diversity", *Ecology Letters*, vol. 3, no. 2, pp. 73–76, 2000.

[LOR 01] LOREAU M., NAEEM S., INCHAUSTI P. *et al.*, "Biodiversity and ecosystem functioning: current knowledge and future challenges", *Science*, vol. 294, pp. 804–808, 2001.

[LUI 12] LUIZ O.J., MADIN J.S., ROBERTSON D.R. *et al.*, "Ecological traits influencing range expansion across large oceanic dispersal barriers: insights from tropical Atlantic reef fishes", *Proceedings of the Royal Society B: Biological Sciences*, vol. 279, no. 1730, pp. 1033–1040, 2012.

[LUI 13] LUIZ O.J., ALLEN A.P., ROBERTSON D.R. *et al.*, "Adult and larval traits as determinants of geographic range size among tropical reef fishes", *Proceedings of the National Academy of Sciences*, vol. 110, no. 41, pp. 16498–16502, 2013.

[MAC 67] MACARTHUR R.H., WILSON E.O., *The Theory of Island Biogeography*, The University of Chicago Press Princeton. New Jersey, 1967.

[MEY 85] MEYER J.L., SCHULTZ E.T., "Migrating haemulid fishes as a source of nutrients and organic matter on coral reefs", *Limnology and Oceanography*, vol. 30, pp. 146–156, 1985.

[MIC 05] MICHELI F., HALPERN B.S., "Low functional redundancy in coastal marine assemblages", *Ecology Letters*, vol. 8, pp. 391–400, 2005.

[MIL 06] MILLIEN V., LYONS S., OLSON L. *et al.*, "Ecotypic variation in the context of global climate change: revisiting the rules", *Ecology Letters*, vol. 9, no. 7, pp. 853–869, 2006.

[MOR 11] MORA C., TITTENSOR D.P., ADL S. *et al.*, "How many species are there on Earth and in the ocean?", *PLoS Biology*, vol. 9, no. 8, p. e1001127, 2011.

[MOU 13] MOUILLOT D., BELLWOOD D.R., BARALOTO C. *et al.*, "Rare species support vulnerable functions in high-diversity ecosystems", *PLoS Biology*, vol. 11, no. 5, p. e1001569, 2013.

[MOU 14] MOUILLOT D., VILLÉGER S., PARRAVICINI V. *et al.*, "Functional over-redundancy and high functional vulnerability in global fish faunas on tropical reefs", *Proceedings of the National Academy of Sciences*, vol. 111, no. 38, pp. 13757–13762, 2014.

[NOR 84] NORMAN M.D., JONES G.P., "Determinants of territory size in the pomacentrid reef fish, *Parma victoriae*", *Oecologia*, vol. 61, pp. 60–69, 1984.

[PAR 13] PARRAVICINI V., KULBICKI M., BELLWOOD D.R., *et al.*, "Global patterns and predictors of tropical reef fish species richness", *Ecography*, vol. 36, no. 12, pp. 1254–1262, 2013.

[PEL 14] PELLISSIER L., LEPRIEUR F., PARRAVICINI V. *et al.*, "Quaternary coral reef refugia preserved fish diversity", *Science*, vol. 344, no. 6187, pp. 1016–1019, 2014.

[PEN 10] PENIN L., MICHONNEAU F., BAIRD A.H. *et al.*, "Early post-settlement mortality and the structure of coral assemblages", *Marine Ecology Progress Series*, vol. 408, pp. 55–64, 2010.

[PEN 13] PENIN L., ADJEROUD M., "Relative importance of recruitment and post-settlement processes in the maintenance of coral assemblages in an insular, fragmented reef system", *Marine Ecology Progress Series*, vol. 473, pp. 149–162, 2013.

[PLA 11] PLAISANCE L., CALEY M.J., BRAINARD R.E. *et al.*, "The diversity of coral reefs: what are we missing?", *PLoS ONE*, vol. 6, no. 10, p. e25026, 2011.

[POW 09] POWELL J.R., PARRENT J.L., HART M.M. *et al.*, "Phylogenetic trait conservatism and the evolution of functional trade-offs in arbuscular mycorrhizal fungi", *Proceedings of the National Academy of Sciences*, vol. 276, pp. 4237–4245, 2009.

[REA 08] REAKA M.L., RODGERS P.J., KUDLA A.U., "Patterns of biodiversity and endemism on Indo-West Pacific coral reefs", *Proceedings of the National Academy of Sciences*, vol. 105, Suppl. 1, pp. 11474–11481, 2008.

[RIC 93] RICHMOND R.H., "Coral reefs: present problems and future concerns resulting from anthropogenic disturbance", *American Zoologist*, vol. 33, pp. 524–536, 1993.

[ROB 76] ROBERTSON D.R., SWEATMAN H.P.A., FLETCHER E.A. *et al.*, "Schooling as a mechanism for circumventing the territoriality of competitors", *Ecology*, vol. 57, pp. 1208–1220, 1976.

[ROB 02] ROBERTS C.M., MCCLEAN C.J., VERON J.E. *et al.*, "Marine biodiversity hotspots and conservation priorities for tropical reefs", *Science*, vol. 295, no. 5558, pp. 1280–1284, 2002.

[ROB 09] ROBERTSON D.R., CRAMER K.L., "Shore fishes and biogeographic subdivisions of the Tropical Eastern Pacific", *Marine Ecology Progress Series*, vol. 380, pp. 1–17, 2009.

[ROG 93] ROGERS C.S., "Hurricanes and coral reefs: the intermediate disturbance hypothesis revisited", *Coral Reefs*, vol. 12, nos. 3–4, pp. 127–137, 1993.

[ROS 88] ROSEN B.R., "Progress, problems and patterns in the biogeography of reef corals and other tropical marine organisms", *Helgoländer Meeresuntersuchungen*, vol. 42, pp. 269–301, 1988.

[ROS 07] ROSENBERG E., KOREN O., RESHEF L. *et al.*, "The role of microorganisms in coral health, disease and evolution", *Nature Reviews in Microbiology*, vol. 5, pp. 355–362, 2007.

[ROX 04] ROXBURGH S.H., SHEA K., WILSON J.B., "The intermediate disturbance hypothesis: patch dynamics and mechanisms of species coexistence", *Ecology*, vol. 85, no. 2, pp. 359–371, 2004.

[SAL 92] SALVAT B., "Coral reefs-a challenging ecosystem for human societies", *Global Environmental Change*, vol. 2, pp. 12–18, 1992.

[SCH 85] SCHUMACHER H., ZIBROWIUS H., "What is hermatypic: a redefinition of ecological groups in corals and other organisms", *Coral Reefs*, vol. 4, pp. 1–9, 1985.

[SCH 00] SCHAUS M.H., VANNI M.J., "Effects of gizzard shad on phytoplankton and nutrient dynamics: role of sediment feeding and fish size", *Ecology*, vol. 81, pp. 1701–1719, 2000.

[SEL 11] SELKOE K.A., TOONEN R.J., "Marine connectivity: a new look at pelagic larval duration and genetic metrics of dispersal", *Marine Ecology Progress Series*, vol. 436, pp. 291–305, 2011.

[SOU 10] SOUTER P., "Hidden genetic diversity in a key model species of coral", *Marine Biology*, vol. 157, no. 4, pp. 875–885, 2010.

[SPA 07] SPALDING M.D., FOX H.E., ALLEN G.R. *et al.*, "Marine ecoregions of the world: a bioregionalization of coastal and shelf areas", *BioScience*, vol. 57, no. 7, pp. 573–583, 2007.

[STÖ 12] STÖHR S., O'HARA T.D., THUY B., "Global diversity of brittle stars (Echinodermata: Ophiuroidea)", *PLoS ONE*, vol. 7, no. 3, p. e31940, 2012.

[TIT 10] TITTENSOR D.P., MORA C., JETZ W. *et al.*, "Global patterns and predictors of marine biodiversity across taxa", *Nature*, vol. 466, no. 7310, pp. 1098–1101, 2010.

[VER 76] VERON J.E.N., PICHON M., *Scleractinia of eastern Australia, vol. I., Families Thamnasteriidae, Pocilloporidae*, Australian Institute Marine Sciences, Monography Series, p. 86, 1976.

[VER 09] VERON J., DEVANTIER L.M., TURAK E. *et al.*, "Delineating the coral triangle", *Galaxea, Journal of Coral Reef Studies*, vol. 11, no. 2, pp. 91–100, 2009.

[VER 13] VERON J., "Overview of the taxonomy of zooxanthellate Scleractinia", *Zoological Journal of the Linnean Society*, vol. 169, pp. 485–508, 2013.

[WIL 08] WILKINSON C., Status of Coral Reefs of the World, Global Coral Reef Monitoring Network and Reef and Rainforest Research Center, Townsville, Australia, 2008.

[WIN 14] WINTERBOTTOM R., HANNER R.H., BURRIDGE M. *et al.*, "A cornucopia of cryptic species-a DNA barcode analysis of the gobiid fish genus Trimma (Percomorpha, Gobiiformes)", *ZooKeys*, vol. 381, p. 79, 2014.

[YAM 12] YAMANO H., SUGIHARA K., WATANABE T. *et al.*, "Coral reefs at 34° N., Japan: Exploring the end of environmental gradients", *Geology*, vol. 40, no. 9, pp. 835–838, 2012.

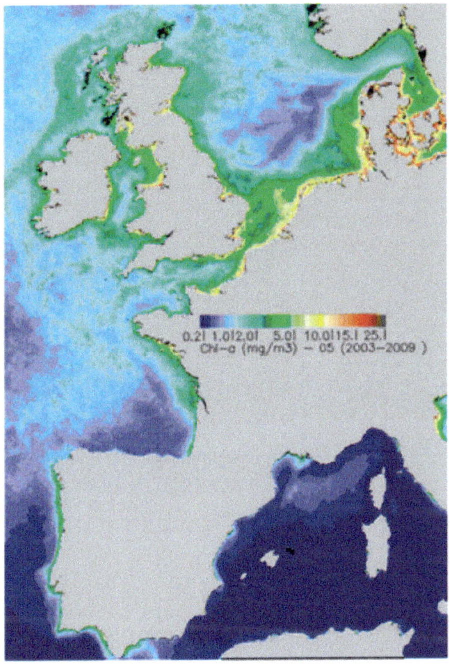

Box 2.1. *Spatial variability in chlorophyll a in the north-east Atlantic ocean, in May, measured using satellite images (according to [GOH 11])*

Figure 3.1. *Morphology of marine magnoliophytes– example of the species Thalassodendron ciliatum (photo C. Hily)*

Figure 3.4. *Seagrass beds, biodiversity hotspots: epibiosis under a canopy of Talassodendron ciliatum (photo C. Hily)*

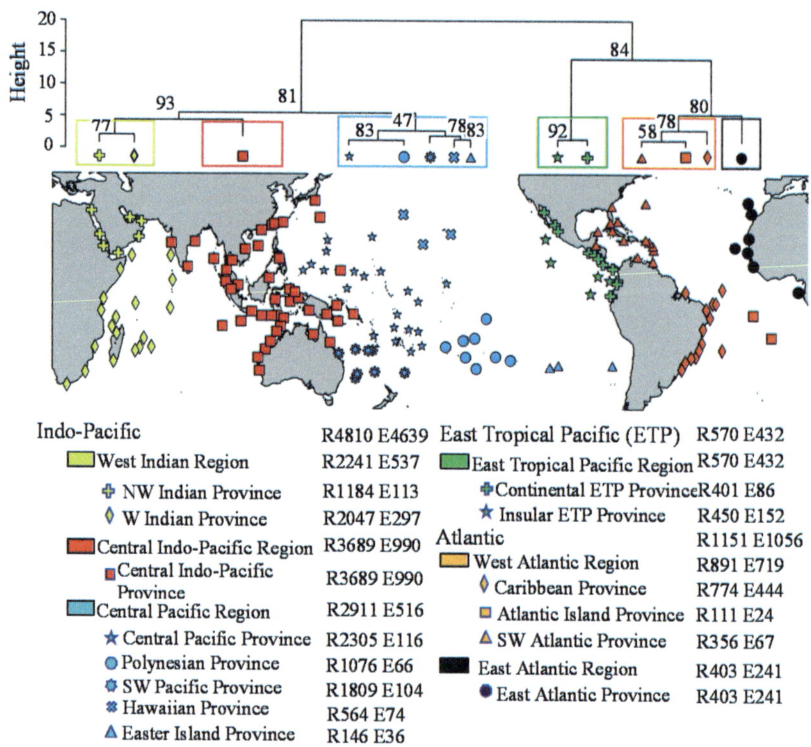

Figure 4.10. *Biogeographic regions defined by the species composition of reef fish [KUL 13]. The R numbers indicate the number of known species for the geographic area and the E numbers indicate the number of endemic species*

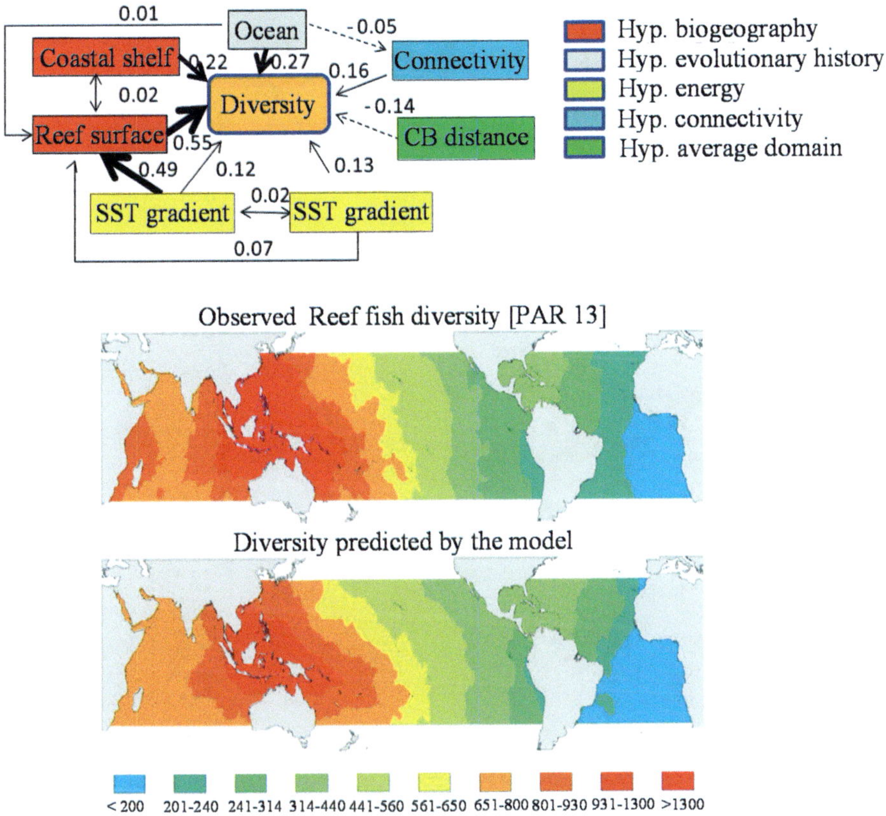

Figure 4.13. *Model (SEM: structural equation model) representing the relative role of different factors to explain the level of reef fish diversity throughout the world [PAR 13]. The maps show, on the one hand, the distribution observed in the diversity of these fish, and on the other hand the diversity predicted by the model. The adjustment is optimal in the zones where the diversity is at its maximum; the most significant differences are observed around the periphery (south-west Indian Ocean; oceanic islands in the East Pacific)*

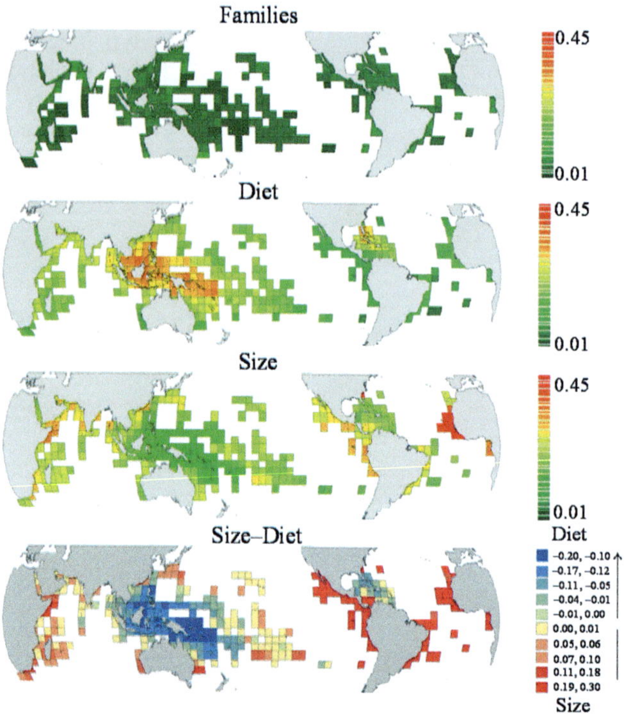

Figure 4.14. *Deviation from a neutral model for the spatial distribution of reef fish for taxonomic composition (at the family level), trophic composition (seven feeding classes) and size (six size groups)*

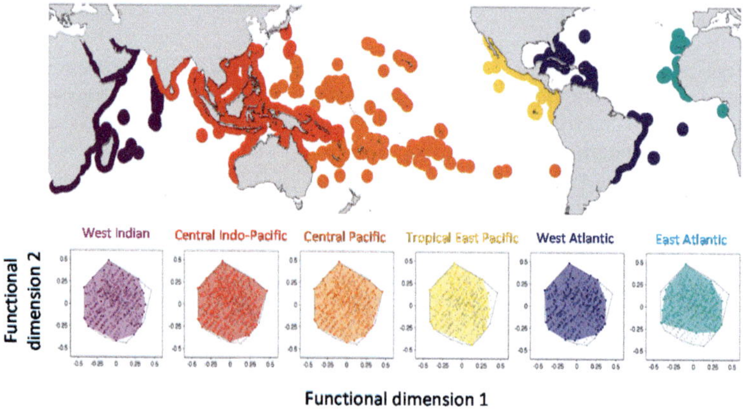

Figure 4.22. *Space of functional volume occupied by coral fish species from six biogeographical provinces of the world [KUL 13] (adapted from [MOU 14])*

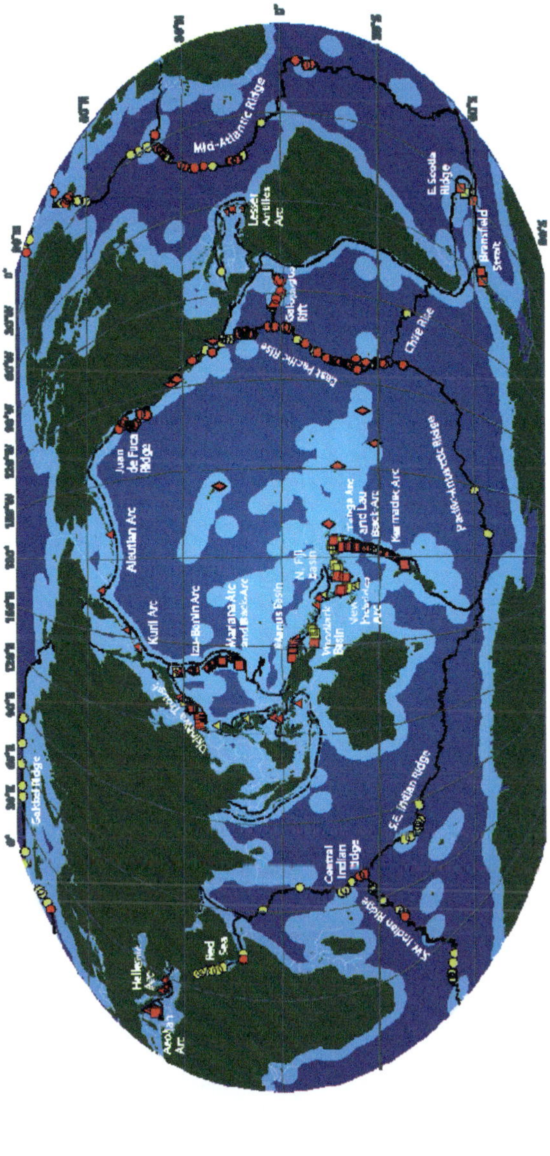

Figure 6.2. *Distribution of active hydrothermal sites (in red) and sites where indices of activity have been identified (in yellow). Continuous lines represent ridges and faults whereas dotted lines represent trenches. Exclusive economic zones are indicated in pale blue (according to [BEA 10])*

Figure 6.9. *The main engineer species of the hydrothermal ecosystem. (a) Riftia pachyptila, EPR (b) Alvinella pompejana, EPR (c) Bathymodiolus azoricus, MAR (d) Rimicaris exoculata, MAR (e) Alviniconcha spp., Lau basin (f) Vulcanolepas sp., Lau basin. (Photos a, b, c, d belong to Ifremer and photos e and f are a courtesy of C. Fisher, Penn State University)*

5

Man and Diversity in the Coral Environment

5.1. Introduction

In Chapter 4, we examined how to use biocomplexity models to try to understand the organization and functioning of a highly diverse system, coral reefs. We are still a long way from fully understanding this system; however, biocomplexity does help structure a set of ideas that support the services provided by such diversity. In fact, coral reefs are a source of many services involving approximately 500 million people in more than 100 countries worldwide. At present, there is no well-defined conceptual framework that can help structure the management of these services at different spatial scales. The responses are currently disparate and they are generally based on a purely local approach, which does not account for the processes functioning at larger scales, or on ideas from other regions where the operating rules may be significantly different.

Let us show two examples that illustrate the current limits. First example: managers rightly want to know what area of maritime space would need to be protected to ensure the permanence of ecosystem services [ATT 97]. The International Union for Conservation of Nature (IUCN) recommended that 20% of the coastline be protected [IUC 92]. Australians have applied this figure of 20% to the Great Barrier Reef (GBR) [DAY 02], a highly diverse system (e.g. > 2,000 species of fish). This figure was used by US lawmakers

Chapter written by Michel KULBICKI, Mehdi ADJEROUD, Laure CARASSOU, Pascale CHABANET, Valeriano PARRAVICINI, Dominique PONTON, Fanny HOULBREQUE and Laurent VIGLIOLA.

intending to apply it to all reef systems in the United States, whether in Florida, which is a system with relatively few species (420 species of fish) and a large proportion of large species, or in Hawaii where the number of species is similar to that in Florida (530 species) but with a high level of endemism, or in Samoa where the diversity is much greater (1,080 species of fish) with a lower proportion of large species. Chapter 4 shows that different processes were operating according to biogeographic regions and diversity levels, therefore, it seems unlikely that setting aside 20% of the reef area will have the same impact in all regions mentioned.

Second example: in Vanuatu (South-West Pacific) before the arrival of Europeans, there were different management systems for reef resources at the village level. These management methods have disappeared over the decades. At present, management at the village level is being reinstated, but each village has its own rules and makes its own decisions regarding the areas, habitats and species to protect, and usually without collaborating with neighboring villages. This leads to a dense network of protected areas, but whose global efficiency remains to be shown. It is likely that simple rules concerning the connectivity between reserves, a simple framework about which species to protect and the combination of habitats to be favored could greatly increase the local and global efficiency of these citizen initiatives.

5.2. *Diversity and ecological services*

5.2.1. *Main ecological services*

Biodiversity provides many services, usually known as "ecosystem services". There are many general classifications for these services, one of the most common being that of De Groot [DEG 02] who split these services into four large groups:

– regulation functions, i.e. the set of processes that help maintain natural systems as well as providing direct or indirect services to man while maintaining a certain air quality, water quality etc.;

– habitat functions which provide habitats that contribute to the conservation of diversity and evolutionary processes;

– production functions, i.e. the production of biomass, usually dietary;

– information functions or even cultural services, with nature representing an essential reference to the human condition.

In the marine environment, the production of biomass is the service best studied and still the focal point of conservation and management of marine zones [WOR 06]. The reef environment does not breach this classification, with the production of biomass still the most characteristic service (e.g. [MUM 08] and [STO 11]). Coral reefs, however, provide many other services, in particular tourism activities, the protection of coastal zones against severe weather conditions or erosion, the presence of organisms that produce substances with a certain pharmacological or industrial interest (e.g. antifouling), the possibility of aquaculture (e.g. pearls, algae and organisms for aquaria), the production of materials (sand, ground coral, etc.). Coral reefs and the communities that they support are also at the center of many cultural aspects.

Most coastal populations exploit their coral reefs by fishing for many organisms, fish representing the largest contribution in weight and economic value. Echinoderms (holothuria and sea urchins), algae, crustaceans and mollusks are also exploited in many localities and may provide significant economic contributions. Fishing is the service best perceived by local populations and is subject to many regulations aiming to make an acceptable economic and sustainable productivity. Reefs provide services associated with the production of biomass, for example nursery or refuge zone for some species. Many reef organisms have developed chemical substances that protect them against predators, pathogens, parasites or environmental effects. These substances can be used in pharmacology, cosmetics or industry. Anti-tumor, antiviral and antimicrobial substances have been discovered in reef organisms and have been tested for use as treatments. Anti-UV agents have also been discovered in these reef organisms and form the basis of products used in sun cream. Antifouling agents have also been found and are being studied for use in the production of paints and varnishes that prevent the incrustation of marine organisms.

Reef-based tourism is blooming with the possibility to visit sites rich in flora and fauna and encounter characteristic species (sharks, manta rays, turtles, dugongs, large groupers, giant Napoleon wrasses, etc.). One advantage of tourism is its sustainable aspect (no direct extraction, creation of protected zones and changes in the way resources are exploited), but it does cause some problems (mooring, overvisiting, the need for water in arid zones or low islands for tourists, fishing of local fish to feed tourists, etc.).

Reefs or adjacent environments that depend on them (sea grass beds, algal beds, coastline and mangroves) can also become aquaculture sites. The organisms most farmed at present are algae and pearl oysters, but cultures may also gradually establish such as in the rearing of fish (in cages or in a semi-open environment), coral culture (for aquaria or repopulation), the extensive farming of organisms including holothuria, clams and other macroinvertebrates. The impact of these activities on reefs is still poorly understood, but certainly not harmless, as recent analyses have shown for pearl farms in Polynesia [AND 12]. The aquarium market is large and blooming. The organisms collected (fish and macrofauna) can have a high-added value (for example, much greater than their value on the local fish market); however, often due to destructive collection methods, this activity puts reefs in danger, whereas a sustainable development would be possible in most of the cases (e.g. [LOV 99]). The development of farming (corals, some invertebrates and fish) should help maintain this service.

Reefs protect to some extent the coast against the effects of swell and erosion of the coastline or even to a lesser extent against exceptional events (cyclones and tsunamis). They also help purify the water, since many filter-feeding organisms and physical processes associated with slowing the water circulation help eliminate or store certain pollutants or suspended particles. Reefs form part of the heritage and therefore often represent an essential part of a culture. Polynesian societies and many Micronesian societies daily refer to their reefs, fauna and flora. The destruction of this cultural reference can have serious consequences on the societal structure of these populations. All services mentioned are renewable if they are accompanied by an adapted management. Reefs can also provide non-renewable services such as providing building materials, sand, etc.

Ideally, we should move toward the harmonized management of these services in a way that optimizes their sustainability. This is not, however, easy since our understanding of how these systems function is still poor and there are many conflicts of interest. The population in coastal zones is increasing in almost all countries with coral reefs, whereas the area and quality of these reefs decrease inexorably. Usually, these populations are poor and do not have the financial resources to implement alternatives, such as tourism, to the direct exploitation of reef resources (fishing). Moreover, increasing coastal populations increases chronic stress caused by pollution and coastal development. In developed countries (e.g. United States, Australia, Japan, France and Taiwan), the nature of these conflicts changes.

For example, should mass tourism be encouraged or tourism with privileged clientele, should recreational fishing replace professional fishing, how can we make the population participate in the co-management of the reef? These management requirements have raised the question of how to quantify these services objectively and comparatively.

It is only relatively recently that economists and managers have become interested in quantifying the different ecosystem services provided by coral reefs. In other words, how much is a reef worth? Is such a quantification possible and if so to what end? Some services, such as the production from fishing, appear to be relatively easy to quantify. To do this, the monetary mass represented by the extracted products must be assessed and the costs must be estimated, for example the price of ships, fishing gear, fuel, port facility, etc. This assessment rapidly becomes complex in most cases since it may be beneficial in some cases and a cost in others. To continue with fishing, the fish captured are no longer accessible to tourists, nor can they provide an ecological function such as controlling the algae-coral equilibrium, which helps protect the reefs and their cultural role or even the level of biodiversity. At present, these economic estimates are, therefore, limited by our understanding of how coral systems function. Moreover, some costs-benefits are very difficult to quantify, such as cultural aspects, or even the settling of fishing populations by maintaining fishing activities has both social advantages and disadvantages involving complex compensatory phenomena. It is, therefore, unrealistic to expect an exact figure for the ecosystem services of a reef. However, using a comparative approach, where reef A provides twice as many services as reef B, this monetization of services is a powerful tool. Some examples will help us better understand this potential.

EXAMPLE 5.1.– Is the creation of a marine protected area (MPA) in a reef environment beneficial to the local economy? An MPA incurs an implementation and monitoring cost, which represents the scientific analyses carried out prior to creating the MPA, scientific monitoring, implementation of legislation and consultative meetings, monitoring by rangers (salaried), surveillance (boats, vehicles and buildings), their maintenance, etc. An MPA also has social costs; some fishermen will be restricted or even banned from accessing the resource and additional trades (e.g. ship repair and sale of fishing equipment) will decrease. Benefits may include better yields for the remaining fishermen, the influx of tourists that visit the MPA, a healthier reef thus better protecting the coastal zone, etc. Not creating an MPA can lead to progressive losses in fishing yields with impoverishment of part of the population, a degradation in the health of the reef, which, eventually, will affect the safekeeping of the coastal zone, its touristic value, etc. Given that degradation is increasingly difficult to reverse beyond a certain degree of disturbance, the benefit of an MPA will decrease if the level of disturbance is allowed to increase to a level beyond which the creation of an MPA is no longer possible (Figure 5.1).

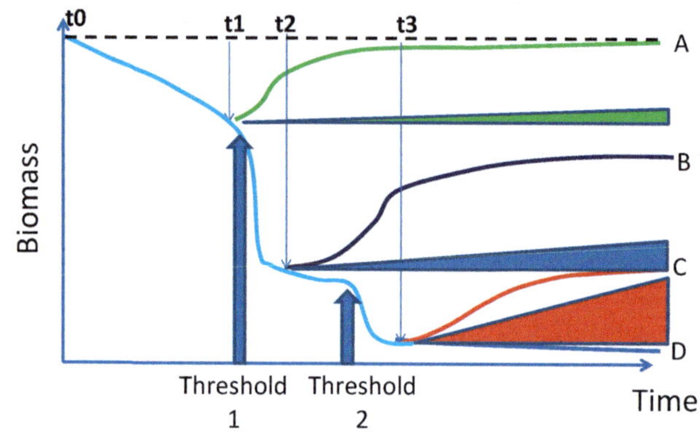

Figure 5.1. *Expected evolution in biomass of reef fish from t_0 (no fishing) to t_1 (low fishing), t_2 (moderate fishing) and t_3 (very intense fishing)*

COMMENTS ON FIGURE 5.1.– At t_1, line A represents the evolution of biomass if fishing was completely stopped or at least effectively controlled. The corresponding triangle above represents the cumulative effort (cost) to control fishing. Beyond t_1, biomass is expected to collapse beyond a certain disturbance threshold. Trying to regulate fishing in t_2 does not allow the return to the level of biomass (line B) at t_0 and the cumulative effort (corresponding triangle) is much greater than that in t_1. Continuing to allow the stocks and the environment to degrade could lead to other thresholds. Trying to regulate fishing at t_3 allows a slight recovery of biomass (line C) because the system no longer has the capacity and the cumulative effort becomes very high (corresponding triangle).

EXAMPLE 5.2.– Building a toll route on a reef *versus* passing through the coastal zone. The route will generate dividends, allowing the circulation to flow more freely, etc. Cutting through a reef zone will reduce costs, but the reef zone considered may have a higher economic, ecological and cultural value than this cost reduction. In this context, a financial assessment of the services as direct products of fishing, the occupation of coastal population (recreational or subsistence fishing), the eventual nursery grounds in this reef for fish, crustaceans and others, the protective role of this reef for the coastline, etc. allow a decision to be made with regard to the outline of this route.

EXAMPLE 5.3.– The cost of an emblematic fish, such as a big grouper, can be calculated from a fishing point of view (at best several euros per kg – no reuse) or from a touristic point of view (diving charged over many years and conservation of progenitors); the difference usually being in favor of a touristic use, however, the beneficiaries are not the same; tourist operators are often not autochthonous and have significant upfront capital that fishermen do not usually have, allowing them to invest in heavy infrastructure (e.g. hotels, diving clubs and leisure activities for tourists).

An important aspect in the analysis of ecosystem services is what is perceived by the beneficiaries. The second example about financial estimates reveals that in our society, financial and monetary aspects often take precedence over well-being and ethical considerations. How these services are perceived by the coastal population according to whether they are city-dwellers or rural will differ significantly. For city-dwellers, the recreational aspect and reef esthetics with the possibility of tourism will likely prevail over fishery products, the role of nursery grounds, coastline protection or quality of life, which on the opposite are essential for a country dweller, especially if they depend on the reef for subsistence. For these populations, the degradation or destruction of a reef is more than just a financial loss, but also a lifestyle loss. Paradoxically, these are often the same populations, as they usually have no alternative, who degrade the reef the most, on which they are so dependent.

5.2.1.1. *The production of biomass*

The production of biomass is probably the most understood ecosystem service provided by reefs and one of the easiest to observe. Reef organisms grow and die. The exploited organisms are removed from the community before they naturally die and this extracted biomass, usually by fishing, represents an outflow from this system. The quantity and quality of this biomass depend on multiple factors, which can be split into intrinsic factors, i.e. specific to the system, and extrinsic factors, which externally regulate. These considerations can be applied to any natural exploited system, and we will, therefore, focus on a particular characteristic of coral reefs, the role of diversity on this production of biomass.

The production of a system first depends on the energy that reaches it and the nutrients available. Coral reefs receive a high quantity of energy because they are situated in tropical zones. However, reefs are often situated in oligotrophic waters, therefore, with relatively low levels of nutrients (see section 5.3.1). Another characteristic of reefs is the fast recycling of nutrients, any nutrient available is rapidly used. The sheer diversity in reef communities facilitates this phenomenon due to "resource partitioning" [ROS 86]: with equal resources, a diverse system possesses organisms that are capable of transforming part of the resource into new products which in turn can be transformed by other organisms; the more diverse organisms there are, the more products will be transformed, which through a cascade effect, leads to an increase in diversity. Let us take the following example: let us assume that on a particular reef there are only herbivorous fish; these will be able to produce X1 $g/m^2/yr$ of fish meat according to the production of algae. If detrivourous fish are introduced, capable of utilizing the waste produced by herbivores, the production increases by X2 $g/m^2/yr$ of detritivorous fish meat, the total production is, therefore, X1+X2. The waste from detritivores can in turn be used by other species and so on. Of course, this partitioning cannot continue indefinitely, many additional events occur (e.g. predator–prey relationships) and the production reaches its maximum with energy and nutrients, but the efficiency of the system tends to increase with diversity. Reich *et al.* [REI 01] give an example of these cascades in a controlled environment. With all other factors being equal, it is, therefore, important to understand the relationships between diversity and production.

An increase in diversity is usually accompanied by a decrease in part of the resource for each species, even if the partitioning of resources slightly mitigates this phenomenon. Therefore, the proportion of small species will increase as the diversity increases [BRO 04, KUL 15]. As the total biomass tends to be limited, an increase in the proportion of small species will usually be accompanied by an increase in abundance (any other factor remaining equal). The resource-partitioning hypothesis suggests that as the diversity increases the production also increases. By combining these different aspects: link between diversity and species size, local diversity and regional diversity, species size and abundance, with the exponential law, which links size and biomass (Figure 5.2), it becomes clear that diversity will affect the biomass. As seen in Chapter 4, the diversity is not distributed uniformly across coral reefs throughout the world. In particular, the diversity is affected by the geological history, island size, degree of isolation of islands and water temperature. These factors will, therefore, have an effect on biomass and production, which can be observed on a given reef. Chapter 4 showed that in the Atlantic and East Pacific, the diversity is lower than in the Indo-Pacific, thus *a priori* the biomass potential should be lower. However, as species in the Atlantic and East Pacific are larger on average than in the Indo-Pacific, we can expect, at an equal diversity and similar environment, a greater biomass in the Atlantic or East Pacific. Similarly, in a given region, a small, isolated island will have a lower diversity per unit area than a large island or continental shelf, and therefore a lower biomass potential. However, the proportion of large species is greater on smaller islands; therefore, at an equal level of diversity, a greater biomass is expected on a small isolated island than on a large island. Note that if a larger part of the biomass is composed on larger species on this small island, the biomass will be much more sensitive to exploitation (for an equal fishing effort, a greater proportion of biomass will be extracted and these will mainly be large species which are generally more susceptible to exploitation).

Several local factors are also involved; fishing is probably the most characteristic. Anything that will locally favor the diversity, such as significant coral cover, will increase the biomass; however, additional species will not necessarily be "interesting" from a resource point of view. Therefore, at an equal level of diversity between a zone with high coral cover and a zone without corals, the zone without corals will usually have more large species and therefore a greater biomass. A brief analysis (Figure 5.3) shows that the same fishing effort will have different effects on the level of biomass, and consequently production, depending on the levels of these factors. This analysis reveals a very important message: the management of reef ecosystem services should take into account the local and regional context; a reef in a diverse region or zone should not be managed in the same way as a reef in a region or zone with fewer species, all other factors being equal.

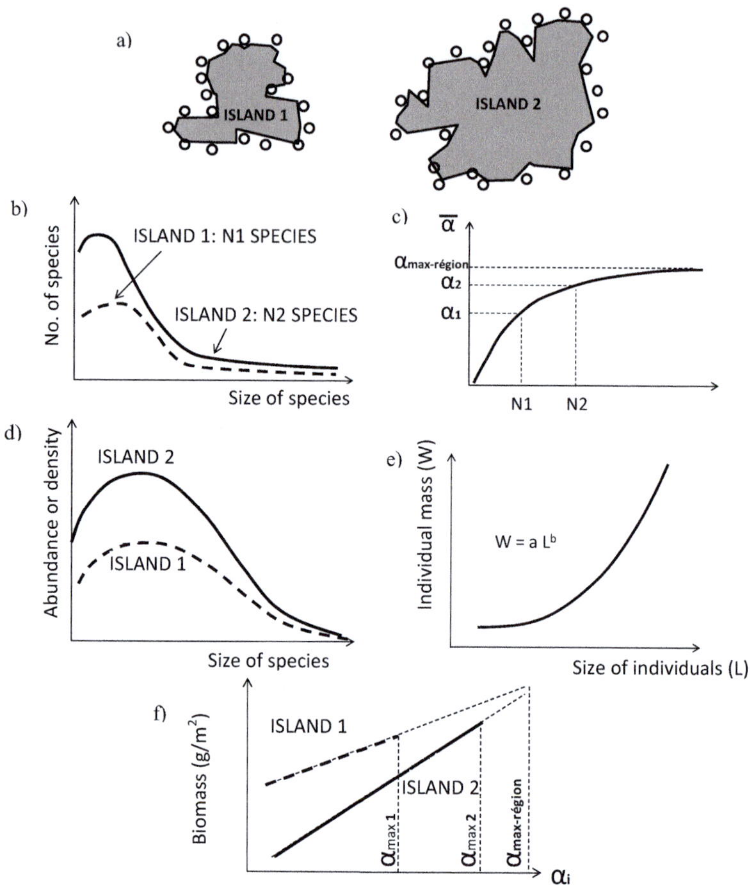

Figure 5.2. *a) Let us assume that two islands 1 and 2 are different in size. Around each island, stations of the same size are regularly sampled (small circles). b) The number of species on island 1 will be lower than on island 2 and the mean size of species on island 1 will be greater than on island 2. c) Both islands belong to the same region; the local diversity (α) observed at a station on one of these islands will be proportional to the regional diversity and number of species known on the island; the number of species observed at a station in this region cannot exceed the* α_{max} *of the region. d) The abundance (or density) of a species is associated with size; the abundance is also associated with the number of resources (fewer on island 1 due to its smaller size), we will, therefore, observe a lower abundance for the same fish size on island 1 and a general decrease in abundance as fish size increases. e) The weight of a fish is a function of its size, according to an exponential law. f) By combining graphs B, C, D and E, the biomass on station i on one of these two islands will be a function of the alpha diversity observed; with species being larger on average on island 1 their biomass will be greater at an equal diversity compared with island 2; the maximal local diversity observed is bounded on each of these two islands (*α_{max1}*;* α_{max2}*)*

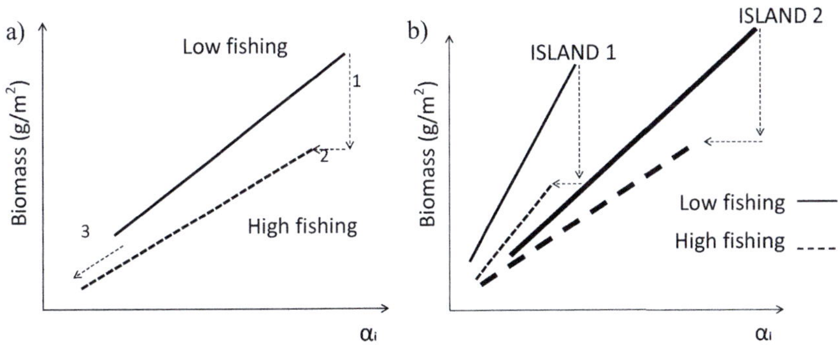

Figure 5.3. *a) Moving from a low level of fishing to a high level of fishing causes a loss of biomass (loss of individuals removed by fishing) (1), but also a depletion of individuals, which causes a decrease in species density (2), this loss of diversity is greater when the initial diversity is low (3). These losses of diversity (2 and 3) in turn cause a decrease in resource partitioning and therefore the amount of biomass attainable, as there is a dual effect of fishing on biomass. b) On island 1 (small), the loss of biomass will be greater at an equal fishing effort, whereas the loss of diversity will be greater on island 2 (large)*

The biomass of fish, or most other reef resources, is quite easy to measure, for example, using direct underwater observations of fish. There is a relationship between production and biomass for a set of organisms [BRO 04], based on the analysis of many species, the production P is approximately equal to biomass $M^{\alpha}\alpha$ which is on average equal to 0.75 with slight deviations essentially associated with species, size and temperature. Biomass, however, is not necessarily a good proxy for production. In fact, at equal biomass small and young fish produce more than larger and older fish of the same species. Therefore, an exploited community of fish, characterized by many small and young fish, may have a greater production than a community that is fished less, with high biomass but whose individuals are large and produce little new biomass. If, however, the level of exploitation (fishing effort) is the same in the zones that are being compared, then biomass is a good proxy for production in a comparative approach. The question asked is, therefore, how can we assess this production to estimate the potential of different locations subject to specific conditions? In the reef environment, this has rarely been attempted [DEP 07].

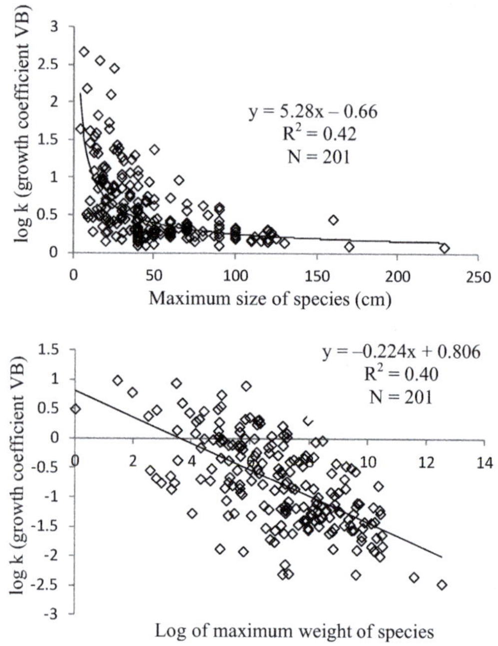

Figure 5.4. *Relationship between the growth coefficient (k) and size (top graph) or log of weight (bottom graph)*

There are two solutions. The first solution is to estimate over a long period, usually 1 year, the increase in biomass due to growth and loss of biomass through death, both natural and due to fishing. This exercise attempts to estimate the mortality, which is very difficult to reliably obtain, even for mortality due to fishing. The second solution is to ignore mortality and only focus on instantaneous growth (usually growth in 1 day). The production obtained is, therefore, an "instantaneous" production, which better reflects the potential of a reef to produce biomass than the biomass actually produced. This formulation does, however, allow the comparison of reefs. What remains is to estimate growth, which is obtained for fish by analyzing otoliths, ossicles located in the inner ear of fish. These otoliths have regular rings of aragonite; a cross-section allows these rings to be counted (just as we would count the rings of a tree trunk to determine its age). Once the periodicity of these deposits is verified (forming either daily or seasonal rings), it becomes possible to link the number of rings to the age and size of the fish and therefore estimate growth [MOR 00]. The analysis of otoliths may now involve algorithms that allow back calculations to obtain growth curves from a small number of otoliths (< 20), whereas previously, a large number of otoliths (several dozens) were required [VIG 09]. With this obstacle overcome, it is now possible to obtain reliable growth curves from less than 20 otoliths/species, which allows the growth of many species to be studied and therefore obtain estimates of production for increasingly complex assemblages. Generally speaking, large species have slower growth rates than smaller species (Figure 5.4). Therefore, at equal biomass, a group of small individuals will produce more than a single large individual. The rules determining the abundance and mean size of individuals will, therefore, significantly influence production.

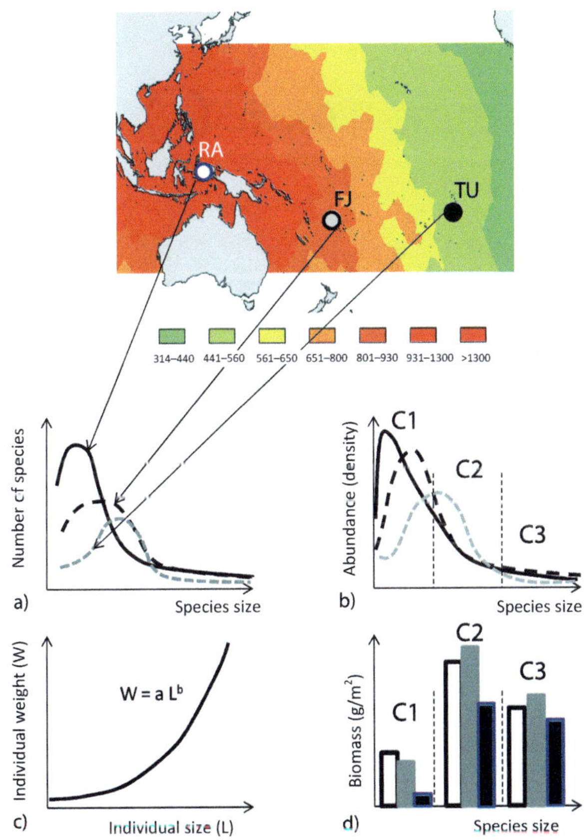

Figure 5.5. *Relationships between biomass, size and biogeographical origin. Three regions, Raja Ampat (RA circle), Fiji (FJ circle) and Tuamotu (TU circle), have declined diversities of reef fish (sources: Raja Ampat: A. Green (pers. com.), Fiji and Tuamotu: IRD database)*

COMMENTS ON FIGURE 5.5.– (a) Relationship between diversity and size of species in three regions, the more the diversity increases the more the average size increases. (b) Relationship between species size and abundance, deduced from A, and abundance increases with diversity (not shown). (c) Relationship between weight and length, which is different for each species. (d) Biomass for three size classes (C1, C2 and C3) depending on the region is generated from the combination of band c.

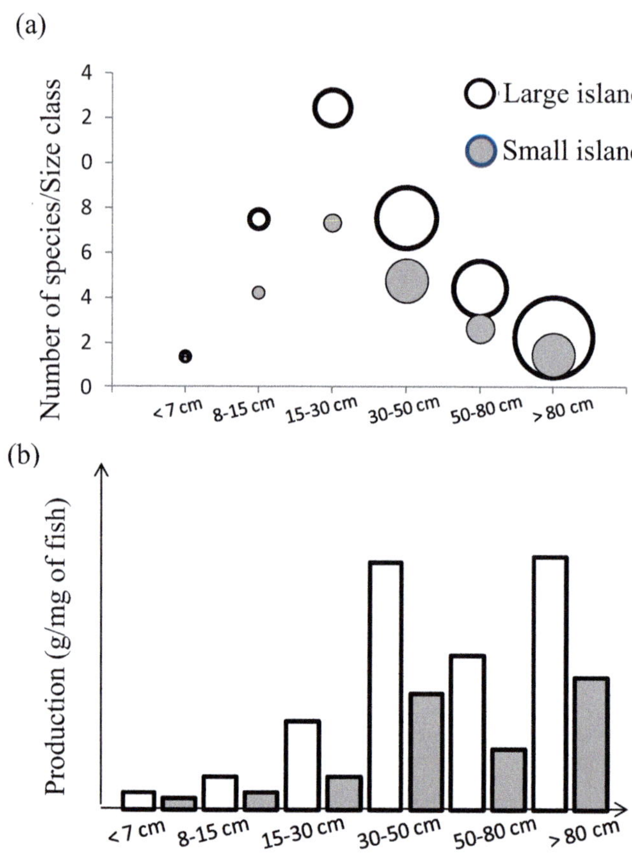

Figure 5.6. *Relationships between size, diversity, biomass and production. Example taken from Tuamotu. a) Ten atolls were classed into two size classes (small < 100 km²; large > 100 km²), each circle represents the biomass for a given size class (abscissa) the level of the circle being given by the diversity of the size class (ordinates); the largest circle represents 50.3 g/m². b) Based on biomass-production relationships according to size, this diagram shows a higher expected production in large atolls*

5.2.1.2. *Functional and phylogenetic redundancy*

Another aspect that has been largely overlooked until now in our approach of services provided by diversity is redundancy (number of species in a group), whether functional or phylogenetic. Ecologists traditionally focus on species diversity. The management of ecosystems is, therefore, largely based on the principle that species diversity must be protected. Recently, this concept of functional and phylogenetic diversity has increasingly been taken into account because theoretical and experimental studies indicate that these diversities play a key role in the functioning of ecosystems [CAD 12, CAD 13]. In particular, in Chapter 4, it was shown how functional diversity structures communities, with a strong association between taxonomic, functional and phylogenetic diversity. In this situation, there are two key questions:

– Are the functional and phylogenetic diversity affected by human pressures?

– Is protecting the species diversity effective to maintain other facets of diversity?

A recent study on the diversity of *Scarinae* (parrotfish, in the family *Labridae*) in 17 countries of the insular tropical Pacific indicated that the human population density has a greater impact on the functional and phylogenetic diversity than on species diversity [DAG 14]. Therefore, in addition to reducing species diversity, Man significantly reduces the diversity of functions ensured by fish populations as well as the richness of their evolutionary history (Figure 5.7).

These results show the significant drop in the levels of functional and phylogenetic diversity, particularly beyond 20 habitants per km^2 of reef, whereas species richness remains largely unaffected across the human pressure gradient (Figure 5.7). Thus, when the human population density reaches 1,700 habitants per km^2 of reef, the impact on the levels of functional and phylogenetic diversity (respectively, –46 and –36%) is greater than on the species richness (–12%). Moreover, these studies emphasize that the number of species is an indicator with low sensitivity with regard to anthropogenic pressure, whereas the other two components of biodiversity are far more affected by human density. This highlights the importance of conserving the whole set of biodiversity components, and not solely species diversity. This also indicates that it is possible to use diversity of functional traits and phylogenetic lineages as reliable and sensitive indicators of the degradation of species communities with regard to management plans.

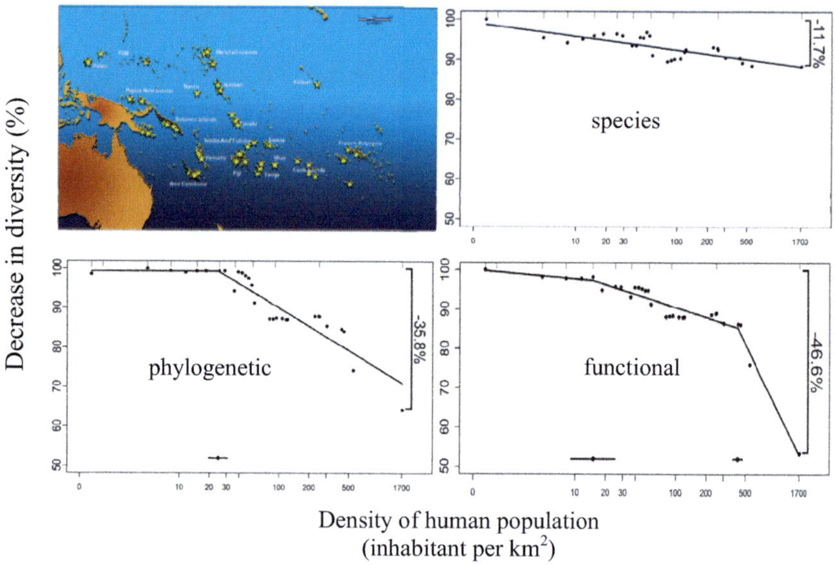

Figure 5.7. *Erosion of species, phylogenetic and functional diversity of parrotfish from coral reefs in 17 insular countries in the tropical Pacific (adapted from [DAG 14])*

5.2.2. *Diversity, robustness, resilience and disturbances*

Until the 1970s, most authors considered that coral ecosystems were in a state of equilibrium, with relatively stable species assemblages, with this stability essentially due to internal forces of trophic relationships and to the constancy of environmental constraints [END 76, GOR 69, JOH 75]. According to those hypotheses, the ecological succession of reefs would proceed toward increasing complexity and diversity with the final stage being climax. The complexity of the reef ecosystem would be precisely adjusted to allow for maximal optimization, and reefs would by highly fragile with low resilience against environmental changes. Following the initial interannual monitoring of reef communities and initial studies on the impact of large-scale disturbances (including the massive die-off of sea urchins in the Caribbean and the bleaching event of 1982, which caused mass coral death in most reefs on the planet [GAR 03, HUG 94, MUM 07]), this vision for coral reefs has been considerably undermined. It is now accepted that the reef ecosystem rarely attains a steady state, and that reefs are greatly and frequently affected by many natural and anthropogenic disturbances, along with chronic and localized stress [CON 78, GRI 74, KAR 93]. In this context, theoretical studies, despite being recently contested by some authors [FOX 13], hypothesize that the high diversity in coral reefs would be maintained by these disturbances, if their intensity and frequency remain "average" (*intermediate disturbance hypothesis* [CON 78]). These disturbances would prevent the ecosystem from reaching its climax and limit the dominance of some species. According to this hypothesis, the ecosystem would be a temporal mosaic, i.e. composed of communities subject to different disturbances and at different stages of evolution and succession. This would result in a greater robustness and resilience to disturbances. At present, most authors consider that the structure and dynamics of reef communities are essentially determined by these sporadic but intense disturbances, which sustainably modify the coral ecosystem (concept of "ecological shock" [WOO 92]).

Over the past four decades, monitoring the health of reefs has shown that most reefs on Earth are significantly affected by diverse natural or anthropogenic changes, such as cyclones and tsunamis, coral bleaching events, die-offs of key species, proliferation of coral predators including the starfish *Acanthaster planci*, hypersedimentation, water pollution or overfishing [BEL 04, DEA 12, HUG 03, HOE 07]. Although some of these disturbances have formed part of the history of reefs for thousands of years, the recent increase in their frequency and intensity, associated with global climate change, has caused serious degradation (e.g. Figure 5.8). Moreover, other types of disturbances, such as ocean acidification, will quite probably affect reefs in the coming years. The most recent global assessments of the health of reefs are particularly worrying and clearly show that reefs are one of the most threatened marine ecosystems. At present, an estimated 20% of reefs have been permanently destroyed, and 35% are in a critical state and seriously endangered [WIL 08].

Most large-scale disturbances cause significant death, sometimes mass death, of corals and many other reef organisms, both directly and indirectly [BEL 06, KAY 12, PRA 14]. Some physical changes cause habitats to degrade, which consequently affects the overall reef biodiversity. Furthermore, when they are intense or frequent, these changes can modify the structure and functioning of reef communities and usually cause disequilibrium of the ecosystem.

These disequilibria occur, for example, by the rapid transition (several years) of an ecosystem dominated by corals to an ecosystem dominated by algae (phase-shift), especially when nutrient concentrations are high and/or populations of herbivores are not abundant enough to control algal proliferation [DON 92, GAR 03, MCM 04]. It may also involve radical changes in the dominance of some species within key functional groups (shift in dominance [SUN 12]). In the long run, it is likely that these disequilibria will reduce the biodiversity and productivity of reefs, as well as the robustness and resilience of reefs.

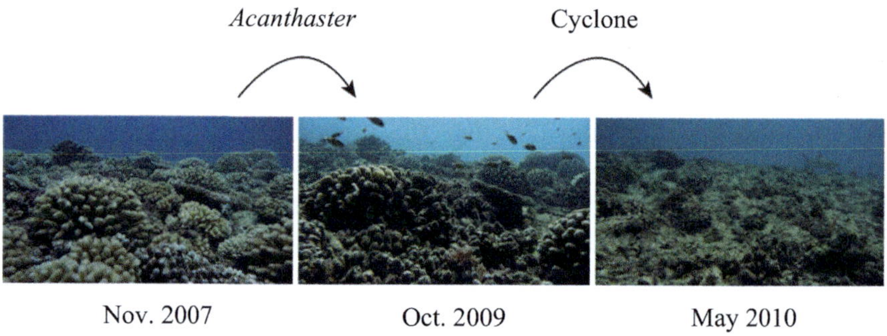

Figure 5.8. *Photos of the outer slope reef of the island of Moorea (French Polynesia), before (photo from November 2007) and after an outbreak of the corallivorous starfish Acanthaster planci (photo from October 2009), and after cyclone Oli in February 2010 (photo from May 2010) (photo credits: M. Kayal)*

COMMENTS ON FIGURE 5.8.– This sequence illustrates the considerable impact of the successive effects of predation by *A. planci*, which left colonies dead, and then the cyclone destroyed most of the remaining colonies, therefore, affecting the three-dimensional structure of the reef.

Following a disturbance that caused mass death of corals and associated fauna, corals recolonize the environment via several mechanisms. First, the substrate is recolonized via the recruitment of larvae from sexual reproduction of surviving coral colonies or neighboring reefs unaffected by the disturbance, revealing the importance of connectivity between metapopulations [ADA 11, ELM 09, PEA 81]. However, if corals only usually reproduce once a year, the recruited juveniles must wait at least 4 years before they become capable of sexually reproduction. The re-establishment of coral communities after a disturbance, therefore, assumes that no major disturbance or local stress intervenes during the recolonization process. Moreover, recolonization can also occur via the growth of colonies still in place that survived the disturbance. However, for slow-growing corals (several mm/cm per yr), the re-establishment of communities requires time, approximately a decade for the most "performing" reefs.

Over the past two decades, one of the main questions in reef ecology concerns the resilience of reefs [HUG 03, NYS 00, NYS 01]. Despite some controversies regarding its definition [MUM 14], most authors agree on defining the ecological resilience of a reef and its communities as its capacity to return to its initial state after a major disturbance that caused the decline of some key functional groups. This not only assumes a return to "pre-disturbance" values of diversity and abundance, but also the re-establishment of the structure and functioning of the disrupted reef ecosystem (Figure 5.9). At present, and despite recent advances, the challenge faced by scientists is to identify indicators that can measure the resilience capacity of reefs, which requires long and laborious interannual monitoring, involving many components of reef communities. Some authors also distinguish robustness and vulnerability of coral ecosystems [MUM 14]. The robustness of a reef can be defined as its capacity to maintain itself in an "acceptable range" of diversity, structure and function, despite the disturbances. With regard to the vulnerability, this is the degree of exposure of a reef to one or several well-defined disturbances (e.g. a reef may be vulnerable to chemical pollution but not to thermal stress).

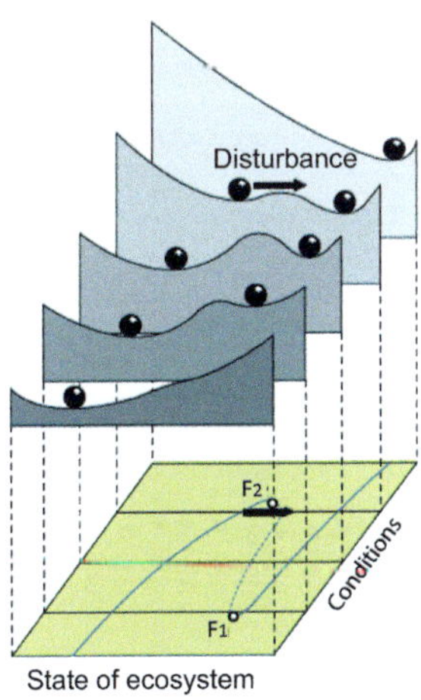

Figure 5.9. *Conceptual diagram illustrating the concept of ecological resilience (according to [HOL 73])*

Although many reefs have been permanently degraded and appear to have low resilience, particularly in the Caribbean and South-East Asia, some reefs have shown stronger resilience. This is the case for coral communities around the island of Palau in the West Pacific, Cocos Islands in the East Pacific and the outer slope (part of the reef directly in contact with the ocean) of the island of Moorea in French Polynesia (Central Pacific) (Figure 5.10). The rapid increase in coral cover following disturbances in 1991 (cyclone and coral bleaching) and the return to initial values after 2001 reveal the capacity of communities to re-establish themselves [ADJ 09]. However, the speed at which coral cover is re-established has rarely been quantified [DON 99, HAL 04]. Nevertheless, it is important to note that this high resilience of corals in Moorea concerns only one of the structural aspects of coral communities, the percentage of total recovery, across all genera. Disturbances, particularly those of 1991, have greatly modified the relative contribution of the main genera (Figure 5.11), with an increase in species of the genus *Porites* at the expense of those of the genus *Pocillopora* (shift in dominance). These results indicate that the higher frequency of disturbances on the outer slope of Moorea has prevented some species from recolonizing bare substrate, and that this recolonization is mainly due to more opportunistic species, such as those from the genus *Acropora*, or those competitively superior and particularly resistant such as massive colonies of the genus *Porites* [BER 06].

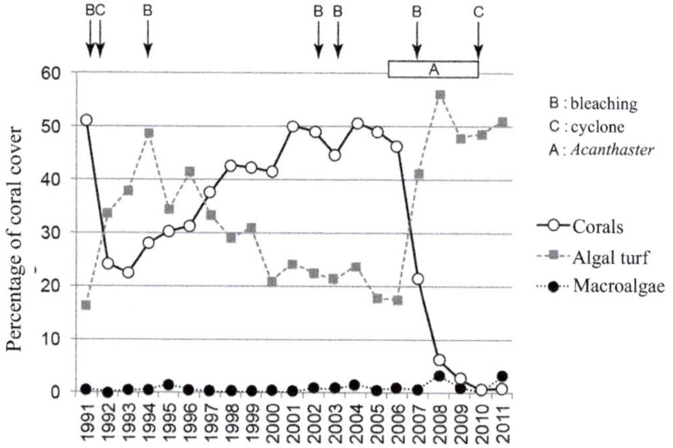

Figure 5.10. *Variation in the percentage of coral cover (across all genera) and algal cover (macroalgae and microalgal turf) on the outer slope of Tiahura (Island of Moorea, French Polynesia), impacted by five coral bleaching events, two cyclones and an outbreak of the corallivorous starfish Acanthaster planci*

COMMENTS ON FIGURE 5.10.– After considerably declining after the 1991 disturbances, living coral cover reached "pre-disturbance" values (~50%) in a decade, thereby showing a certain degree of resilience.

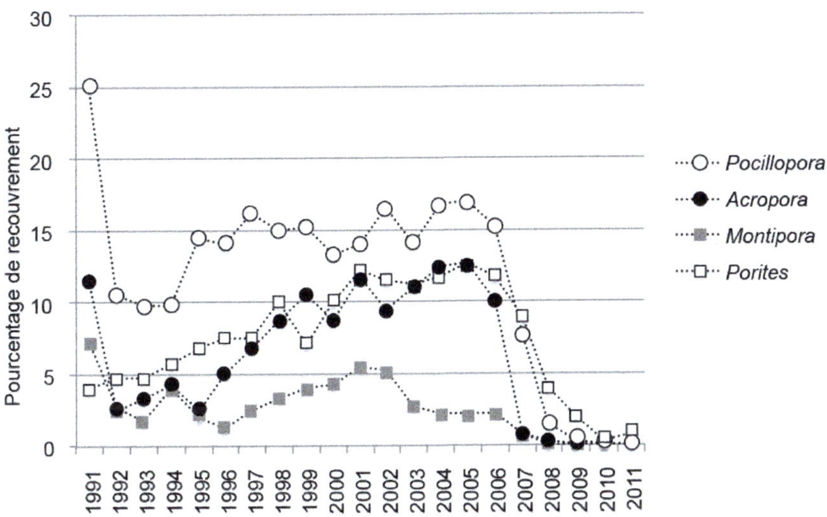

Figure 5.11. *Variation in percentage of coral cover of four main coral genera (which represent more than 70% of total cover) on the outer slope of Tiahura, showing a change in their relative contributions within assemblages*

Up until 2006, coral assemblages around Moorea had managed to recover after a dozen major disturbances within a decade. However, the damage caused by a recent outbreak of the starfish *A. planci* (2006–2009), which slowly decimated coral populations on the outer slopes (drop in coral cover to < 5%), and after cyclone Oli (February 2010) which considerably affected the three-dimensional structure of the reef, provided the final blow to these coral communities that had just recovered from the disturbances in the 1990s [KAY 12]. With this level of degradation, unknown till now, the recovery of coral communities and associated fauna is still hypothetical. The recent history of the Tiahura reef reveals at which point coral ecosystems are still fragile, even for those showing high robustness and resilience in the past.

With regard to the present coral reef crisis [BEL 04], many initiatives have been taken to implement studies on the evolution of reef status and health, on the impact of disturbances and on the mechanisms that erode the biodiversity and those that maintain it. This information is not only required to determine the vulnerability and resilience of reefs, but it is also essential to guide conservation measures that aim to maximize their recovery potential after a large disturbance or local stress. Although these conservation measures at the local or regional scale cannot prevent a cyclone or thermal stress in an ocean, a reduction in local stress (e.g. to limit algal proliferation and degradation of habitats essential for the installation of corals) can quite possibly increase the resilience of reefs [CAR 09, GRA 13,]. Maintaining the diversity, in particular the functional diversity of key species, is a crucial element for the robustness and resilience of reefs [ELM 03, ROG 13].

For many scientists, the outlook is bleak in the coming decades for most reefs, even in more optimistic scenarios of global changes [ANT 11, MCC 12, MOR 13]. In this context, it is urgent to identify the most robust/resilient reefs and understand the mechanisms behind these capacities to recover [HUG 10]. It will also involve implementing conservation measures as efficiently as possible [ALM 09, MOR 06, MUM 08, RAU 12]. In addition, the change in reef health will also depend on the ability of their main reef-builders, i.e. corals, to adapt [HOE 12, HOW 12, MCC 14]. Research on this topic has only, however, just begun, and other, such as ocean acidification, are a threat in the near future [HOE 10, PAN 11].

5.3. Local versus global threats: what are local threats and what are their effects?

Many factors can disrupt the equilibrium of coral ecosystems. Local anthropogenic pressures are one of the greatest threats to coral reefs. More than 450 million people live in proximity to coral reefs and, faced with an increasing demographic pressure, reefs have degraded extremely quickly over the past 50 years [HUG 03, WIL 08]. So, at present, 19% of coral reefs are considered to have been permanently lost and just over 35% are seriously endangered [CAR 08, WIL 08]. Human activities are particularly responsible for the increased mechanical erosion of soils, domestic and agricultural pollution of reef water and the destruction of part of reefs, during multiple dredging operations, increasing the sedimentary load of water. Some natural phenomena, such as successive cyclones and population bloom of the starfish *A. planci*, add to the degradation of reefs and increase reef mortality.

5.3.1. *Eutrophication*

Reef waters are usually oligotrophic and contain low levels of inorganic nutrients. The concentrations of nitrate are approximately between 0.1 and 0.5 µM/l, 0.2 and 0.5 µM/l of ammonium and less than 0.3 µM/l of phosphorus [FUR 91]. Unfortunately, over the past 50 years, an increase in nutrient concentrations has been observed in many reef ecosystems (phosphorus > 0.3 µM/l, inorganic nitrogen > 1–2 µM/l [JOH 79, TOM 87]). This excess of nutrients is essentially due to the discharge of wastewater and runoff. This eutrophication is one of the most harmful forms of pollution for coral reefs. Laboratory and *in situ* experiments [GEN 95, MUS 89, TOM 87] have shown that corals and coral ecosystems were widely affected by these high concentrations of nutrients. In the ecosystem, high nutrient content stimulates algal growth. The proliferation of macroalgae leads to competition with corals for space, whereas simultaneously, phytoplankton blooms cause an increase in turbidity of the water meaning there is less light available for corals [GEN 95, RIE 91]. The survival and installation of coral larvae are also compromised [HUN 92]. At the cellular level, high amounts of nutrients cause an increase in density of zooxanthellae and chlorophyll within these zooxanthellae [DUB 96, MUS 89]. The levels of photosynthesis are, therefore, reduced due to less carbon available [DUB 90], with most carbon acquired by photosynthesis being respired by algae instead of being transferred to animals [FAL 93, MUS 89]. High levels of nutrients disrupt the symbiosis between zooxanthellae and the host organism. Many studies have also reported modifications to calcification and growth of corals when concentrations of nutrients reach a certain level. The reef community [SMI 78] also experiences low calcification rates measured in a lagoon in the Canton Atoll due to high phosphate levels in reef waters. Similarly, Kinsey and Domm [KIN 74] measured a clear decrease (50%) in the calcification of communities following the fertilization of some areas in the GBR [KIN 79]. At the level of the organism, many studies have clearly shown that coral growth can be decreased by 50% in high levels of nitrogenous elements [MAR 96] with the strongest decrease in growth rates recorded during a simultaneous input of nitrogen and phosphate [FER 00].

These high levels of nutrients can have very different effects on other parts of coral reefs. The increase in plankton production encourages the growth of plankton feeders, whether fixed organisms (lamellibranches) or mobile organisms (fish). This increases the number of prey for higher trophic levels; however, this increase may be accompanied by habitat loss, due to the decrease in coral; many small plankton-feeding fish (e.g. *Chromis spp.*, *Pseudanthias spp.* and *Apogonidae*) depend directly on the quality and quantity of refuges offered by corals. Competition between corals and macroalgae, as already mentioned, can have serious consequences. Some macroalgae will benefit from these nutrient levels, but if there are sufficient herbivore populations (invertebrates, fish, turtles and dugongs) the production of macroalgae will probably be absorbed by these herbivores, providing more services, in particular an increase in biomass. In practice, most zones with significant levels of nutrients are also zones disturbed by various human activities, including fishing, and herbivore populations are frequently fished. Consequently, beyond a certain level of nutrients, herbivores are unable to control the growth of macroalgae, which have a harmful effect on corals following the occupation by macroalgae of rocky substrates required for the installation of new coral colonies. The paroxysm is a sustainable disequilibrium of the relationship between algae and corals, as observed in the Caribbean in the 1990s, with the sudden disappearance of sea urchins, due to disease, which are essential grazers of benthic algae, together with the poor status of herbivorous fish populations due to overfishing. The consequence of this was a sustainable dominance of macroalgae cover, which caused significant changes in all reef communities in the Caribbean.

In some regions, another indirect effect of high nutrient content is the stimulation of demographic explosions of the starfish *A. planci*, which feeds on coral. For example, a fourfold increase in nutrient input in the GBR in Australia during the rainy season was correlated with phytoplankton blooms, allowing the simultaneous development of zooplankton populations and *Acanthaster* larvae, which benefit from additional prey [BRO 05]. This is true for reefs in the GBR but does not explain the bloom of *Acanthaster* observed, for example, on isolated islands. Here, it is worth noting that for many reef components, the destruction of corals by *Acanthaster* has different consequences to the destruction observed during other types of disturbances. Current research reveals that in the case where corals are destroyed by cyclones, fish populations were less affected than after being destroyed by *Acanthaster*. This may be due to the types of coral affected the percentage destroyed as well as the depth reached by these destructions.

5.3.2. *Pollution by chemical products and hydrocarbons*

5.3.2.1. *Metals*

The problems caused by metal pollution (typically, copper, zinc, manganese, aluminum and iron) affect many reefs throughout the world (e.g. Costa Rica, Red Sea and Indonesia) [GUZ 92, REI 12]. They are considered as persistent pollutants that remain unchanged in the environment for many years. The main input of metals is from storm water runoff, industrial discharge, mining exploitations and atmospheric contaminants, in the particulate or dissolved form [REI 94]. Exhaust gases, sewage sludge and dredging as well as the use of "antifouling" paint for boats also significantly increase these concentrations. Studies on the effects of these metals on corals can be resumed as follows: a decrease in their reproductive capacity [REI 05], a decrease in recruitment and survival of larvae [REI 00], loss of zooxanthellae within their tissues [SAB 09], changes in their rate of photosynthesis as well as a decrease in their calcification and growth rates [HAR 89, HOU 12] and finally an increase in mortality [SAB 09].

5.3.2.2. *Organic pollutants*

Unlike metals, organic pollutants are more difficult to analyze and detect in the marine environment. Among these organic pollutants, herbicides are considered as the most toxic compounds to reefs, equally affecting zooxanthellae within coral tissues, seagrasses and macroalgae. Many other biocides have also appeared on reefs due to agricultural runoff and more recently as "antifouling" paint, following the ban of tributyltin (TBT). Among the biocides, irgarol® has been found as a very effective compound in this new generation of paints but has also been found to be particularly toxic to photosynthesis in zooxanthellae, inhibiting mechanisms of photosystem II (enzymatic complex involved in photosynthesis) for concentrations as low as 50 ng L^{-1} (50 ppt). Many studies carried out on the Australian GBR have also provided evidence that many pesticides were discharged from the Queensland coast. The concentrations of some of these pesticides were also found to be toxic to corals, destroying part of the reef, which questioned the GBR as a UNESCO World Heritage Site [HUT 05].

5.3.2.3. *Hydrocarbons*

Many studies have been carried out on the impact of hydrocarbons on coral reefs [GUZ 91, LOY 80]. Dominant corals in chronically polluted zones have been found to have a high mortality as well as a lower reproductive potential and reduced ability of larvae to establish themselves [LOY80]. A case of significant hydrocarbon discharge in Panama revealed that the impacts of these oil products could affect up to 8 m in depth and that the coral cover and species diversity were lower in the sites affected [GUZ 91]. Many studies have also revealed that dispersing agents, used during coastal clean up operations, accentuate the harmful effects of these hydrocarbons on corals [KNA 83].

Research into the effect of these three types of substances on other reef components is still rare. As with nutrients, they have indirect effects due to the destruction or changes occurring to the structure of the habitat. These substances rarely cause death directly, but they do, however, have very harmful effects on many terms. Heavy metals are renowned for their toxicity to many invertebrates, particularly in the larval stage (e.g. lamellibranches and sea urchins). Most pesticides act on the reproductive capacities, behavior and assimilation, particularly of lipids, in fish. The sublethal effects are known to affect the intrinsic rate of increase in populations through the combination of a decrease in number and size of eggs, fecundation rate, larval survival rate, juvenile growth rate and consequently size and age at maturity. This results in weaker populations that are more vulnerable to predation, competition and disease, and ultimately a loss of ecosystem service.

5.3.3. *Impacts of coastal development*

Backfilling and dredging are two of the most harmful activities for coral reefs [ERF 12]. Indeed, during these activities, all or part of a reef can become buried by sediment plumes that travel over several kilometers. The backfilling of fringing reefs increases building areas that have become rare on many insular zones. This is the case, for example, in Singapore, where the fringing reef was completely backfilled and merged with the coast. The destruction of fringing reefs or coral pinnacles situated inside a lagoon is also a widespread practice, allowing the construction of ports or mooring sites.

The sediment plumes formed by these activities are particularly harmful to corals. In fact, the increased turbidity reduces light penetration, thereby reducing the photosynthetic capacities of corals and therefore their energy intake [DOD 74, FAB 05]. Unfortunately, in parallel, corals use large amounts of mucus in order to clean this sedimentary layer deposited on their tissues, which paradoxically increases their energy demand. Sedimentation seriously affects coral reefs [ROG 90]. It reduces the growth and survival or a wide range of coral species although the intensity of responses varies according to the species and sediment type [RIE 95, STA 92, STA 93]. The recruitment and survival of young coral lifeforms are particularly affected and have larval installation rates of almost zero on surfaces covered by sediment and a much lower tolerance to sedimentation in recruited corals than in adult corals [FAB 05]. Reef waters are usually extremely clear, with less than 5 mg of sediment l^{-1}. A concentration of 5–20 mg $liter^{-1}$ of sediment is considered as a stress factor, and a value of 20 mg $liter^{-1}$ is above the tolerance level of many species [BRO 97, ROG 90].

These physical disturbances of the environment have other consequences. For example, most fish and many invertebrates both directly and indirectly benefit from coastal habitats, which provide refuge for juveniles, breeding or feeding grounds or even the production of aging species will in turn feed reefs during migrations.

5.3.4. *Coral diseases*

The number of coral diseases and their distribution have increased considerably over the past decades. Most pathogens: bacteria, fungi, protozoa, trematodes, ciliates and cyanobacteria have been associated with one or several coral diseases. Most diseases appear to be caused by a combination of several of these pathogens and are named according to their appearance: white band disease, white pox disease, red band disease, black band disease and purple band disease. Links between coral disease and deterioration of reef waters have often been found.

Until present, more than three quarters of coral diseases have been identified in the Caribbean [ROS 04]. In the 1970s, white band disease proliferated and decimated branched corals belonging to the genus *Acropora* in these reefs, reducing coral cover by 80–90%. Whereas some colonies seemed to recover, white pox disease emerged in 1996. This highly contagious disease is caused by a coliform bacteria *Serratia marcescens* associated with wastewater discharge [SUT 04], and also contributed to the decline of Caribbean reefs.

At present, coral diseases are also problematic in the Indo-Pacific [WIL 04], with the appearance of some diseases such as white band diseases.

The intensification of human activities along the coast, overfishing and the increase in sea temperature all contribute to the development of coral diseases on reefs, reducing coral cover as well as reef diversity and productivity.

5.4. What are the combined effects of local and global threats on corals?

Global climate change, which manifests in many ways, two of the most significant threats to coral ecosystems being warming and ocean acidification (e.g. [HOE 10]). Approximately 25% of CO_2 emitted into the atmosphere is absorbed by oceans, which modifies the chemistry of seawater: reducing the pH (by 0.1 unit of pH throughout the 20th Century), concentrations of carbonate and the saturation of aragonite (Ω_{arag}). It is currently estimated that by the end of the century, calcification of scleratinian corals should decrease from 17 to 37%, due to the decrease in carbonate ions [KLE 99]. Until now, the effects of these changes in climate have been studied in an isolated manner, assuming that coral reefs were located in protected zones, away from human influence. As discussed in the previous section, this assumption is far from reality since most reefs will face direct forms of pollution such as sedimentation due to coastal development, high nutrient input as well as metal pollution and toxic substances [SHE 09]. The multitude of stress factors acting on reefs is unprecedented in the history of humanity and studying their synergy is still in its early stages [DEA 09].

Local anthropogenic impacts (fishing, habitat destruction and pollution) that fragment or decrease the size of populations do not solely have immediate ecological effects, they will also reduce the capacity of coral reefs to adapt to warmer and more acidic conditions predicted in the years to come [HUG 03, PAN 03]. For example, corals that regularly face poor water quality due to wastewater runoff are less resistant to thermal stress, such that when they are exposed to suboptimal temperatures (>28°C), they become more susceptible to bleaching [WIL 08, WOO 09]. The input of dissolved inorganic nitrogen stimulates the growth of zooxanthellae, which breaks the stability and functioning of coral/zooxanthellae endosymbiosis. Similarly, overfishing, which can affect the density of herbivorous fish, can in some circumstances cause algae to dominate reefs [DAR 10]. Generally speaking, where the impact of fishing is minimized, diverse and productive ecosystems appear more robust and resilient and recover quicker from bleaching events or other disturbances [HUG 07]. In general, any decrease in coral calcification caused by an increase in temperature, any decrease in coral cover due to beaching or any decrease in coral heterotrophic nutrition (reducing the energy available for calcification) could accelerate the decline in reef calcification over the next century [ANT 08, DEP 11, JUR 10, MAR 01, MAR 03, REY 03].

Global changes also have direct and indirect influences on many other reef organisms, invertebrates as well as vertebrates. Few of these organisms can tolerate temperatures above 33°C, and if they cannot find favorable habitats in deeper and cooler zones, they may disappear from zones affected by these high temperatures. Acidification can also affect calcification phenomena in many invertebrates (e.g. mollusks and echinoderms) and consequently, their predators, primarily fish. Increases in temperatures in subtropical zones can cause reef species to migrate toward higher latitudes to the detriment of autochthonous species. One such phenomenon is particularly worrying in the Mediterranean with the invasion of the east Mediterranean basin by species from the Red Sea via the Suez Canal. These species, known as Lessepsian species, sometimes represent over half of the fish caught by coastal artisanal fishing in these regions.

An important issue is the need to predict how species distribution ranges modify according to global changes, in order to predict changes in diversity. This also goes for regions whose temperature will become too hot for some species and therefore cause losses in diversity and service, and also in regions whose temperature, previously too cold, will become adequate and allow a greater diversity. Most current models are based on two concepts: first, climate change scenarios, which generate maps of surface temperature, acidity, currents, etc. for future periods; second, ecological envelopes for the species tested. These envelopes are usually based on the minima and maxima of environmental conditions where species are currently observed. Although there is still no credible alternative, these models have several loopholes; uncertainties with regard to scenarios that are associated with predictive models whose accuracy may be questioned, especially in the long term. Another flaw is found with ecological envelopes: a species facing a change in its environment may or may not adapt and we have few methods of knowing whether they do or not. Adaptation means that the species manages to live in new conditions, unknown in its current distribution range (*niche expansion*). Non-adaptation can take several forms. A species may not survive in a new environment, or it may not be able to settle in new territories where conditions are similar to conditions in its current environment (*niche unfilling*).

Another important point to consider is the heterogeneity of responses by communities to all these changes and disturbances. As already seen, highly diverse communities will have a greater robustness, resilience and stability when faced with disturbances as there is a certain insurance that essential functions will be maintained due to the functional redundancy (several species insure similar ecological functions). Responses may vary greatly according to the traits they possess. For example, if we compare an assembly of reef fish in the Pacific and Caribbean faced with two disturbances: fishing and changes in temperature. When subject to fishing, there is a greater proportion of large species in the Caribbean, which should, therefore, be more vulnerable. When subject to higher temperatures, these same large species will be more tolerant to both direct variations in temperature as well as habitat loss due to the decrease in coral cover, which can occur because of warming. However, in the Pacific, smaller species often have a narrower tolerance to changes in temperature, and what is more, these small species are much more dependent on corals and a decrease in corals would undoubtedly affect a large proportion of these small species.

5.5. Functions and diversity

5.5.1. *What to protect?*

The consensus is that reefs must be protected. Reefs are, however, very complex entities and it is rarely possible to protect reef areas large enough so that the whole system is preserved. It is, therefore, often necessary to select zones to protect according to precise objectives; different zones are chosen to be protected according to the target: rare or emblematic species, a type of organism (e.g. a field of sea fans), a particular habitat (e.g. nursery or spawning area), biomass (fishery-based protection) or diversity (e.g. tourism-based protection). Protection involves prioritizing what is fragile and vulnerable, hence the need to define vulnerability. As already mentioned, vulnerability may concern a species, a functional group or a community. There are many ways to define vulnerability, but there are three main concepts, threat, vulnerability and risk, that interact. A species is vulnerable to a threat, for example a species of fish may be sensitive to fishing, but not sensitive to changes in coral cover or increases in temperature. Depending on the level of threat and vulnerability, the species incurs a higher or lower risk. The most frequent risk considered is an increase in mortality, but other risks include a decrease in fecundity or growth and a change in behavior (e.g. a species no longer migrates).

Due to their high diversity and functional complexity, coral reefs do not lend themselves well to a protection targeted to a restricted number of species. Of course, there are specific regulations for some species, usually emblematic species such as turtles, manta rays, dugongs, sharks and some large fish. There are also regulations to restrict the number of species captured commercially or recreationally to maintain a sustainable production. Most current regulations focus on creating MPAs. The protection assured by these MPAs depends not only on:

– associated regulations, for example there are "full" MPAs that a restricted number of people can episodically access, whereas other MPAs authorize certain activities: fishing, boat mooring, touristic activities, etc.;

– the area and spatial organization of the MPA. For example, it is possible to have a single large MPA, or a group of small MPAs which may or may not be connected, exclusively reef MPA or MPA including a reef but also adjacent environments such as mangroves, seagrass beds, algal beds and soft bottoms. These options are discussed in section 5.5.2.3.

As mentioned in the previous sections, each species ensures one or several ecological functions and therefore helps maintain the system. Many functions are either ensured by several species, or do not have a major impact on the system. These functions are, therefore, considered as not very vulnerable and require less protection than functions ensured by a single species or whose impact is essential for the system to function properly. The areas where these functions are located should be identified in order to try to protect them. Recent analyses (Figure 5.12) show that on a large scale, there is not necessarily any overlap between levels of threat, species protection and distribution and vulnerable functions. In particular, these analyses suggest that reefs in the Atlantic will be in greater danger than those in the Indo-Pacific as many species are endemic and many functional groups are assured only by a single species. These initial analyses do not, however, take into account abundance and consider threats overall, in particular threats or functions specific to a single species are not taken into account. Nevertheless, this approach does show that in a biocomplex system, such as coral reefs, the politics for protection should not be the same everywhere and it is necessary to implement large-scale collaboration.

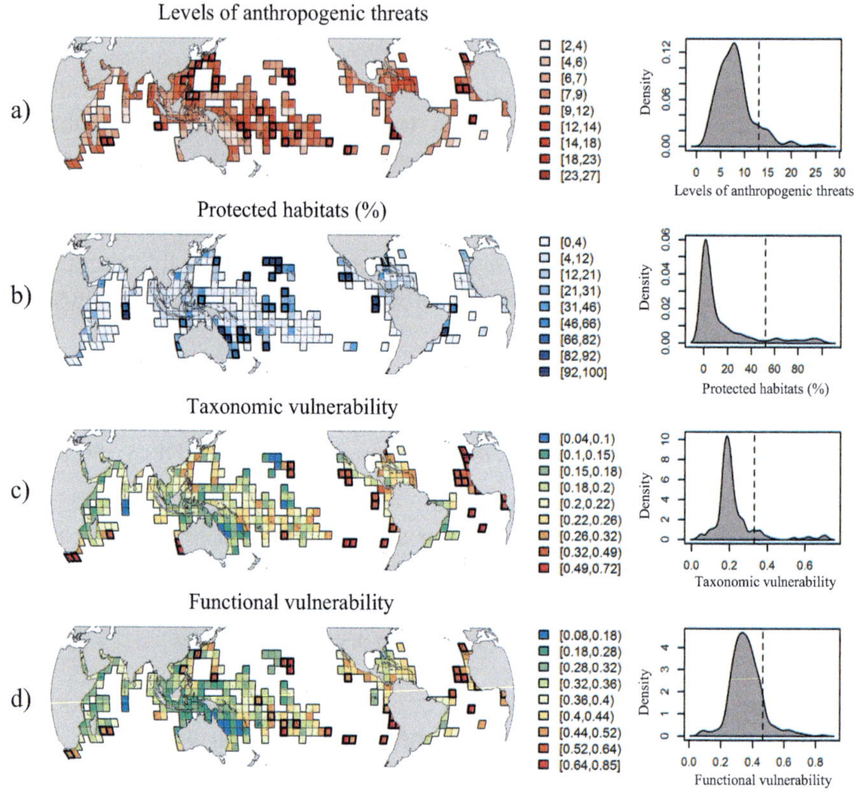

Figure 5.12. *World maps a) of the levels of threats (according to [HAL 08]); b) the proportion of protected habitats (source: IUCN); c) taxonomic vulnerability and d) functional vulnerability of reef fish throughout the world. The graphs to the right indicate the frequency (according to [PAR14])*

Another aspect that is often overlooked when implementing protection measures is the consideration of coastal populations dependent on reef resources. The world is changing very quickly and in many reef areas, the local populations are forced to integrate into an open market economy, which often causes conflicts over the use of natural resources including those from the reefs. The right to access a resource [JOH 02], the maintenance of traditional conservation strategies [CIN 05], the introduction of new exploitation methods [SUU 12] and targeting resources still not exploited by local populations [NYS 00, WIL 08] are among the questions raised by the entry into the globalized economy.

5.5.2. *Protection means*

Man has long been trying to establish rules to preserve the services provided by coral reefs. In what follows, we will only focus on fishing, in some cases with a broader scope, in particular allowing the preservation of the environment.

5.5.2.1. *Traditional tools*

In coral reef regions, local populations have developed local management rules to maintain a satisfactory level of resources. The most common were "taboo" or banned; the access to a reef resource was usually restricted to coastal populations, which ensured their implementation and monitoring. These could include many rules, including gear restrictions, seasonal and/or spatial closures, specific species protection or restriction of people authorized to fish or consume fish products.

In many reef regions, many methods were used, often specifically adapted to certain species or local conditions. A common method was to use fish traps to stop the passage of fish, the materials being made of stone walls, poles and nets; these barriers are used to channel fish into a reduced space where they are then captured (by drying out with the tide, using nets, with harpoons, etc.). The location, extent and timing of these traps all followed a complex set of rules, with them being used collectively or by specialized groups. Another common method is the use of poison, which was mainly used on coral plateaus in zones with little water (for example, at low tide). The use of these substances was often reserved for women. This type of fishing could only take place at certain times of the year, in order not to deplete sources of poison (some coastal trees and holothuria) or prevent too many juveniles from dying during periods of recruitment. Other methods were reserved for special events, such as stone fishing in French Polynesia, which requires lengthy preparation and is implemented by the whole community. Fishing was often subject to zoning, with fishing of all species banned in certain areas throughout the whole year, other zones where only some species were banned or even zones where fishing was banned at only some times of the year, either for all species, or for some species or type of fish (e.g. ban on taking very large fish or very small fish). Such a zoning existed even after the arrival of Europeans (e.g. on the islands of Ouvea and Tiga in New Caledonia) or was reinstated (e.g. village reserves in Vanuatu). Banning fishing at certain times of the year was to protect species during migrations, spawning or the arrival of juveniles. Some species protected from fishing, particularly sharks, were often subject to strong taboos, for example, in French Polynesia or on Ouvea (New Caledonia). Other species were subject to temporary bans, of could only be consumed by part of the population. In addition, some particularly prized species (e.g. *Nasounicorn* is in the Loyalty Islands) were only consumed by the elite. Other species were only captured during special events (e.g. it seems that dugongs were only caught in New Caledonia during important ceremonies). Not everyone had the right to fish, and some types of fishing or some species were reserved for specific clans. Women were generally not allowed to fish from boats and men could not collect from reefs. Some fishing techniques were reserved for specific people, for example fishing *Chanoschanos* on the island of Pines (New Caledonia) requires heavy nets and expertise on how to detect the fish and maneuver boats and was, therefore, only done by specialized clans. It is difficult to determine the efficiency of these techniques as it has never been formally measured, but based on current knowledge, it is likely that many of them were based on accurate observations of the environment and were fairly efficient, as many of these societies depended directly on their marine resources, with poor management being a synonym of food shortages. As we will see, many modern rules were directly inspired by these ancestral methods.

5.5.2.2. *Early western rules*

Most countries with reefs have been colonized and have shifted from traditional rules to western rules, often with a juxtaposition of both. The first rules introduced by newcomers were inspired by what was in place in Europe, essentially gear constraints, the seasons and sizes of captures and often on rare or emblematic species (e.g. marine turtles, dugongs and marine birds). However, protecting certain zones was not a rule, with only local rules (e.g. reproduction zones and concentration zones during migration) being applied.

The use of rules based on population dynamics, which was very popular between 1960 and 2000 in western countries, was barely carried out in the coral environment due to the many species and gears, as well as the high fragmentation of fishing territories, which made the application of these rules impossible. There have been many attempts, particularly in the Caribbean [MUN 83] and several articles (e.g. [WAR 03]) proposed to use as a baseline the maximum sustainable yield (MSY) or to assess the level of catches per country according to this principle [NEW 07]. These studies rightly set limits for what was *a priori* considered reasonable to fish, but there were many conceptual flaws. The first downfall was the multispecific character of fisheries in the coral environment, with each species depending on many interactions with the environment and other species, which is not considered in a realistic manner in population dynamic models. The second downfall was that estimating the load capacity of these systems depends on both local factors (level of fishing, habitat complexity, nutrient input, etc.) and regional factors such as size and degree of isolation of islands and biogeographic regions. The multiplicity of gears and the often non-targeted nature of captures are also a worry when applying these models. Finally, the recruitment of coral fish (and many other commercial species such as giant clams and holothuria) occurs as oceanic larvae with no link to the size of parent stocks, which means that stock–recruit relationships, very important in population dynamics, cannot be used. A final point is the amount of information that must be collected on their source of fishing to supply this type of model, with this amount usually being unrealistic in many countries with coral reefs. The general downfall for this approach in temperate fisheries (e.g. cod in Canada [MYE 97]) has led to a certain disinterest in these concepts, despite some points being interesting, particularly some warning signs, such as ratios between maximum size, size at first reproduction and average size, or even the existence of no return thresholds (at least in the short term).

5.5.2.3. *New approaches*

The main change to coral reef regulations was the establishment of MPAs. This measure was first applied in the tropics in the 1990s, before moving onto temperate zones, even if there were several marine parks (modest size) prior to 1990 in Europe and North America. The similarity between the concept of MPA and taboo areas is striking and we may wonder to what extent the former inspired the latter. The number of MPAs has increased very rapidly and this concept has diversified considerably and been subject to many studies [CIC 05]. Under the term "MPA", there are many different regulations, from integral reserves where all human activity, besides surveillance and scientific research, is prohibited, to zones where some types of fishing and different types of activities (diving, hotels, water sports, etc.) are permitted. The size and spatial structure of MPAs are also highly variable, whether several hectares to several thousand km^2, an MPA in one piece with a single regulation to complex zonings with a range of regulations allowing fully protected zones to mix with zones where some activities are permitted. The popularity of MPAs can be explained by several reasons:

– these measures are fairly easy to implement, even if applying the rules is quite difficult ("paper" MPA where regulations are not applied due to lack of personnel, logistical reasons or corruption);

– the effects of protection are easy to detect;

– it is possible to set up a dialogue with stakeholders (participatory management), the underlying concepts of MPA are easy to explain in populations;

– the political message is easily transmitted (e.g. actual protected area, compared with proposed total surface area to protect).

This does not mean that MPAs are a solution to all problems or that they are effective every time. Some of the main problems encountered are:

– conflicts of use: the aim of MPAs is usually to improve the state of resources and their environment. By blocking access to the most productive zones or those most attractive to fishermen or other users, this may generate conflicts of use. Improving a resource within an MPA may not be offset by the increased pressure exerted on the remaining fishing zones available outside the MPA. Usually, the professions that benefit the most from MPAs are businesses linked to tourism, and fishermen excluded from these zones do not usually have the means or knowledge to take advantage of the new situation;

– effective application of regulations: as already mentioned, for an MPA to be effective, regulations must first be applied, which requires human resources and significant logistics, which countries with the greatest need for MPA cannot properly provide. There are solutions to solve this, participatory management being one of them. This type of management involves the participation of local populations in the definition and application of regulations, with the need for communication between administration and local populations. We will see that local self-management (e.g. "micro-MPA" in the South Pacific) is derived from these participatory management concepts and may represent one of the responses required to maintain the ecosystem services and reef biodiversity;

– monitoring the state of the system: implementing an MPA is the first step, but it is important to ensure that the results expected are in fact there. It is, therefore, useful to implement monitoring plans, which often arise from the concept of *Before and After Control Impact Paired Series* (BACIPS) whose principle is as follows: study areas are defined within and outside an MPA before it is created, with these zones having similar environmental and ecological characteristics; samples are taken in these test areas before and after the MPA was implemented. As well as the design of this monitoring, the objects studied and the methods implemented are the most critical. The key question is "what to preserve?" and the follow-up must show that MPAs are effective. The protocols and competencies will vary depending on whether one is focusing on commercial species, coral cover or water quality. One of the main problems encountered during this follow-up is, besides interpretation and analyses, the acquisition of high-quality data. The high diversity of reef communities usually requires a high qualification level, which is often not available or is too expensive, hence the research into simplified solutions. For example, monitoring coral fish requires a lot of practice before being able to correctly collect samples during diving expeditions from sites with hundreds of species. One solution is to restrict the monitoring to indicator species that are easy to identify and quantify and whose characteristics accurately reflect the MPA performance in order to use less qualified personnel, who can be easily trained and where possible from local coastal populations. Until present, the indicators available do not accurately take into account the actual characteristics of communities, essentially due to the biocomplexity of the systems analyzed;

– what area to protect, according to which criteria, in order to protect what? At present, there is no concrete theory to objectively determine the best criteria for the creation of MPA. The following sections will provide some details about these fundamental issues in the creation of MPA.

MPAs are often created in areas that are considered to have a particular interest, for example with a high diversity, or that are still well preserved, or that contain ecological activities of interest, for example breeding zones. It is also common that these zones have a particular heritage, cultural or even religious importance and that the decision to create an MPA, its location and size are all political acts, with science being taken into account afterward. There are also re-established rules that are not necessarily scientifically proven, one of the main rules being the need to protect 20% of coastal zones [IUC 92]. This rule has no ecological sense as it does not specify which zones should be included in this 20%; also, 20% may be necessary in one region but insufficient or too great in other regions, because as we saw in

Chapter 4, the potential of the services depends on both local and regional factors, which vary from one region to another. At present, using MPAs usually appears to follow a precautionary principle until a better alternative is found, rather than an action based on well-established rules. Without returning to the validity of MPA, which has clearly proven its usefulness, it is wise to ponder about what we know and what we do not know about MPAs and how this can influence the creation of new MPAs or the use of existing MPAs. It has been found that MPAs, when they are correctly implemented, can improve levels of stocks (in quantity and quality), especially for species with a home range in line with the size of these MPAs. MPAs can help increase the size of the brood stock and therefore the potential recruitment, but not necessarily local recruitment (due to source and sink phenomena, the larvae produced do not necessary colonize their site of origin, and the survival of larvae does not depend on numbers). MPAs help local diversity to increase and improve the quality of the environment, which is often a source of tourism. However, the choice of area and fragmentation of MPA is still open for debate. Is it better to create one large MPA in the same biotope or several small MPAs across several biotopes? What is the minimum useful area? There is no single answer; each case must be considered along with a range of local and regional factors. To show this, there is no single law, we will take two examples.

EXAMPLE 5.4.– What relative area is required for an effective MPA for commercial fish? On a small isolated island, we now know that the proportion of large species is greater than what is observed on large islands or unisolated islands. For the proper conservation of large species, which have home ranges, large reserves adapted to this action radius are required. Therefore, on a small island, the percentage of habitats to protect should be considerably greater than on a large island, ultimately with MPA being unsuitable on the smallest islands if fishing is to be preserved. Other protection measures will, therefore, need to be used (for example, quotas, size limits, etc.).

EXAMPLE 5.5.– Is it better to have one large MPA or five small MPAs? Both choices can be justified, depending on what we want to protect and the configuration of the system to be protected. In a complex system, adjacent habitats such as mangroves, seagrass beds, algal beds, estuaries etc. are essential for many processes and therefore they may be worth including in the MPA, but this assumes counterparts too (e.g. less reef area protected, greater human and financial resources and protection of zones whose residents do not always see the value, which leads to a lower acceptance of the MPA). Creating five smaller reserves covering these habitats rather than one large reef reserve does not account for another important factor when maintaining fish assemblages: connectivity, and its opposite, fragmentation. If in our example, the five small MPAs are too far apart from each other, the benefit of interactions between habitats may be lost. However, a large reserve integrating these different habitats will maybe provide a better result than an MPA of the same size with only one reef. This simple example shows at what point the complexity of reef systems imposes constraints on their management and there is, therefore, no single or ideal solution.

This leads us to consider two current opposite trends, "micro-MPA" and regional MPA (also known as transboundary MPA). In some countries in the Pacific, local communities have implemented small reserves under their direct control, usually with significant participation by the local community. For instance, in Vanuatu, more than 200 micro-MPAs, often restricted to several hectares, have been implemented in the past few years. The government tried to ensure a certain cohesion with flexible legislation allowing these zones to be officially recognized, which is an important step to ensure that other communities comply. The effects of these micro-MPAs are difficult to quantify, apart from species whose movement is restricted, such as mollusks and echinoderms. The undeniable outcome is the awareness of the importance of participatory management and limited natural resources. This approach is still individualistic, in that each village decides for itself, without consulting with its neighbors to optimize protected areas, but even in societies like ours, this behavior is also found with little consulting between groups of fishermen on the Mediterranean coast in the zoning of protected areas. Quite contrary to these micro-MPAs are very large MPAs, which are sometimes transboundary MPAs. The aim of these MPAs is to protect the whole system by encompassing enough biotopes and large enough areas to ensure the whole system is covered. In particular, this involves the connectivity between zones. It is likely that there are source zones and sink zones in terms of recruitment or migration. In this configuration, it becomes important to identify and protect source zones. An example currently being created is the Coral Sea protected area that will unite Australia, New Caledonia and to a lesser extent New Zealand, in an area of several million km^2, which will include a rather complex system of zones and uses. There will be sanctuaries, particularly for mammals and marine birds, and zones partly open to human activities. This type of MPA aims to protect whole areas of biodiversity to assure a large enough space for species that require large territories (e.g. mammals, birds, sharks and turtles), and ensure large-scale connectivity, which is currently found between the GBR, many ocean reefs in the Coral Sea and New Caledonia. There is also a political side to these MPAs, since they allow countries to more easily obtain protective quotas expressed in international treaties. These MPAs exceed simple marine reserves, but their implementation still raises many legislative, logistical and scientific concerns. At present, these MPAs are possible only in certain regions, for both political and pragmatic reasons, but it is likely that this concept will continue to develop.

This section on protection tools has only touched on the aspects of biocomplexity. Due to their complexity, coral systems raise universal questions and provide an insight into what could be used in other situations. In addition, participatory management and science, taking into account the connectivity, the transfers between ecosystem services (e.g. fishing-tourism shift), accounting for groups of species rather than a species-by-species approach, are all concepts that have benefitted from actions carried out on coral reefs throughout the world.

5.5.2.4. *From collection to culture?*

Artisanal fishing of organisms from coral reefs makes a living for hundreds of millions of people in developing countries [MCM 97, WAR 03]. Assuring the maintenance of these fisheries is, therefore, a priority, but the sustainability of many of them is much more insecure as the demand for seafood increases in the years to come [NEW 07]. Thus, in 2030, only a small proportion of countries and territories in the Pacific will be able to feed their population with the recommended quantities of seafood while properly managing their coastal resources [BEL 09b]. In this context, aquaculture is presented by many as a solution to respond to the increasing demand for seafood while limiting the environmental impacts [POM 06].

Traditionally, in addition to the direct use of marine resources, young stages of some species, such as the mangrove crab *Scylla serrata* [QUI 10], the lobster *Panulirushomar* us [JON 11] and the fish *Chanos* [NEL 07], are captured in the natural environment and then kept in captivity to allow them to grow to adulthood. At present, production in captivity is based increasingly on controlling the complete lifecycle of a species (Figure 5.13). Other types of aquaculture are initially based on controlling reproduction and survival of larval stages to ensure breeding completely cooped up such as for the cod *Epinepheluscoioides*[1], or also for reseeding natural populations such as for the holothurian *Stichopus japonicus* [SAK 11].

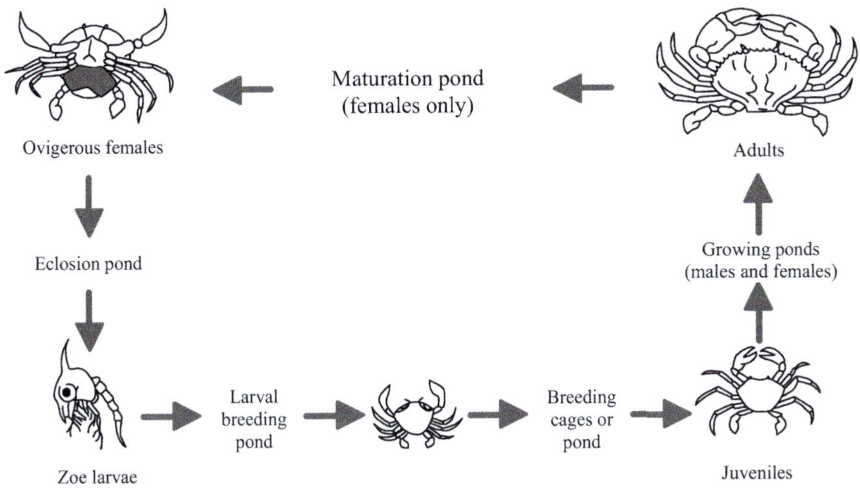

Figure 5.13. *Production cycle for Scylla serrata illustrating the different stages of breeding (adapted from [QUI 10]) – see [SHE 11] for a detailed description of each stage*

Unlike most other exploited ecosystems, in coral environments, the animals with the highest added value are those collected and kept alive, in particular for the aquarium trade [WAR 03]. Unfortunately, the capture techniques used are often highly destructive and the inappropriate animal housing practices and transport conditions rarely allow the sustainable use of these resources [BEL 09a, REK 07, SAD 02]. Since the full lifecycle is controlled in captivity for only a very restricted number of species of interest to tropical aquariums [WAR 03, WOO 01], the development of other practices is necessary to make this sustainable.

In this context, capturing young fish in their natural environment to rear them for the aquarium market and aquaculture has been proposed for a viable and sustainable method since the early 2000s [DUF 02, HAI 00]. Now, well known under the label post-larvae capture and culture (PCC)

1 www.fao.org/fishery/culturedspecies/Epinephelus_coioides/en (accessed on 7 March 2013).

[LEC 07,][2], this method is based on the idea of collecting crustaceans or fish at the end of their larval stage, just before they settle as juveniles to benthic habitats in the coral environment (Figure 5.14). The individuals captured are, therefore, saved from the high mortality they would naturally face when they first arrive on coral reef habitats at the end of their larval stage, and they easily acclimate to farming conditions [DOH 04, PLA 01]. Larvae of crustaceans and fish can be captured using different types of gear [BEL 09a, LEC 07]: crest nets placed on the barrier reef, hoa nets[3], different types of light traps (see photos in Moana Initiative 2007 and [MAL 08]) or simply pierced coconuts to capture *puerulus* lobster larvae. The larvae captured are then reared until they reach the size required for entering the aquarium market, for consumption, or to be used for restocking purposes.

Although the PCC principle is promising and its benefits are clear to habitants in developing countries as well as for reef habitats [BEL 09a], this approach still has some obstacles to overcome with the help of research and development. Therefore, despite increasing knowledge of post-larvae (e.g. [CAR 08]), predicting the seasonal and variability of larval abundances of some target species is complex. These predictions require a modeling approach based on long-term monitoring data, an approach often out of financial and logistical reach of most commercial organizations, particularly in developing countries [BEL 09a]. However, with the exception of pierced trunks to capture *puerulus* lobster larvae, the gears used to capture post-larvae are usually not very selective. Large quantities of commercially interesting organisms are, therefore, captured [HAI 00]. In the near future, it is likely that the selectivity of light traps will improve by using particular wavelengths, and or by adding to the emitted light other types of attractors such as sounds and chemical molecules obtained from the environment. Finally, identifying captured post-larvae is still a tedious task. Therefore, despite the recent publication of several guides [COL 13, JUN 07, MAA 06], of the 44,750 post-larvae captured between 2011 and 2013 in La Réunion during the program POLARUN[4], only 58% have been identified to the species level, 25% were identified only to the family level and 14% could

2 www.moanainitiative.org/home.html.

3 Shallow channels at the boundary of some coral atolls.

4 POLARUN: project studying the post-larval colonization of fish on reefs in La Reunion and their value potential, associating the Natural Marine Reserve of La Reunion, IRD, IFREMER and ARDA.

not be identified at all, even after several weeks of rearing [OCE 13]. In this field, approaches based on molecular tools (barcode) from reference individuals, coupled with morphometric approaches, which allow a large number of individuals to be analyzed at a low cost, appear to be particularly promising [PON 13]. With regard to predictive models of larval abundances, as well as tools that help identify larvae upon their capture, recent research in the Indian Ocean should help address some of the challenges that still exist for the sustainable and economically viable development of PCC in the reef environment.

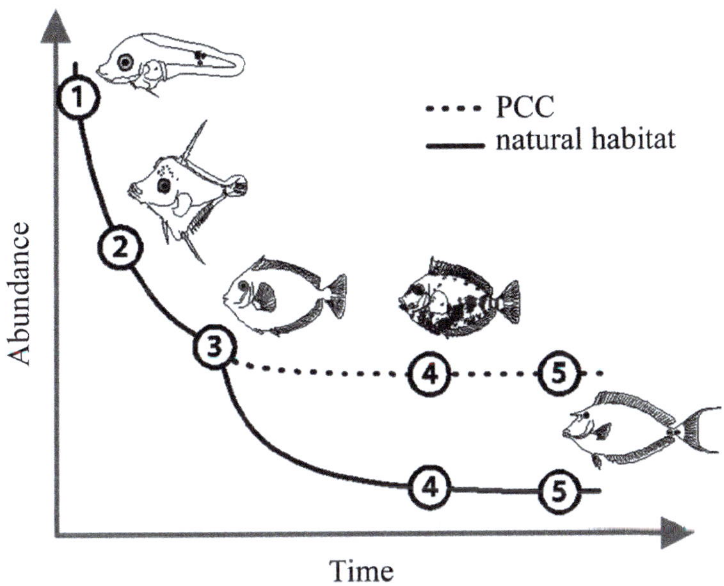

Figure 5.14. *Diagram showing the principle of post-larval capture and culture (PCC)*

COMMENTS ON FIGURE 5.14.– Pelagic larval stages (1 and 2: 2 and 5 mm, respectively) are subject to a sharp decline in numbers over time. Upon settlement, post-larvae (3: approximately 15 mm) are captured at a more mature age (dotted line) to avoid significant decreases in numbers in the natural environment (continuous line). Juveniles from PCC (4: approximately 60 mm) are, therefore, more abundant than in nature, and if they are farmed, the adults (5) will, therefore, be more abundant. The sizes indicated on this diagram correspond to *Nasounicornis* (*Acanthuridae*); this principle is identical for fish species and crustaceans destined for the aquarium trade.

5.6. Conclusion

5.6.1. *Reefs as a model?*

The previous sections have shown that coral reefs are one of the most diverse ecosystems in the world, and probably one of the most complex in terms of functioning. It was also found that besides this abundant complexity, these ecosystems are highly structured, both spatially and temporally, largely associated with their evolutionary history and high environmental constraints. It is likely that most other highly diverse systems function based on similar principles. However, everything that was observed on reefs clearly does not apply to the terrestrial environment, for example recruitment and colonization are controlled on land by different traits, probably much more diverse than in the reef environment. The question this current chapter poses is to whether reefs can be a model with regard to ecosystem services, their users, their reactions when facing disturbances and their methods of protection.

Some services ensured by coral reefs are highly specific, for example they offer sheltered lagoon areas. However, many other services, such as biomass production, functional or phylogenetic redundancy, cultural services, etc. are universal in nature, even if each system has its own specific characteristics. For example, interest in the link between diversity and production emerged next to the work by Tilman [TIL 96] which was carried out on grasslands but whose scope is almost universal. This work and those which followed suffered from a major drawback as they were experimental studies (reductionist in the sense of [MIC 01]). Coral reefs offer the possibility to extend this type of approach to "real" situations as they are based on observations of systems *in situ*. Many threats to reefs are specific, but others, such as global changes and even the destruction of habitats or overexploitation, are common to current complex systems. From this point of view, reefs represent a particularly interesting field of study since this ecosystem is highly spatially structured with the possibility of a comparative approach rarely available in other environments. Therefore, it is possible to compare islands of different sizes, of different degrees of isolation, with different degrees of disturbance, etc. while maintaining a base structure common to all sites studied, a rare situation in other systems. For example, if we want to model global warming, we would have access to sites whose coral cover will likely strongly decrease and other sites where the diversity and abundance of coral will likely increase, with a sequence of consequences with regard to biodiversity, processes and services, which we are already able to model rather precisely. Another example for the use of MPA, there is a range of MPAs on coral reefs, which to our knowledge is unrivalled in the marine environment (and perhaps in tropical terrestrial environments). Many questions regarding the implementation (size, number, connectivity, etc.) of MPAs or reserves in general can be analyzed using MPAs in the coral environment, with increasing levels of diversity (taxonomic, functional and landscape). Such comparative analyses are very difficult elsewhere because of the lack of such a range of conditions.

Although coral reefs are very useful when trying to understand complex systems and propose solutions to manage their services, we must bear in mind that it cannot always be applied to other complex systems. Coral reefs have many unique characteristics, particularly in terms of habitats, traits associated with recruitment and colonization of environments as well as a rather peculiar evolutionary history.

5.6.2. *Do coral reefs have a future?*

Coral reefs are one of the most threatened ecosystems on the planet. Coral reefs are particularly vulnerable due to their dependence on corals as habitat builders. Corals are very sensitive to both local disturbances (predators, pollution, physical destruction, disease, etc.) and global disturbances (warming and acidification) and their interaction. A significant part of coral reefs has already disappeared over the 20th Century and this is far from being over. The greatest worry is that almost all coral reefs are threatened. *A priori*, we can expect a greater resilience and robustness in zones with the highest diversity, but this is still a working hypothesis, since this robustness is not infinite and the moment will come where the system disintegrates with little chance of recovering in the near future. It is, however, worth mentioning that in regions of high diversity the endemism is also usually lower, meaning that there may still be biodiversity reservoirs in these regions, but the spatial and temporal dimensions required for such reservoirs to remain efficient are overlooked. As in an increasing number of situations in ecology, the response is increasingly regional or global.

There are also many isolated coral ecosystems, still relatively untouched by local disturbances. Mora *et al.* [MOR 11] state that it is likely that these refuges will still be fairly unaffected in the coming decades as these zones are far from human centers of development. However, this isolation will not necessarily protect these islets against global changes, in particular ocean warming and acidification. Coral bleaching events have already been observed in isolated zones (see *Reef Base*)[5]. Due to their isolation, these zones may find it difficult to recover since recolonization can only be initiated from remote locations.

Global warming causes a change in the spatial distribution of species in warm waters, which may colonize subtropical, or even temperate zones. In Japan, reef-building corals have drifted northward. This provides hope with regard to the capacity of coral systems to move as global warming continues. Another glimmer of hope lies in the capacity of species to adapt to a gradual change in environmental conditions. Some corals have also shown some resistance to heat stress, which could increase. These hopes must, therefore, be qualified. On the one hand, we do not know what proportion of species is capable of rapidly changing their geological distribution or adapt to new environmental conditions. On the other hand, the speed of these migrations or adaptations will not necessarily be enough, given the unprecedented speed of these global changes.

5.7. Bibliography

[ADA 11] ADAM T.C., SCHMITT R.J., HOLBROOK S.J. *et al.*, "Herbivory, connectivity, and ecosystem resilience: response of a coral reef to large-scale perturbation", *PloS ONE*, vol. 6, no. 8, p. e23717, 2011.

[ADJ 09] ADJEROUD M., MICHONNEAU F., EDMUNDS P.J. *et al.*, "Recurrent disturbances, recovery trajectories, and resilience of coral assemblages on a south Central Pacific reef", *Coral Reefs*, vol. 28, pp. 775–780, 2009.

[ALM 09] ALMANY G.R., CONNOLLY S.R., HEATH D.D. *et al.*, "Connectivity, biodiversity conservation and the design of marine reserve networks for coral reefs", *Coral Reefs*, vol. 28, pp. 339–351, 2009.

[AND 12] ANDRÉFOUËT S., CHARPY L., LO-YAT A. *et al.*, "Recent research for pearl oyster aquaculture management in French Polynesia", *Marine Pollution Bulletin*, vol. 65, no. 10, pp. 407–414, 2012.

5 "Reef Base" is an online data-base and GIS on coral reefs. Information on coral bleaching can be found at: http://www.reefbase.org/global_database/default.aspx?section=t4.

[ANT 08] ANTHONY K.R.N., KLINE D.I., DIAZ-PULLIDO G. *et al.*, "Ocean acidification causes bleaching and productivity loss in coral reef builders", *Proceedings of the National Academy of Sciences of the United States of America*, vol. 105, no. 45, pp. 17442–17446, 2008.

[ANT 11] ANTHONY K., KLEYPAS J.A., GATTUSO J.P., "Coral reefs modify their seawater carbon chemistry – implications for impacts of ocean acidification", *Global Change Biology*, vol. 17, no. 12, pp. 3655–3666, 2011.

[ATT 97] ATTWOOD C.G., HARRIS J.M., WILLIAMS A.J., "International experience of marine protected areas and their relevance to South Africa", *South African Journal of Marine Science*, vol. 18, no. 1, pp. 311–332, 1997.

[BEL 04] BELLWOOD D.R., HUGHES T.P., FOLKE C., *et al.*, "Confronting the coral reef crisis", *Nature*, vol. 429, pp. 827–833, 2004.

[BEL 06] BELLWOOD D.R., HOEY A.S., ACKERMAN J.L. *et al.*, "Coral bleaching, reef fish community phase shifts and the resilience of coral reefs", *Global Change Biology*, vol. 12, pp. 1587–1594, 2006.

[BEL 09a] BELL J.D., CLUA E., HAIR C.A. *et al.*, "The capture and culture of post-larval fish and invertebrates for the marine ornamental trade", *Reviews in Fisheries Science*, vol. 17, pp. 223–240, 2009.

[BEL 09b] Bell J.D., KRONEN M., VUNISEA A. *et al.*, "Planning the use of fish for food security in the Pacific", *Marine Policy*, vol. 33, pp. 64–76, 2009.

[BER 06] BERUMEN M.L., PRATCHETT M.S., "Recovery without resilience: persistent disturbance and long-term shifts in the structure of fish and coral communities at Tiahura Reef, Moorea", *Coral Reefs*, vol. 25, pp. 647–653, 2006.

[BRO 97] BROWN B.E., "Coral bleaching: causes and consequences", *Coral Reefs*, vol. 16, pp. S129–S138, 1997.

[BRO 04] BROWN J.H., GILLOOLY J.F., ALLEN A.P. *et al.*, "Toward a metabolic theory of ecology", *Ecology*, vol. 85, pp. 1771–1789, 2004.

[BRO 05] BRODIE J., FABRICIUS K., DE'ATH G. *et al.*, "Are increased nutrient inputs responsible for more outbreaks of crown-of-thorns-starfish? An appraisal of the evidence", *Marine Pollution Bulletin*, vol. 51, nos. 1–4, pp. 266–278, 2005.

[CAD 12] CADOTTE M.W., DINNAGE R., TILMAN D., "Phylogenetic diversity promotes ecosystem stability", *Ecology*, vol. 93, pp. 223–233, 2012.

[CAD 13] CADOTTE M.W., "Experimental evidence that evolutionarily diverse assemblages result in higher productivity", *Proceedings of the National Academy of Sciences of the United States of America*, vol. 110, pp. 8996–9000, 2013.

[CAR 08] CARASSOU L., Les assemblages de larves de poissons dans le lagon de Nouvelle-Calédonie: structure spatio-temporelle et relations avec les facteurs abiotiques et biotiques de l'environnement, Thesis, Ecole Pratique des Hautes Etudes, Perpignan, p. 292, 2008.

[CAR 09] CARILLI J.E., NORRIS R.D., BLACK B.A. *et al.*, "Local stressors reduce coral resilience to bleaching", *PloS ONE*, vol. 4, no. 7, p. e6324, 2009.

[CIC 05] CICIN-SAIN B., BELFIORE S., "Linking marine protected areas to integrated coastal and ocean management: a review of theory and practice", *Ocean Coastal Management*, vol. 48, no. 11, pp. 847–868, 2005.

[CIN 05] CINNER J.E., MARNANE M.J., MCCLANAHAN T.I.M., "Conservation and community benefits from traditional coral reef management at Ahus Island., Papua New Guinea", *Conservation Biology*, vol. 19, no. 6, pp. 1714–1723, 2005.

[COL 13] COLLET A., BOSC P., POTHIN K. *et al.*, Guide d'identification des post-larves de poissons des récifs coralliens: île de la Réunion, 2013.

[CON 78] CONNELL J.H., "Diversity in tropical rain forests and coral reefs", *Science*, vol. 199, no. 4335, pp. 1302–1310, 1978.

[DAG 14] D'AGATA S., MOUILLOT D., KULBICKI M. *et al.*, "Human-mediated loss of phylogenetic and functional diversity in coral reef fishes", *Current Biology*, available at http://dx.doi.org/10.1016/j.cub.2014.01.049, 2014.

[DAR 10] DARLING E.S., MCCLANAHAN T.R., COTE I.M., "Combined effects of two stressors on Kenyan coral reefs are additive or antagonistic, not synergistic", *Conservation Letters*, vol. 3, no. 2, pp. 122–130, 2010.

[DAY 02] DAY J.C., "Zoning – lessons from the Great Barrier Reef marine park", *Ocean Coastal Management*, vol. 45, no. 2, pp. 139–156, 2002.

[DE 02] DE GROOT R.S., WILSON M.A., BOUMANS R.M., "A typology for the classification., description and valuation of ecosystem functions., goods and services", *Ecological Economics*, vol. 41, no. 3, pp. 393–408, 2002.

[DE 09] DE'ATH G., LOUGH J.M., FABRICIUS K.E., "Declining coral calcification on the Great Barrier Reef", *Science*, vol. 323, no. 5910, pp. 116–119, 2009.

[DEP 11] DE PUTRON S.J., MCCORKLE D.C., COHEN A.L. *et al.*, "The impact of seawater saturation state and bicarbonate ion concentration on calcification by new recruits of two Atlantic corals", *Coral Reefs*, vol. 30, pp. 321–328, 2011.

[DE 12] DE'ATH G., FABRICIUS K.E., SWEATMAN H. *et al.*, "The 27-year decline of coral cover on the Great Barrier Reef and its causes", *Proceedings of the National Academy of Science of the United States of America*, vol. 109, pp. 17995–17999, 2012.

[DEP 07] DEPCZYNSKI M., FULTON C.J., MARNANE M.J. *et al.*, "Life history patterns shape energy allocation among fishes on coral reefs", *Oecologia*, vol. 153, no. 1, pp. 111–120, 2007.

[DOD 74] DODGE R.E., ALLER R.A., THOMSON J., "Coral growth related to resuspension of bottom sediments", *Nature*, vol. 247, pp. 574–577, 1974.

[DOH 04] DOHERTY P.J., DUFOUR V., GALZIN R. *et al.*, "High mortality during settlement is a population bottleneck for a tropical surgeonfish", *Ecology*, vol. 85, pp. 2422–2428, 2004.

[DON 92] DONE T.J., "Phase shifts in coral reef communities and their ecological significance", *Hydrobiologia*, vol. 247, pp. 121–132, 1992.

[DON 99] DONE T.J., "Coral community adaptability to environmental change at the scales of regions, reefs and reef zones", *American Zoologist*, vol. 39, pp. 66–79, 1999.

[DUB 90] DUBINSKY Z., STAMBLER N., BEN-ZION M. *et al.*, "The effect of external nutrient resources on the optical properties and photosynthetic efficiency of Stylophora pistillata", *Proceedings of the Royal Society of London B*, vol. 239, pp. 231–246, 1990.

[DUB 96] DUBINSKY Z., STAMBLER N., "Marine pollution and coral reefs", *Global Change Biology*, vol. 2, no. 6, pp. 511–526, 1996.

[DUF 02] DUFOUR V., "Reef fish post-larvae collection and rearing programme for the aquarium market", *SPC Live Reef Fish Information Bulletin*, vol. 10, pp. 31–32, 2002.

[ELM 09] ELMHIRST T., CONNOLLY S.R., HUGHES T.P., "Connectivity, regime shifts and resilience of coral reefs", *Coral Reefs*, vol. 28, pp. 949–957, 2009.

[ELM 03] ELMQVIST T., FOLKE C., NYSTRÖM M. *et al.*, "Response diversity, ecosystem change, and resilience", *Frontiers in Ecology and the Environment*, vol. 1, pp. 488–494, 2003.

[END 76] ENDEAN R., "Destruction and recovery of coral reef communities" in JONES O.A., ENDEAN R. (eds.), *Biology and Geology of Coral Reefs*, Academic Press, New York, pp. 215–254, 1976.

[ERF 12] ERFTEMEIJER P.L.A., RIEGL B., HOEKSEMA B.W. *et al.*, "Environmental impacts of dredging and other sediment disturbances on corals: a review", *Marine Pollution Bulletin*, vol. 64, no. 9, pp. 1737–1765, 2012.

[FAB 05] FABRICIUS K.E., "Effects of terrestrial runoff on the ecology of corals and coral reefs: review and synthesis", *Marine Pollution Bulletin*, vol. 50, no. 2, pp. 125–146, 2005.

[FAL 93] FALKOWSKI P.G., DUBINSKY Z., MUSCATINE L. *et al.*, "Population control in symbiotic corals", *Bioscience*, vol. 40, pp. 608–611, 1993.

[FER 00] FERRIER-PAGÈS C., GATTUSO J.P., DALLOT S. *et al.*, "Effect of nutrient enrichment on growth and photosynthesis of the zooxanthellate coral Stylophora pistillata", *Coral Reefs*, vol. 19, pp. 103–112, 2000.

[FOX 13] FOX J.W., "The intermediate disturbance hypothesis should be abandoned", *Trends in Ecology and Evolution*, vol. 28, nos. 86–92, 2013.

[FUR 91] FURNAS M.J., "Net *in situ* growth rates of phytoplancton in an oligotrophic, tropical shelf ecosystem", *Limnology and Oceanography*, vol. 36, no. 1, pp. 13–29, 1991.

[GAR 03] GARDNER T.A., CÔTÉ I.M., GILL J.A. *et al.*, "Long-term region-wide declines in Caribbean corals", *Science*, vol. 301, pp. 958–960, 2003.

[GEN 95] GENIN A., LAZAR B., BRENNER S., "Vertical mixing and coral death in the Red Sea following the eruption of Mount Pinatubo", *Nature*, vol. 377, pp. 507–510, 1995.

[GOR 69] GOREAU T.J., "Post pleistocene urban renewal in coral reefs", *Micronesica*, vol. 5, pp. 323–326, 1969.

[GRA 13] GRAHAM N.A.J., BELLWOOD B.R., CINNER J.E. *et al.*, "Managing resilience to reverse phase shifts in coral reefs", *Frontiers in Ecology and the Environment*, vol. 11, pp. 541–548, 2013.

[GRI 74] GRIGG R.W., MARAGOS J.E., "Recolonization of hermatypic corals on submerged lava flows in Hawaii", *Ecology*, vol. 55, pp. 387–395, 1974.

[GUZ 91] GUZMÁN H.M., JACKSON J.B.C., WELL E., "Short-term ecological consequences of a major oil spill on Panamanian subtidal reef corals", *Coral Reefs*, vol. 10, nos. 1–12, 1991.

[GUZ 92] GUZMÁN H.M., JIMÉNEZ C.E., "Contamination of coral reefs by heavy metals along the Caribbean coast of Central America (Costa Rica and Panama)", *Marine Pollution Bulletin*, vol. 24, no. 11, pp. 554–561, 1992.

[GUZ 93] GUZMÁN H.M., HOLST I., "Effects of chronic oil-sediment pollution on the reproduction of the Caribbean reef coral Siderastrea siderea", *Marine Pollution Bulletin*, vol. 26, no. 5, pp. 276–282, 1993.

[HAI 00] HAIR C., DOHERTY P.J., BELL J.D. *et al.*, "Capture and culture of presettlement coral reef fishes in Solomon Islands", *Proceedings of the 9th International Coral Reef Symposium,* Bali, pp. 819–827, 2000.

[HAL 04] HALFORD A., CHEAL A.J., RYAN D. *et al.*, "Resilience to large-scale disturbance in coral and fish assemblages on the Great Barrier Reef", *Ecology*, vol. 85, pp. 1892–1905, 2004.

[HAL 08] HALPERN B.S., WALBRIDGE S., SELKOE K.A. *et al.*, "A global map of human impact on marine ecosystems", *Science*, vol. 319, pp. 948–952, 2008.

[HAR 89] HARLAND A.D., BROWN B.E., "Metal tolerance in the scleractinian coral *Porites lutea*", *Marine Pollution Bulletin*, vol. 20, no. 7, pp. 353–357, 1989.

[HOE 07] HOEGH-GULDBERG O., MUMBY P.J., HOOTEN A.J. *et al.*, "Coral reefs under rapid climate change and ocean acidification", *Science*, vol. 318, pp. 1737–1742, 2007.

[HOE 10] HOEGH-GULDBERG O., BRUNO J.F., "The impact of climate change on the world's marine ecosystems", *Science*, vol. 328, pp. 1523–1528, 2010.

[HOE 12] HOEGH-GULDBERG O., "The adaptation of coral reefs to climate change: Is the Red Queen being outpaced?", *Scientia Marina,* vol. 76, pp. 403–408, 2012.

[HOL 73] HOLLING C.S., "Resilience and stability of ecological systems", *Annual Review of Ecology and Systematics*, vol. 4, pp. 1–23, 1973.

[HOU 12] HOULBRÈQUE F., RODOLFO-METALPA R., JEFFREE R. *et al.*, "Effects of increased pCO2 on zinc uptake and calcification in the tropical coral Stylophora pistillata", *Coral Reefs*, vol. 31, no. 1, pp. 101–109, 2012.

[HOW 12] HOWELLS E.J., BELTRAN V.H., LARSEN N.W., *et al.*, "Coral thermal tolerance shaped by local adaptation of photosymbionts", *Nature Climate Change*, vol. 2, pp. 116–120, 2012.

[HUG 94] HUGHES T.P., "Catastrophes, phase shifts, and large scale degradation of a Caribbean coral reef", *Science*, vol. 265, pp. 1547–1551, 1994.

[HUG 03] HUGHES T.P., BAIRD A.H., BELLWOOD D.R. *et al.*, "Climate change, human impacts, and the resilience of coral reefs", *Science*, vol. 301, pp. 929–933, 2003.

[HUG 07] HUGHES T.P., RODRIGUES M.J., BELLWOOD D.R. *et al.*, "Phase shifts, herbivory and the resilience of coral reefs to climate change", *Current Biology*, vol. 17, no. 4, pp. 360–365, 2007.

[HUG 10] HUGHES T.P., GRAHAM N.A.J., JACKSON J.B.C. *et al.*, "Rising to the challenge of sustaining coral reef resilience", *Trends in Ecology and Evolution*, vol. 25, pp. 633–642, 2010.

[HUN 92] HUNTE W., WITTENBERG M., "Effects of eutrophication and sedimentation on juvenile corals", *Marine Biology*, vol. 114, pp. 625–631, 1992.

[HUT 05] HUTCHINGS P., HAYNES D., "Marine Pollution Bulletin special edition editorial", *Marine Pollution Bulletin*, vol. 51, pp. 1–2, 2005.

[IUC 92] IUCN Caracas action plan. In Plenary session and symposium papers of the 4th World Congress on national Parks and Protected Areas, Caracas, Venezuela Gland, Switzerland, World Conservation Union (IUCN), pp. 301–310, 1992.

[JOH 75] JOHANNES R.E., "Pollution and degradation of coral reef communities", in FERGUSON-WOOD E.J., JOHANNES R.E. (eds.), *Tropical Marine Pollution*, Elsevier Scientific Publishing, New York, Elsevier Oceanography Series, vol. 12, pp. 15–51, 1975.

[JOH 79] JOHANNES R.E., KIMMERER W., KINZIE R. *et al.*, The impacts of human activities on Tarawa lagoon, Report to Government of Gilbert Islands, 1979.

[JOH 02] JOHANNES R.E., "The renaissance of community-based marine resource management in Oceania", *Annual Review of Ecology and Systematics*, pp. 317–340, 2002.

[JON 11] JONES C., Panulirus homarus, *Cultured Aquatic Species Information Programme*, FAO Fisheries and Aquaculture Department, Rome, 2011. Available at http://www.fao.org/fishery/culturedspecies/Panulirus_homarus/en [accessed on 7th March 2013].

[JUN 07] JUNCKER M., Jeunes poissons coralliens de Wallis et du Pacifique central: guide d'identification, Service Territorial de l'Environnement de Wallis et Futuna, 2007.

[JUR 10] JURY C.P., WHITEHEAD R.F., SZMANT A.M., "Effects of variations in carbonate chemistry on the calcification rates of Madracis auretenra (=Madracis mirabilis sensu Wells, 1973): bicarbonate concentrations best", *Global Change Biology*, vol. 16, no. 5, pp. 1632–1644, 2010.

[KAR 93] KARLSON R.H., HURD L.E., "Disturbance, coral reef communities, and changing ecological paradigms", *Coral Reefs*, vol. 12, pp. 117–125, 1993.

[KAY 12] KAYAL M., VERCELLONI J., LISON DE LOMA T. *et al.*, "Predator Crown-of-Thorns Starfish (Acanthaster planci) outbreak, mass mortality of corals, and cascading effects on reef fish and benthic communities", *PloS ONE*, vol. 7, no. 10, p. e47363, 2012.

[KEN 98] KENNISH R., *Pollution Impacts on Marine Biotic Communities*, CRC Press, Boca Raton, FL, 1998.

[KIN 74] KINSEY D.W., DOMM A., "Effects of fertilization on a coral reef environment – primary production studies", *Proceedings of 2nd International Symposium Coral Reefs*, vol. 1, pp. 49–66, 1974.

[KIN 79] KINSEY D.W., DAVIES P.J., "Effects of elevated nitrogen and phosphorus on coral reef growth", *Limnology Oceanography*, vol. 24, pp. 935–940, 1979.

[KLE 99] KLEYPAS J.A., BUDDEMEIER R.W., ARCHER D. *et al.*, "Geochemical consequences of increased atmospheric carbon dioxide on coral reefs", *Science*, vol. 284, no. 5411, pp. 118–120, 1999.

[KNA 83] KNAP A.H., SLEETER T.D., DODGE R.E. *et al.*, "The effects of oils spills and dispersant use on corals: a review and multidisciplinary experimental approach", *Oil and Petrochemical Pollution*, vol. 1, no. 3, pp. 157–169, 1983.

[KUL 15] KULBICKI M., PARRAVICINI V., MOUILLOT D., "Patterns and processes linked to body size", in MORA C. (ed.), *Ecology of Fishes on Coral Reefs*, Cambridge University Press, 2015.

[LEC 07] LECAILLON G., LOURIÉ S.M., "Current status of marine post-larval collection: existing tools, initial results., market opportunities and prospects", *SPC Live Reef Fish Information Bulletin*, vol. 17, pp. 3–9, 2007.

[LOV 99] LOVELL E.R., TUMURI M., Provisional environmental impact assessment for the extraction of coral reef products for the marine aquarium and curio trade in Fiji, Report prepared for the Fisheries Division, Government of Fiji, p. 116, 1999.

[LOY 80] LOYA Y., RINKEVICH B., "Effects of oil pollution on coral reef communities", *Marine Ecology Progress Series*, vol. 3, pp. 167–180, 1980.

[MAA 06] MAAMAATUAIAHUTAPU M., REMOISSENET G., GALZIN R., *Guide d'identification des larves de poissons récifaux de Polynésie française*, Editions Téthys, Tavaro, 2006.

[MAL 08] MALPOT E., GALZIN R., REMOISSENET G., "Using coral reef fish larvae: synopsis of work conducted in French Polynesia", *SPC Live Reef Fish Information Bulletin*, vol. 18, pp. 29–38, 2008.

[MAR 96] MARUBINI F., DAVIES P.S., "Nitrate increases zooxanthellae population density and reduces skeletogenesis in corals", *Marine Biology*, vol. 127, pp. 319–328, 1996.

[MAR 01] MARUBINI F., BARNETT H., LANGDON C., *et al.*, "Dependence of calcification on light and carbonate ion concentration for the hermatypic coral Porites compressa", *Marine Ecology*, vol. 220, pp. 153–162, 2001.

[MAR 03] MARUBINI F., FERRIER-PAGÈS C., CUIF J.P., "Suppression of skeletal growth in scleractinian corals by decreasing ambient carbonate –ion concentration: a cross-family comparison", *Proceedings of the Royal Society of London B: Biological Sciences*, vol. 270, no. 1511, pp. 179–184, 2003.

[MCC 12] McCulloch M., Falter J., Trotter J. *et al.*, "Coral resilience to ocean acidification and global warming through pH up-regulation", *Nature Climate Change*, vol. 2, pp. 623–627, 2012.

[MCC 14] McClanahan T.R., Graham N.A.J., Darling E.S., "Coral reefs in a crystal ball: predicting the future from the vulnerability of corals and reef fishes to multiple stressors", *Current Opinion in Environmental Sustainability*, vol. 7, pp. 59–64, 2014.

[MCM 97] McManus J.W., "Tropical marine fisheries and the future of coral reefs: a brief review with emphasis on Southeast Asia", *Coral Reefs*, vol. 16, pp. S121–S127, 1997.

[MCM 04] McManus J.W., Polsenberg J.F., "Coral-algal phase shifts on coral reefs: ecological and environmental aspects", *Progress in Oceanography*, vol. 60, pp. 263–279, 2004.

[MIC 01] Michener W.K., Baerwald T.J., Firth P. *et al.*, "Defining and unraveling biocomplexity", *BioScience*, vol. 51, no. 12, pp. 1018–1023, 2001.

[MOR 00] Morales-Nin B., "Review of the growth regulation processes of otolith daily increment formation", *Fisheries Research*, vol. 46, no. 1, pp. 53–67, 2000.

[MOR 06] Mora C., Andréfouët S., Costello M.J. *et al.*, "Coral reefs and the global network of marine protected areas", *Science*, vol. 312, pp. 1750–1751, 2006.

[MOR 11] Mora C., Aburto-Oropeza O., Bocos A.A. *et al.*, "Global human footprint on the linkage between biodiversity and ecosystem functioning in reef fishes", *PLoS Biology*, vol. 9, no. 4, p. e1000606, 2011.

[MOR 13] Mora C., Frazier A.G., Longman R.J. *et al.*, "The projected timing of climate departure from recent variability", *Nature*, vol. 502, pp. 183–187, 2013.

[MUM 07] Mumby P.J., Hastings A., Edwards H.J., "Thresholds and the resilience of Caribbean coral reefs", *Nature*, vol. 450, pp. 98–101, 2007.

[MUM 08a] Mumby P.J., Steneck R.S., "Coral reef management and conservation in light of rapidly evolving ecological paradigms", *Trends in Ecology and Evolution*, vol. 23, pp. 555–563, 2008.

[MUM 08b] Mumby P.J., Broad K., Brumbaugh D.R. *et al.*, "Coral reef habitats as surrogates of species, ecological functions, and ecosystem services", *Conservation Biology*, vol. 22, no. 4, pp. 941–951, 2008.

[MUM 14] Mumby P.J., Chollett I., Bozec Y.M. *et al.*, "Ecological resilience, robustness and vulnerability: how do these concepts benefit ecosystem management?", *Current Opinion in Environmental Sustainability*, vol. 7, pp. 22–27, 2014.

[MUN 83] MUNRO J.L., (ed.), *Caribbean Coral Reef Fishery Resources (No. 7)*, WorldFish, 1983.

[MUS 89] MUSCATINE L., FALLOWSKI P.G., DUBINSKY Z. *et al.*, "– The effect of external nutrient resources on the population dynamics of zooxanthellae in a reef coral", *Proceedings of the Royal Society of London B*, vol. 236, no. 1284, pp. 311–324, 1989.

[MYE 97] MYERS R.A., HUTCHINGS J.A., BARROWMAN N.J., "Why do fish stocks collapse? The example of cod in Atlantic Canada", *Ecological Applications*, vol. 7, no. 1, pp. 91–106, 1997.

[NEL 07] NELSON A.L., MARYGRACE C.Q., *Chanos Chanos. Cultured Aquatic Species Information Programme*, FAO Fisheries and Aquaculture Department, Rome, 2007. Available at http://www.fao.org/fishery/culturedspecies/Chanos_chanos/fr [accessed on 7th March 2013].

[NEW 07] NEWTON K., CÔTÉ I., PILLING G.M. *et al.*, "Current and future sustainability of island coral reef fisheries", *Current Biology*, vol. 17, pp. 655–658, 2007.

[NYS 00] NYSTRÖM M., FOLKE C., MOBERG F., "Coral reef disturbance and resilience in a human-dominated environment", *Trends in Ecology Evolution*, vol. 15, no. 10, pp. 413–417, 2000.

[NYS 01] NYSTRÖM M., FOLKE C., "Spatial resilience of coral reefs", *Ecosystems*, vol. 4, pp. 406–417, 2001.

[OCE 13] OCEA CONSULT, Collecte de post-larves de poissons récifaux, Rapport final du programme POLARUN: Étude de la colonisation post-larvaire des poissons sur les récifs de La Réunion et potentialités de valorisation, p. 73, 2013.

[PAN 03] PANDOLFI J.M., BRADBURY R.H., SALA E. *et al.*, "Global trajectories of the long-term decline of coral reef ecosystems", *Science*, vol. 301, no. 5635, pp. 955–958, 2003.

[PAN 11] PANDOLFI J.M., CONNOLLY S.R., MARSHALL D.J. *et al.*, "Projecting coral reef futures under global warming and ocean acidification", *Science*, vol. 333, pp. 418–422, 2011.

[PAR 14] PARRAVICINI V., VILLÉGER S., MCCLANAHAN T.R. *et al.*, "Global mismatch between species richness and vulnerability of reef fish assemblages", *Ecology Letters*, vol. 17, no. 9, pp. 1101–1110, 2014.

[PEA 81] PEARSON R.G., "Recovery and recolonization of coral reefs", *Marine Ecology Progress Series*, vol. 4, pp. 105–122, 1981.

[PLA 01] PLANES S., LECAILLON G., "Caging experiment to examine mortality during metamorphosis of coral reef fish larvae", *Coral Reefs*, vol. 20, pp. 211–218, 2001.

[POL 84] POLOVINA J.J., "Model of a coral reef ecosystem", *Coral Reefs*, vol. 3, no. 1, pp. 1–11, 1984.

[POM 06] POMEROY R.S., PARKS J.E., BALBOA C.M., "Farming the reef: is aquaculture a solution for reducing fishing pressure on coral reefs?", *Marine Policy*, vol. 30, no. 2, pp. 111–130, 2006.

[PON 13] PONTON D., CARASSOU-BOUILLERET L., RAILLARD S. *et al.*, "Geometric morphometrics as a tool for identifying emperor fish (Lethrinidae) larvae and juveniles", *Journal of Fish Biology*, vol. 83, no. 1, pp. 14–27, 2013.

[PRA 14] PRATCHETT M.S., HOEY A.S., WILSON S.K., "Reef degradation and the loss of critical ecosystem goods and services provided by coral reef fishes", *Current Opinion in Environmental Sustainability*, vol. 7, pp. 37–43, 2014.

[QUI 10] QUINITIO E.T., Scylla serrata. Cultured Aquatic Species Information Programme, FAO Fisheries and Aquaculture Department, Rome, 2010. Available at http://www.fao.org/fishery/culturedspecies/Scylla_serrata/en [accessed on March 2013].

[RAU 12] RAU G.H., MCLEOD E.L., HOEGH-GULDBERG O., "The need for new ocean conservation strategies in high-carbon dioxide world", *Nature Climate Change*, vol. 2, pp. 720–724, 2012.

[REI 94] REICHELT A.J., JONES G.B., "Trace metals as tracers of dredging activity in Cleveland Bay – field and laboratory studies", *Marine and Freshwater Research*, vol. 45, no. 7, pp. 1237–1257, 1994.

[REI 01] REICH P.B., KNOPS J., TILMAN D. *et al.*, "Plant diversity enhances ecosystem responses to elevated CO_2 and nitrogen deposition", *Nature*, vol. 410, no. 6830, pp. 809–810, 2001.

[REI 00] REICHELT-BRUSHETT A.J., HARRISON P.L., "The effect of copper on the settlement success of larvae from the scleractinian coral Acropora tenuis", *Marine Pollution Bulletin*, vol. 41, nos. 7–12, pp. 385–391, 2000.

[REI 05] REICHELT-BRUSHETT A.J., HARRISON P.L., "The effect of selected trace metals on the fertilization success of several scleractinian coral species", *Coral Reefs*, vol. 24, pp. 524–534, 2005.

[REI 12] REICHELT-BRUSHETT A.J., "Risk assessment and ecotoxicology: limitations and recommendations for ocean disposal of mine waste in the Coral Triangle", *Oceanography*, vol. 25, no. 4, pp. 40–51, 2012.

[REK 07] REKSODIHARDJO-LILLEY G., LILLEY R., "Towards a sustainable marine aquarium trade: An Indonesian perspective", *SPC Live Reef Fish Information Bulletin*, vol. 17, pp. 11–19, 2007.

[REY 03] REYNAUD S., LECLERCQ N., ROMAINE-LIOUD S., "Interacting effects of pCO_2 partial pressure and temperature on photosynthesis and calcification in a scleractinian coral", *Global Change Biology*, vol. 9, no. 11, pp. 1660–1668, 2003.

[RIE 91] RIEGL B., VELIMIROV B., "How many damaged corals in Red Sea systems? A quantitative survey. Coelenterate biology: recent research on Cnidaria and Ctenophora", *Developments in Hydrobiology*, vol. 66, pp. 249–256, 1991.

[RIE 95] RIEGL B., BRANCH G.M., "Effects of sediment on the energy budgets of four scleractinian alcyonacean (Lamouroux 1816) corals", *Journal of Experimental Marine Biology and Ecology*, vol. 186, pp. 259–275, 1995.

[ROG 90] ROGERS C.S., "Responses of coral reefs and reef organisms to sedimentation", *Marine Ecology Progress Series*, vol. 62, pp. 185–202, 1990.

[ROG 13] ROGERS C.S., "Coral reef resilience through biodiversity", *ISRN Oceanography*, p. 739034, 2013.

[ROS 86] ROSS S.T., "Resource partitioning in fish assemblages: a review of field studies", *Copeia*, pp. 352–388, 1986.

[ROS 04] ROSENBERG E., LOYA Y., *Coral Health and Disease*, Springer-Verlag, Berlin, p. 488, 2004.

[SAB 09] SABDONO A., "Heavy metal levels and their potential toxic effect on coral galaxea fascicularis is from Java sea, Indonesia", *Research Journal of Environmental Sciences*, vol. 3, no. 1, pp. 96–102, 2009.

[SAD 02] SADOVY Y.J., VINCENT A.C.J., "Ecological issues and the trades in live reef fishes", in SALE P.F. (ed.), *Coral Reef Fishes: Dynamics and Diversity in a Complex System*, Academic Press, San Diego, pp. 391–420, 2002.

[SAK 11] SAKAI Y., Stichopus japonicus Cultured Aquatic Species Information Programme, FAO Fisheries and Aquaculture Department, Rome, 2011. Available at http://www.fao.org/fishery/culturedspecies/Stichopus_japonicus/en [accessed on 7 March 2013].

[SHE 09] SHEPPARD C.R.C., DAVY S.K., PILLING G.M., *The Biology of Coral Reefs*, Oxford University Press, New York, 2009.

[SHE 11] SHELLEY C., LOVATELLI A., Mud crab aquaculture – A practical manual, FAO Fisheries and Aquaculture Technical Paper, No. 567. Rome, FAO, p. 78, 2011. Available at http://www.fao.org/docrep/015/ba0110e/ba0110e.pdf [accessed on 7 March 2013].

[SMI 78] SMITH S.V., KINSEY D.W., "Calcification and organic carbon metabolism as indicated by carbon dioxide", *Monographs on Oceanographic Methodology*, vol. 5, pp. 409–484, 1978.

[STA 92] STAFFORD-SMITH M.G., ORMOND R.F.G., "Sediment-rejection mechanisms of 42 species of Australian scleractinian corals", *Australian Journal of Marine & Freshwater Research*, vol. 43, no. 4, pp. 683–705, 1992.

[STA 93] STAFFORD-SMITH M.G., "Sediment-rejection efficiency of 22 species of Australian scleractinian corals", *Marine Biology*, vol. 115, pp. 229–243, 1993.

[STO 11] STOECKL N., HICKS C.C., MILLS M. *et al.*, The economic value of ecosystem services in the Great Barrier Reef: our state of knowledge", *Annals of the New York Academy of Sciences*, vol. 1219, no. 1, pp. 113–133, 2011.

[SUN 12] SUNDSTROM S.M., ALLEN C.R., BARICHIEVY C., "Species, functional groups., and thresholds in ecological resilience", *Conservation Biology*, vol. 26, pp. 305–314, 2012.

[SUT 04] SUTHERLAND K.P., RITCHIE K.B., "White pox disease of the caribbean elkhorn coral Acropora palmata", *Coral Health and Disease*, pp. 289–300, 2004.

[SUU 12] SUURONEN P., CHOPIN F., GLASS C. *et al.*, Low impact and fuel efficient fishing–looking beyond the horizon", *Fisheries Research*, vol. 119, pp. 135–146, 2012.

[TIL 96] TILMAN D., DOWNING J.A., "Biodiversity and stability in grasslands", *Ecosystem Management*, Springer New York, pp. 3–7, 1996.

[TOM 87] TOMASCIK T., SANDER F., "Effects of eutrophication on reef building corals. II: structure of scleractinian coral communities on fringing reefs, Barbados, West Indies", *Marine Biology*, vol. 95, pp. 53–75, 1987.

[VIG 09] VIGLIOLA L., MEEKAN M. G., "The back-calculation of fish growth from otoliths", *Tropical Fish Otoliths: Information for Assessment, Management and Ecology*, Springer Netherlands, pp. 174–211, 2009.

[WAR 03a] WARBITZ C., TAYLOR M., GREEN E. *et al.*, *From Ocean to Aquarium*, UNEP – World Conservation Monitoring Centre, Cambridge, p. 65, 2003.

[WAR 03b] WARREN-RHODES K., SADOVY Y., CESAR H., "Marine ecosystem appropriation in the Indo-Pacific: a case study of the live reef fish food trade", *Ambio*, vol. 32, pp. 481–488, 2003.

[WIL 04] WILLIAMS J.D., MCDERMOTT J.J., "Hermit crab biocoenoses: a worldwide review of the diversity and natural history of hermit crab associates", *Journal of Experimental Marine Biology and Ecology*, vol. 305, no. 1, pp. 1–128, 2004.

[WIL 08] WILKINSON C., *Status of Coral Reefs of the World*, Global Coral Reef Monitoring Network and Reef and Rainforest Research Center, Townsville, Australia, p. 296, 2008.

[WOO 92] WOODLEY J.D., "The incidence of hurricanes on the north coast of Jamaica since 1870: are the classic reef descriptions atypical?", *Hydrobiologia*, vol. 247, pp. 133–138, 1992.

[WOO 01] WOOD E.M., *Collection of Coral Reef Fish for Aquaria: Global Trade, Conservation Issues and Management Strategies*, Marine Conservation Society, Ross-on-Wye, UK, p. 80, 2001.

[WOO 09] WOOLDRIDGE S.A., DONE T.J., "Improved water quality can ameliorate effects of climate change on corals", *Ecological Applications*, vol. 19, pp. 1492–1499, 2009.

[WOR 06] WORM B., BARBIER E.B., BEAUMONT N. *et al.*, "Impacts of biodiversity loss on ocean ecosystem services", *Science*, vol. 314, no. 5800, pp. 787–790, 2006.

6

Hydrothermal Vents: Oases at Depth

6.1. Introduction to deep-sea ecosystems

The deep ocean, often called the abyss, covers 66% of the planet. Schematically, it is the marine zone where there is not enough light to allow plants to grow. In fact, the lightened part of the ocean is shallow and does not exceed a few hundred meters; the rest of the ocean can reach much greater depths (11 km in the Mariana Trench) and is in complete darkness except for slight glows of bioluminescence. In the absence of local organic production, the organisms living at depth subsist due to organic matter produced in surface waters. This organic matter, in the form of decomposed matter, feces and molts, is clumped together by secretions from gelatinous organisms, sediments at depth and is degraded under the action of heterotrophic microbial communities, or is being recycled by pelagic organisms. Stratification of the water column, depth, surface production and seasonal variations as well as distance from continental shelves directly affect the flow of organic matter that reaches deep sediments.

Generally speaking, the further we move away from the coast, the greater the depth, the lower the organic input and consequently, the lower the benthic biomass. Production in the abyssis approximately a hundredth of the primary production at the surface and benthic biomass expressed in fresh weight rarely exceeds 2 g/m^2. Only some populations of filter-feeding

---

Chapter written by Jozée SARRAZIN and Daniel DESBRUYÈRES.

organisms such as sponges and corals have significant biomass, concentrated in zones where deep currents resuspend sedimented organic matter. Moreover, surface plankton blooms occur which lead to the significant accumulation of organic matter at the water/sediment interface. This rapid enrichment of the environment allows for the rapid development of opportunistic fauna (small invertebrates and foraminifera) which disappears as the resource is depleted. Similarly, as the dead carcasses of large pelagic vertebrates (large fish, turtles and marine mammals) and plants (large algae and sunken wood) arrive at depth, they attract diverse populations of necrophages and xylophages (fish, mollusks and crustaceans) that rapidly use this food source and enrich the environment with their feces. The shortage of food, low temperatures (except for the Mediterranean Sea) and high hydrostatic pressure slow down biological processes such as growth and respiration.

Although the biomass, density and intensity of biological processes are low compared with coastal zones, the species diversity is high. A study of a sample of benthic macrofauna revealed that the population is composed of many species, each present in low numbers. Moreover, the similarity in faunal composition between samples taken in the same area is low, revealing the high biodiversity at a local scale. From these samples, the number of new species new to science is high, representing more than half of the sample. For example, in a total surface area of 21 m^2, situated 2,100 m offshore of the New Jersey coast, an American team identified 798 species from a total of 90,677 individuals sampled, without the depletion curve reaching a plateau [GRA 92]. Extrapolating this curve would give a figure of 11,800 species of benthic macrofauna in this community.

Although many articles on the origin of deep-sea biodiversity, often contradictory, have been published, it is still difficult to draw a clear conclusion. Some of the oldest hypotheses suggest that the deep-sea environment is stable and in equilibrium and therefore not selective, leading to an unconstrained diversification and allowing the exploitation of different food sources. However, some researchers have recently suggested that the deep-sea environment is composed of a mosaic of small environments in disequilibrium caused by bioturbation and hydrodynamism (see Figure 6.1). Each of these environments would be colonized by successions of species responding to local conditions.

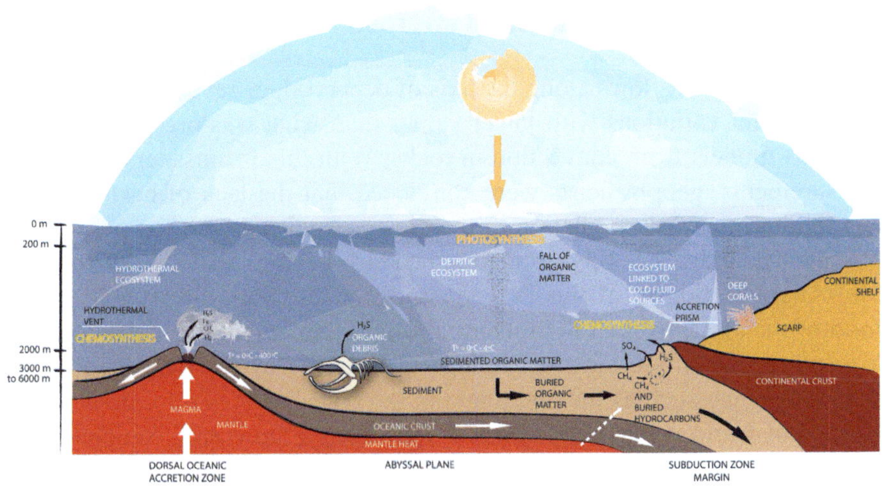

Figure 6.1. *Cross-section of the ocean illustrating the different ecosystems of the deep sea (Capsule Graphik 2014)*

Since the middle of the 19th Century, scientists have questioned the position of deep-sea animals in the evolution of marine fauna, in the hope that they might discover the missing link, i.e. the species or group that may have survived mass extinctions in this protected and stable environment. Currently, despite the discovery of primitive forms such as the mollusk *Neopilina*, many others reveal that the proportion of forms with ancient characteristics is much greater in the coastal and bathyal environment than in the abyss. Therefore, we have contrasting theories. The first theory considers that the abyssal fauna has evolved separately; the second theory proposes that this fauna is the result of successive migrations of coastal or bathyal species, particularly at high latitudes where temperature barriers are absent. In the history of the oceans, periods of anoxia at depth (e.g. between 90 and 50 million years ago), or the sharp change in temperature 16 million years ago, could have caused extinctions or significant variations in abyssal fauna, which would have been locally recolonized by fauna from continental margins.

At the beginning of the 1970s, our general vision of the deep-sea was that it was a relatively homogeneous ecosystem, with low benthic biomass, slow biological turnover, but high diversity of fauna. It was supplied by the fluxes of organic matter from photic zones of the ocean (see Figure 6.1).

6.2. Discovery of hydrothermal sources

Measuring heat flows along the axis of ocean ridges reveals the existence of significant variations with lower values than what was predicted by heat transport models from convection in rocks. With this inconsistency between measurements, geophysicists were convinced that the lack of heat could be due to hydrothermal circulation in the crust, associated with the percolation of seawater. Indices of this circulation had also been revealed in 1974 on the Mid-Atlantic Ridge[1], in the Franco-American program FAMOUS, in the form of metal deposits of hydrothermal origin. American geologists decided to explore not only a slow-spreading ridge as the MAR, but also a faster spreading ridge such as the one found in the Galapagos[2]. To locate potential sources of warm water, in May 1976, they used a "towcam" equipped with cameras and different sensors including temperature probes, towed several meters above the ridge situated at a depth of 2,500 m. After several days, from the photos they noticed a coincidence between the presence of temperature anomalies and concentrations of white areas on black basalts. After analysis, these were considered to be large bivalve shells.

In February 1977, scientists arrived at the Galapagos onboard two oceanography vessels: "Lulu", a submersible support vessel for the submersible "Alvin" and the "Knorr". Alvin led the first dive of the campaign on 17 February 1977 at a depth of 2,500 m on the site found the previous year (00°47' N and 86°09'W). Approximately 15 min after reaching the seafloor, the temperature probe on the submarine detected a temperature anomaly of several tenths of a degree and suddenly the landscape changed: through the portholes, astonished observers saw large white clams, aligned in faults between pillow lavas, anemones, mussel assemblages and orange mineral deposits [LON 77].

Here, in this mineral desert, a flourishing oasis of several hundred meters squared with diffuse emissions of warm water between 15 and 25°C was found; the temperature at these depths usually being around 2°C. Quickly, several dives were planned and approximately 10 km away new sites were discovered, home to unique and colorful life forms. The first observations were fascinating and the sites were given poetic names such as "Rose Garden", "Clambake" or even "Garden of Eden". Large population of invertebrates was dominated by large

---

1 MAR: Mid-Atlantic Ridge.

2 GSC: Galapagos Spreading Center.

bivalves such as the giant clam *Calyptogena magnifica*[3] and the mussel *Bathymodiolus thermophilus* as well as the giant tubeworm *Riftia pachyptila*, which formed bushes of more than 1 m tall, composed of several hundred tubular individuals [HES 83].

Different chemical and isotopic measurements estimated that the pure hydrothermal fluid should reach much higher temperatures, probably between 300 and 400°C; geophysicists were disappointed because the GSC fluids were very diluted. A new exploration program called RITA was, therefore, launched in 1978, in cooperation with French and Mexican teams, located on the East Pacific Rise[4] between Rivera and Tamayo faults. In 1978, the French submarine Cyana discovered, at a depth of 2,645 m, large sulfide structures of several meters tall and "cemeteries" of white bivalves with dissolved shells. The following year, the submersible Alvin brought American and French geophysicists to the same zone (20°50'N and 109°06'W). After eight unsuccessful dives, the American team reported the presence of a chimney emitting black smoke. The first temperature measurements were unsuccessful, but then a temperature of 350°C was measured from this black fluid that was being emitted. The fauna at these sites was comparable to that at GSC sites, apart from the absence of mussels and the presence of a polychaete worm on the chimney walls, called the "Pompeii worm" (*Alvinella pompejana*) due to the hydrothermal ash covering its habitat.

The discovery of these areas of very high biomass surrounding hydrothermal fluid emissions located in the middle of a mineral desert and populated by unique organisms led the biologists to question the mechanisms of biological production in the absence of sunlight.

6.3. Geology and geochemistry of hydrothermal systems

6.3.1. *Formation of hydrothermal vents*

The distribution of hydrothermal emissions depends on geological and geochemical processes. These warm water emissions are formed in the different geodynamic contexts (slow and fast spreading ridges, back-arc basins and volcanic arcs), at depths varying from several meters to more than

3 For the taxonomic references see [DES 06a].

4 EPR: *East Pacific Rise* (with two subdivisions SEPR and NEPR, separated by the Amercian plate).

5,000 m, at the boundary between two tectonic plates where new oceanic crust is formed. These plates move at speeds varying from several millimeters per year at slow ridges (such as the MAR) to 90, or even 180 mm per year at fast ridges (as in the south of EPR). Unlike ridges that are the site of new oceanic crust formation, back-arc basins are formed in subduction zones. Located between the coasts and ocean trenches, these basins are also home to hydrothermal phenomena. The best known for their hydrothermal activity are the North Fiji Basin, the Okinawa area, the Lau Basin and the Manus Basin in the western Pacific and the East Scotia basin in the South Atlantic Ocean. Volcanic arcs can also contain active sites (e.g. South Sandwich arc in the Atlantic and Kermadec arc in the Pacific). A total of approximately 200 hydrothermal sites have been identified to date and several indices suggest the presence of several other active sites (see Figure 6.2), not including parts of the world that have never been explored, particularly at the poles.

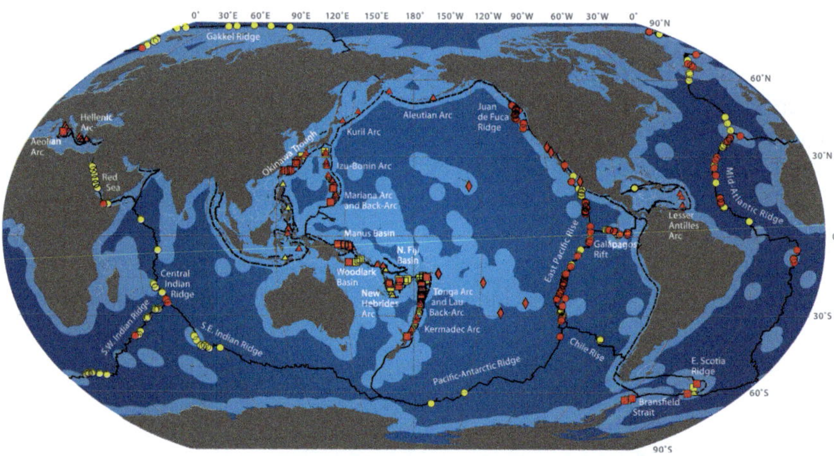

Figure 6.2. *Distribution of active hydrothermal sites (in red) and sites where indices of activity have been identified (in yellow).Continuous lines represent ridges and faults whereas dotted lines represent trenches. Exclusive economic zones are indicated in pale blue (according to [BEA 10]). For a color version of the figure, see color section*

Hydrothermal activity occurs due to the movement of tectonic plates and volcanic activity [FOU 12]. These intense geological processes favor a weakening of the seafloor and the creation of fissures in which cold and well-oxygenated seawater (between 0 and 4°C) will penetrate (see Figure 6.3).

This seawater then reacts with the underlying rocks and undergoes modifications. The final composition of this "hydrothermal fluid" will depend on the composition of these rocks, the depth at which the seawater will percolate and subsurface temperatures. These zones of formation of the oceanic crust are characterized by the presence of shallower (only several kilometers) magma chambers than elsewhere in the ocean, which allows heating of the percolating fluids. These magma chambers contain magma with a melting point of 1,200°C. Intense reactions cause the weathering of rocks and magnesium in the fluid to bind with new metals. This hot, acidic and rich in reduced compounds "hydrothermal fluid" is less dense and will rapidly rise to the surface of the seafloor to form a jet of hot water similar to a geyser called a "hydrothermal vent". The metals contained in the hydrothermal fluids precipitate in contact with the cold surrounding seawater to form polymetallic sulfides which accumulate and form structures reaching up several meters high. Hydrothermal plumes, formed by the dilution of hydrothermal fluid with seawater, rich in particles and characterized by temperature anomalies, can rise up to several hundred meters above active sites, until reaching neutral buoyancy. They then spread laterally according to the currents, favoring the dispersion of metals over tens of kilometers. They, therefore, significantly contribute to the input of chemical elements in the deep ocean.

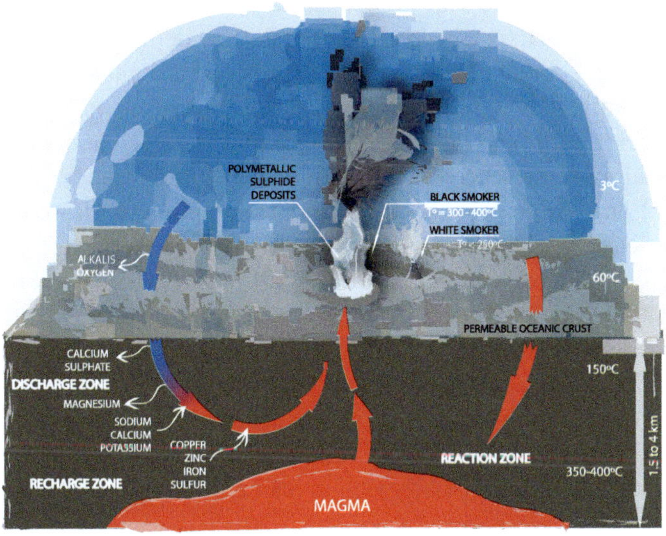

Figure 6.3. *Diagram illustrating the circulation of fluids and the formation of hydrothermal vents (Capsule Graphik 2014)*

6.3.2. *Chemical composition of hydrothermal fluids*

All hydrothermal fluids are hot, acidic, free of oxygen and rich in sulfides and various metals such as iron, manganese, zinc, copper and also, in smaller proportions, cadmium, lead, cobalt, nickel, gold and silver. They are also rich in alkali metals (lithium, potassium, rubidium and cesium) and silica, which come from the leaching of basalts. Despite their similarities, these fluids have chemical compositions that vary from one area to another and even from one vent to another at the same site [CHA 00]. The final composition will depend on factors such as the geodynamic environment, water depth, temperature, the maturity of the deposits and the nature of the substrata affected by hydrothermal circulation. The reaction rate of seawater with this substratum is much faster at higher temperatures.

One particular chemical reaction, called serpentinization, occurs at several points along slow-spreading ridges. In this case, seawater reacts with the outcropping mantle rock, producing fluids extremely rich in hydrogen and methane. On the MAR, five active sites (Rainbow, Logatchev 1 and 2, and Ashadze 1 and 2) have been found on this mantle substrata. The final chemical composition of the fluids results from a pressure and temperature-dependent phenomena, which can lead to the process of phase separation. Thus, the critical point at which the seawater separates into two phases is at 400°C, at a pressure of 300 bar (corresponding to 3,000 m depth). These two phases include a dense, saline solution rich in metals and a lighter fraction, rich in dissolved volatile gases. This phase separation process is responsible for large changes in salinity observed in hydrothermal fluids. There are also hydrothermal systems where the oceanic crust is buried under a thick layer of sediment due to the proximity to the coast and/or pelagic sedimentation. This is the case in the Guaymas Basin (Gulf of California), the Escanaba Trough (Gorda ridge) and the Middle Valley site (Juan de Fuca Ridge) located in the Pacific. There, the heat is trapped and fluids interact with sediment leading to the precipitation of sulfides and the formation of large deposits. The chemical composition of the fluids will then depend on that of the sediments and also on the importance of the reactions between these fluids, the surrounding basalts and buried sulfides. As these sediments trap metals, these sites are particularly rich in lead and arsenic [FOU 12].

The synthesis of organic molecules in hydrothermal fluids [HOL 01] raises the question of what role hydrothermal vents may have played in the emergence of life on Earth. Indeed, this environment contains the essential characteristics for the development of prebiotic syntheses: heat, water and the presence of basic chemical compounds. In addition, these deep-sea ecosystems were most likely protected from extreme conditions at the surface (meteor showers and ultraviolet radiation).

Several types of hydrothermal emissions exist, according to the degree of mixing between the vent fluids and cold seawater. Black smokers are the most spectacular. They result from a pure hydrothermal fluid, which does not undergo any dilution prior to being emitted. They can reach temperatures above 350°C. These emissions are characterized by dense plumes of black smoke created by the precipitation of metal sulfides in contact with cold seawater. The vent diameter, which can reach 10 cm, promotes the release of high-speed fluids at several meters per second. As the sulfide structure ages, the flow is reduced, the ducts become clogged and the fluid continues to percolate through porous minerals, sometimes forming very special black beehive structures. When the hydrothermal emissions last for thousands of years in one place, they can generate the formation of impressive sulfide mounds. Gray or white smokers emit fluids with temperatures ranging between 150 and 270°C, which are more diluted and less rich in chemical elements. Their color is related to silica, anhydrite and barite that precipitate during mixing of hydrothermal fluids and seawater. Finally, the dilution of the hydrothermal fluids with seawater is even greater in the case of diffuse emissions, with temperatures below 50°C. The fluid formed creates a shimmering effect linked to the difference in density with the cold surrounding seawater. It is in this diluted hydrothermal fluid, at moderate temperatures, that the hydrothermal fauna is found.

The chemical composition of hydrothermal fluids also varies over time: while it can remain stable for many years at some sites, it can also change slowly or even suddenly during volcanic or major tectonic events. Thus, the composition of the fluids on fast-spreading ridges appears to change faster than on slow-spreading ridges, where it would tend to remain more or less stable. The lifespan of a site can vary from tens to tens of thousands of years.

6.4. Microbial chemosynthesis

In the first months after the discovery of hydrothermal vents, the possible primary production of organic matter by autotrophic microbes was suggested

[LON 77]. Producing organic matter requires an energy source, a carbon source and a source of electrons. In the case of photosynthesis, the energy source is light, CO_2 is the source of carbon and water is the source of electrons (see Figure 6.4). Although water and CO_2 are abundant in deep environments, the light required for photosynthesis is absent. Microbial chemosynthesis (or chemolithoautotrophy) is, therefore, an organic synthesis pathway, first discovered in 1887 by Sergei Winogradsky, a Russian microbiologist. This researcher, who studied soil biology, found out that filamentous bacteria were able to synthesize their cellular carbon from inorganic compounds, sulfides and carbon dioxide. This pathway analogous to photosynthesis differs in the origin of energy used to operate the CO_2 condensation cycles [JAN 79].

In the case of aerobic chemosynthesis, oxygen is the most common electron acceptor and the electron donor is usually a form of reduced sulfur (sulfide, elemental sulfur and thiosulfate), but also iron, manganese and ammonium. At hydrothermal vents, the most abundant compound in the fluids is hydrogen sulfide or dihydrogen sulfide (H_2S) but methane, iron and hydrogen can also be used [SCM 08]. Chemosynthesis cannot, as it has often been claimed, be separated from solar energy, since it uses oxygen, a product of photosynthesis. When oxygen is absent, the main electron donor is hydrogen which is often present in large quantities in hydrothermal fluids; nitrates, sulfates, sulfur and CO_2 acting as electron acceptors. In chemosynthesis, carbon fixation usually occurs in the Calvin–Benson cycle. However, in certain bacteria or archaea, other pathways may be used. Methanotrophy may also be a primary biosynthetic pathway if methane is synthesized abiotically. It can be used as a carbon source and an electron donor.

Although the role of chemosynthetic production in the hydrothermal ecosystem is generally acknowledged, some authors [KAR 95] state that its importance has not been clearly proven, and other possibilities should be considered, such as the recycling of organic matter within hydrothermal circulation and the formation of hydrocarbons or fatty acids, or the abiotic thermogenesis of organic compounds which could be used by heterotrophic microorganisms. The hydrothermal environment is inhabited by microorganisms (bacteria and archaea), with highly diverse metabolisms that populate almost all microenvironments including the turbulent mixing zone between warm fluids and cold seawater, water column, and biological and mineral surfaces. Filamentous bacteria can, particularly on soft sediments, develop in thick mats.

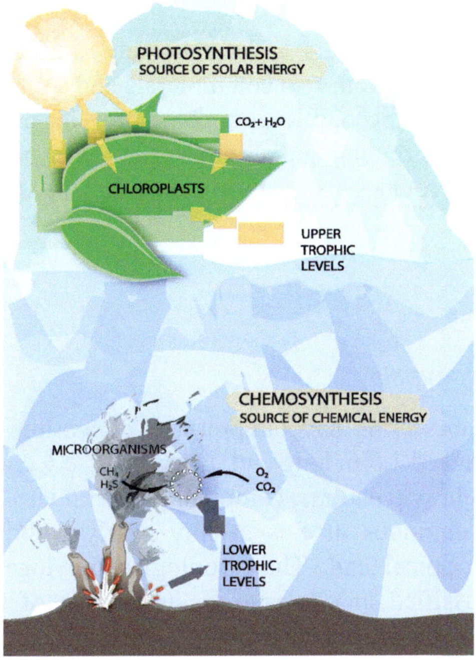

Figure 6.4. *Photosynthesis and chemosynthesis: two processes producing an organic matter that uses different energy sources (Capsule Graphik 2014)*

The discovery of hydrothermal oases was quickly followed by the discovery of cold seep zones (reviewed by Sibuet and Olu-Le [SIB 02]) on continental margins. Microbial production in this case is primarily based on the use of methane of biological origin as a carbon source and an electron donor. Besides direct methanotrophy, consortia of bacteria and archaea present in the sediments can transform methane into sulfides [BOE 00], which are used in chains of chemoautotrophic microbes. Similar populations to those observed in hydrothermal ecosystems have been found. Significant organic inputs (large vertebrate carcasses and sunken wood), which provide food for necrophages and xylophages, are also the site of anaerobic microbial degradation producing sulfides from sulfates in seawater. These sulfides will form the basis of chemosynthesis. Similar populations of organisms to those found in hydrothermal or deep ecosystems also develop around these large inputs of organic matter.

6.5. Symbioses and trophic chains

The vent food web appears to be relatively simple: a local microbial production using energy from reduced minerals and a set of filter feeders and grazers, which form the second level become prey for carnivores, crabs and fish. Studying the intriguing giant tubeworm *Riftia pachyptila* and other engineer species provides key information to the functioning of this ecosystem.

6.5.1. *Symbioses*

6.5.1.1. *Riftia pachyptila*

These giant tubeworms have impressive dimensions (see Figure 6.5). Their tubes can exceed 1.5 m long and measure up to 6 cm in diameter. They are found as upright populations, comprising several tens to several hundred individuals. The chitinous tube is usually pearly white and a red plume emerges from its apical end. Their zoological phylogeny has long been debated. First classified in *Obturata*, a subphylum of the pogonophores, which include worms with a vestimentum (dorsal chamber of the body) and in the phylum vestimentifera, distinct from pogonophores, they are now classified with other members of pogonophores in a family of annelid polychaetes, *Siboglinidae*, similar to *Sabellidae*. Studying their anatomy reveals some interesting features. The highly vasculated, lamellar gill plume ends in a type of operculum, the obturaculum, which closes the tube when the animal retracts. The muscular vestimentum, situated between the tentacular crown and trunk, has two wing-like folds forming a chamber in which a pair of genital openings is located. The elongated sack-like trunk terminates in an opistosome with bristles anchoring in the tube. However, the most unexpected feature in an animal of this size is the absence of a gut. This absence, found in all adults of this species, obviously raises the question of their feeding mode. Dissection reveals that most of the trunks are occupied by a highly vascularized solid tissue, which was initially hypothesized, in a closely related species (*Lamellibrachia Barhami*), to having role in the formation of sperm. Closer analysis of this trophosome via transmission electron microscopy soon revealed that its cells were inhabited by symbiotic bacteria located inside vacuoles and were interspersed with elemental sulfur crystals. These bacterial populations may represent up to a quarter of the volume of this nutritive tissue.

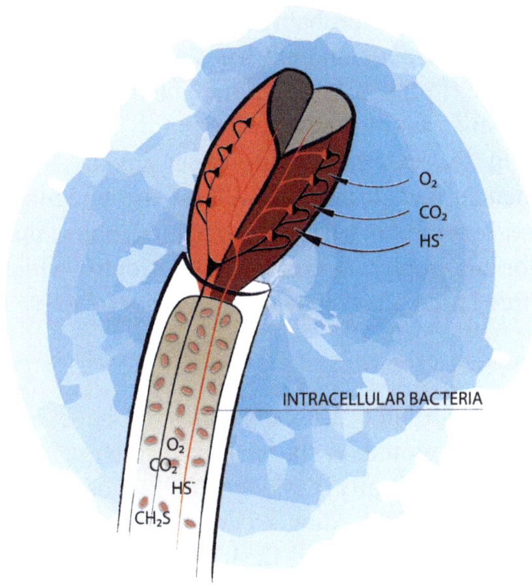

Figure 6.5. *Endosymbiosis in the siboglinid worm Riftia pachyptila (Capsule Graphik 2014)*

Biochemists have revealed that this trophosome harbors high activity of RubisCo, a diagnostic enzyme in the Calvin–Benson cycle, a biochemical cycle used in autotrophy and a characteristic enzyme in sulfur compound cycles. In addition, studying natural isotope ratios of carbon demonstrated that the worm tissues were not made of carbon from photosynthesis [FIS 90]. Recent studies on proteomes have also shown that these symbionts were able to use a different metabolic pathway of carbon fixation (rTCA) which requires less energy than the Calvin–Benson cycle. Together, these observations show that the tubeworm exploits the organic matter produced by its symbiotic bacteria found inside its bacteriocytes. The worm transfers the nutrients needed for the chemosynthesis bacteria via the bloodstream and in exchange, the bacteria provide it with the organic compounds necessary for its growth, metabolism and reproduction. To acquire the oxygen and sulfides required by bacteria, the worm's plume must be in a turbulent mixing zone where rich hydrothermal fluids alternate with oxygenated seawater [CHI 92].

This type of strict symbiosis is very productive. Sequential observations made after an eruption showed a tube growth of about 80 cm per year for a new *Riftia* colony, which is probably a record for a marine invertebrate. Research into embryonic development has shown that *Riftia* eggs give birth to trochophore larvae. In the next stage, the gut is formed from the oralarea. The bacteria are still lacking, while the gut is already functional. Free bacteria present in the external environment invade the worms and the trophosome is then formed from the walls of the midgut. Recent genomic analyses of *Riftia*'s symbionts (γ-proteobacteria) show that they have the genes required for autotrophy, as well as heterotrophy, which were probably used during the free larval phase in open water, out of active sites.

6.5.1.2. *Calyptogena magnifica*

Giant tubeworms are not the only ones to have a very productive symbiosis with chemosynthetic bacteria. The large clam *Calyptogena magnifica* belongs to the family *Vesicomyidae*, whose members are mostly burrowing bivalves. Often exceeding 25 cm at its greatest length, it has thick gills, very red and highly vasculated flesh, and a large muscular foot, which can be inserted deep into the cracks through which warm hydrothermal fluid is released [ARP 84]. The gut is reduced and it is likely that it is little or not functional. The hypertrophied gill immersed in seawater drawn in through the siphon located opposite to the foot [CHI 91]. Studies on carbon stable isotopes show that this animal does not use, or only slightly uses, carbon from photosynthesis. Enzymatic analyses reveal the strong autotrophic activity of Rubisco in gill tissues. Finally, histological studies confirmed the presence of symbiotic bacteria in the vacuoles of the bacteriocytes located on the outer edge of the gill filaments and sulfur crystals in the intercellular spaces. Sulfides are absorbed through the foot, whereas oxygen is stored and transported by haemoglobins in the blood. Various measurements of shells indicate a growth rate varying between 0.5 and 6 cm/year, with a maximum size reached after 20 years. These growth rates are not exceptional but are generally higher than those measured in coastal species. Unlike in *Riftia* tubeworms, the transmission of symbionts occurs during reproduction inside the maternal oocyte.

6.5.1.3. *Bathymodiolus thermophilus*

Bathymodiolus thermophilus mussels are found in the Eastern Pacific and belong to *Mytilidae*. These true mussels aggregate on the basalts and also on each other using their byssus. They reach a maximum size of approximately 20 cm. They are located in low diffusion zone areas and they can even be observed in places where no hydrothermal activity is present, highlighting the high ecological plasticity of this species. Mussels contribute to the horizontal dispersion of fluid, which diffuses through cracks over several meters, allowing a lateral extension of their habitat [JOH 94]. They have a functional gastrointestinal tract, virtually similar to that of their coastal counterparts. The labial palps are used to concentrate the particles collected by their syphons, and their ciliated esophagus allows transport toward the stomach. The animal has the ability to very efficiently filter microorganisms from seawater. However, even if the gut is functional, the gill is thicker and four or five times larger than in other *Mytilidae*. Histological studies of the gill lamellae reveal that their lateral sides are mainly occupied by bacteriocytes which have large vacuoles full of chemosynthetic bacteria [CHI 92]. Elemental sulfur and lysosomes, which are membrane sacs containing the enzymes necessary to digest bacteria, are observed in these bacteriocytes. The mussels do not have protein molecules in their blood that transport oxygen or sulfides and therefore depend on the diffusion of gases from the environment into their palleal liquid. The toxicity of sulfides is controlled by their rapid oxidation into thiosulphates, which will be used in chemosynthesis. Due to the flexibility of their diets, from symbiosis to filtration, mussels are formidable competitors of other hydrothermal species.

6.5.1.4. *Alvinella pompejana*

Pompeii worms, colonizing the active chimneys of the EPR, differ completely in their morphologies from everything that was previously seen in polychaete annelids. Although they share some characters similar with *Ampharetidae*, it was soon clear that a new family, *Alvinellidae*, needed to be created among the *Terebellomorpha* [DES 98]. Unlike other families in this group, there is no clear separation between the thorax and abdomen, and a complex anchoring system is present on anterior segments. Two different species, *Alvinella pompejana* and *A. caudata*, were observed during the first dives at vents, both being extracted from organic tubes forming an intricate network on the walls of active chimneys. Their maximum sizes are approximately 150 mm. They appear to withstand high temperatures for metazoan organisms. Their circulatory system and pigments are well adapted to life in hypoxic environments with the presence of a very large gill surface area and circulating haemoglobin with a very high oxygen affinity and a coelomic liquid rich in red blood cells in order to withstand the internal extreme conditions of the tube. These worms have lamellar gills, a functional gastrointestinal tract and retractile tentacles in the oral cavity as in

Ampharetidae. However, the most surprising character of these animals is the presence of abundant white filaments, which are located in the intersegmental areas and form a very abundant dorsal felting. These are actually abundant colonies of filamentous bacteria embedded in the epidermal secretions of the worm. The inner tube is covered with large diverse microbial populations. Bacterial epibioses are common in marine invertebrates, but they are rarely present on all harvested individuals and are not systematically associated with specialized morphological structures. However, if the epibiont population in the worm is dominated by four strains of filamentous bacteria, it is clear that it must be much more diverse and function in a more complex way. Genetic analyses revealed the existence of a filamentous enzyme, ATP citrate-lyase, characteristic of the rTCA cycle for CO_2 fixation. Two non-exclusive hypotheses have been proposed to explain this epibiosis. The first hypothesis states that there is a complex feeding relationship, by enrichment of the internal environment of the tube into organic matter, the worm mainly feeds from bacteria in the tube. This assumption has not been clearly confirmed by labeling experiments on the seabed. The second hypothesis involves bacteria in the detoxification process inside the tube. Such an association is observed, for example in rice where commensal bacteria on the roots transform sulfides into less toxic sulfites. We must await the results of ongoing experiments to validate these hypotheses. Pompeii worms colonize one of the most hostile environments on our planet, which is also a microbial habitat where primary production is significant.

6.5.1.5. *Rimicaris exoculata*

The deep-sea shrimp *Rimicaris exoculata* colonizes all known active sites in the MAR. They can form dense swarms of thousands of individuals close to the emission zones. The digestive contents of *A. exoculata* are composed of mineral particles from chimney surface and the isotope ratios of muscle tissue indicate that they feed on organic chemosynthetic material. Inside their dilated branchial chamber, some oral appendages are enlarged with long lamellae that beat in the thoracic cavity to ventilate the gills. This appendix has long setae on which filamentous bacteria grow, they are similar to those in *Alvinella pompejana*. Further research into microbial populations reveals that other microbial strains are present in the communities of the gill chamber. Experience in *in situ* simulated conditions [PON 13] shows that commensal bacteria in the gill chamber exhibit a chemoautotrophic metabolism and that nutritional transfers between these bacteria and the shrimp probably occur throughout the teguments. Research into genes involved in the carbon cycle

[HÜE 11] reveals the existence of a complex community of microorganisms that interact in this symbiosis. Two groups of dominant filamentous bacteria belong to sulfur-oxidizing bacteria: the ε-proteobacteria using the rTCA pathway and γ-proteobacteria using the Calvin–Benson pathway for autotrophic synthesis. δ-proteobacteria, also present, may reduce sulfur compounds (sulfates, elemental sulfur and thiosulfate) into sulfides, thus making it available for the other two groups of bacteria.

6.5.2. *Trophic network*

Apart from the presence of these large symbiotic species, forming the majority of the biomass, the trophic network of hydrothermal ecosystems is similar to those observed in other marine ecosystems. In addition to primary autotrophic producers (here, microorganisms), there are several primary and secondary heterotrophic consumers which include suspension feeders, grazers, deposivores, predators and necrophages. Besides endosymbioses and epibioses, several primary production compartments are present either in suspension in the water, fixed on different surfaces, living in the vent plume and that derived from surface processes [VAN 00] (see Figure 6.6).

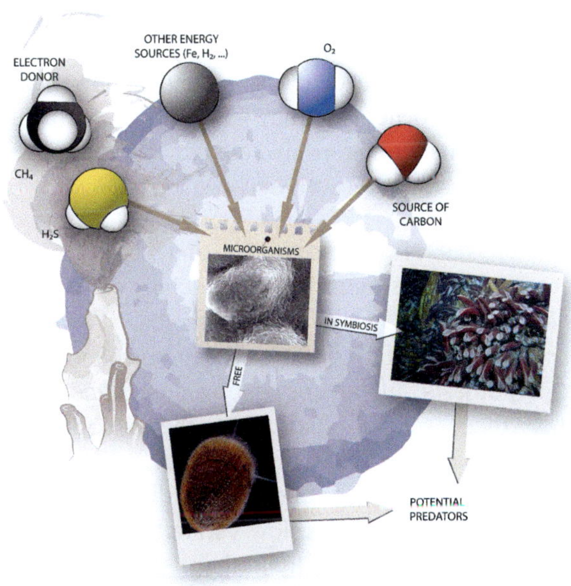

Figure 6.6. *Simplification of the hydrothermal trophic network (Capsule Graphik 2014)*

Given the large animal biomass, the microbial production at hydrothermal vents is undoubtedly very large. Some sites, such as the Southern Trough Hydrothermal Field in the Guaymas Basin, are covered with dense microbial mats that alone form a microhabitat. These mats are also present, but to a lesser extent, on hydrothermal organisms. Microbial production in suspension can support large populations of filter feeding organisms such as barnacles, serpulids and anemones. This is the case on the SEPR ridge where barnacle fields can cover large areas or in Trans-Atlantic Geotraverse (TAG) and Ashadze sites on the MAR, which are surrounded by dense anemone fields.

The first model of hydrothermal food chain was proposed by Hessler and Smithey [HES 83] for the Rose Garden site on the GSC. There, primary producers include symbionts and free-living microorganisms. The primary consumers consist of engineer taxa living in symbioses: mytilids, vesicomyids and siboglinids. These hosts would be consumed, depending on the taxa, by carnivorous gastropods, crabs, shrimp and squat lobsters. However, the free-living microorganisms would be consumed by grazing limpets, polychaetes and small crustaceans (amphipods and leptostraca) or filtered by serpulids, anemones, siphonophores, enteropneusta and small crustaceans. A parasitic or commensal polynoid worm lives in the mantle of mussels and other parasites, particularly copepods, were identified. Finally, deep-sea predators, such as octopus and fish, take advantage of the abundant organic matter near active sites. This first model has benefited from new methods of analysis that have helped consolidate certain assumptions and also review some interactions in the food web.

The isotopic analyses of carbon ($\delta^{13}C$) and ($\delta^{15}N$) provide important information on carbon sources and the trophic levels of species, allowing the reconstruction of the food web. Initial studies showed that $\delta^{13}C$ in hydrothermal species was low compared to surrounding abyssal taxa, supporting the hypothesis of an organic autochthonous carbon source [RAU 79]. Since then, several studies on the isotopic composition of hydrothermal species have revealed that despite the low biodiversity, the structure of the food web may be complex. Generalist species (high variability in $\delta^{13}C$) coexist with specialist species (low variability in $\delta^{13}C$), supporting the hypothesis of resource partitioning in hydrothermal habitats [LEV 06]. These analyses also showed that the organisms with endosymbiotic bacteria were almost not consumed by other species, playing a less significant nutritional role than initially predicted [VAN 94]. This mass symbiotic production is likely reused by numerous scavengers in this environment. It seems that the hydrothermal food chain is based more on organic carbon fixed by free-living microorganisms, which are more accessible. The dominance of deposivore species compared to suspensivores suggests that attached microorganisms play a leading role in the nutrition of hydrothermal fauna. Research on the MAR revealed the presence of two independent trophic groups: one consisting of endosymbiotic mussels *Bathymodiolus azoricus* and their associated polychaetes (*Branchipolynoe seepensis*) and the second grouping the heterotrophic fauna [DEB 09]. The study also showed that the chemical composition of hydrothermal fluids influences the structure of the microbial community, therefore, affecting the carbon sources available. Abiotic conditions even cause changes in the proportion of symbionts in mussels [HAL 08].

Although still rarely used in these ecosystems, the analysis of fatty acids, which are specific biomarkers, can be useful to refine the results obtain through isotopic analyses. In addition, they help to differentiate the contributions from different microorganisms [COL 07]. In the hydrothermal environment, these biomarkers were used to measure the contribution of thiotrophic and methanotrophic bacteria in mussels with double symbioses [PON 98] and to assess the role of phytodetritus. The presence of lipid biomarkers of phytoplankton was detected, but in very low abundances among *Bathymodiolus thermophilus*. However, experiments have shown that the carbon source was insufficient to maintain good physiological conditions. Finally, these biomarkers have shown the low contribution of carbon from photosynthesis in the hydrothermal trophic network.

The role of hydrothermal production on surrounding ecosystems is difficult to quantify. The material from chemosynthetic sources is exported by non-native predators of these ecosystems [MIC 02] and by the hydrothermal plume that can carry on fluids over long distances (>km). Abundant communities of crustaceans (copepods and amphipods) have been observed in the water column above hydrothermal diffuse zones. Only 3% of the total carbon that flows into deep benthic ecosystems would be derived from chemosynthetic production [JAN 95].

The characterization of the trophic network of hydrothermal ecosystems is still incomplete. Some compartments such as those of meiofauna and microorganisms are often absent from trophic studies. They do, however, represent important links toward higher trophic levels and significantly contribute to the complexity of the hydrothermal trophic chain. The functional role of different microbial communities in the transformation and degradation of organic matter remains to be defined.

6.6. Distribution of fauna at different spatial scales

6.6.1. *Ridges of the East Pacific*

Typical communities in the EPR and GSC, dominated by the *Siboglinidae* tubeworms and *Alvinellidae*, Bathymodiolin mussels and giant clams, range from the mouth of the Gulf of California to the Easter Island microplate with slight variations in composition, due to the temporal succession of populations and the isolation effect caused by transform faults. The dispersion of species, which occurs little-by-little along a linear ridge, is in fact interrupted by large breaks (faults or continental plates). More than 15 hydrothermal fields have been located and visited between 21°N and 23°S. Despite their large latitudinal range, their depths vary only slightly, between 2,500 and 2,800 m. The high regional diversity found on the EPR is linked to several factors including its geological age, high spreading rates, a high frequency of natural disturbance and the presence of large diffuse flow zones on the basaltic seafloor. One study shows that the community structure of *Bathymodiolus thermophilus* is similar between two fields separated by 1 km [VAN 02]; these two fields sharing more than 75% of their species. However, also with a high diversity, the structure of communities at a third site, situated 130 km away, showed little similarity with these other sites; this difference is not due to distance but rather due to the stage of activity of this senescent site.

On the same field, species distribution appears to be linked to the heterogeneity of the habitat. Active structures and diffuse emission zones are often colonized by different types of assemblages. A high level of similarity between fauna was observed between communities present in the same type of habitat on the EPR [GOV 05], the Juan de Fuca ridge (NEPR, [TSU 01]) and the MAR [VAN 00]. In the Guaymas Basin, four faunal assemblages were distinguished: two assemblages were associated with soft substrata (microbial mats and vesicomyids) and two with hard substrata on active structures (siboglinids and alvinellids). Many similarities were recorded between vent and seep common habitats in this region, suggesting a greater effect of the physico-chemical conditions in the absence of biogeographical barriers [POR 15].

6.6.2. *Ridges of the Northeast Pacific*

After 1977, other oceanic ridges were rapidly explored by geologists and biologists giving a more comprehensive view of the biogeography of hydrothermal ecosystems. The exploration of the Juan de Fuca in 1983, and that of the Explorer and Gorda ridges, revealed that the hydrothermal fauna was not cosmopolitan [TUN 85]. These Northeast Pacific ridges[5] have been isolated from the EPR by the American continental plate for 28 million years, i.e. at the end of the Oligocene [TUN 96]. Consequently, few species are present on both sides of this plate, but many zoological and functional similarities are observed between the populations. Although few species are common to both (less than five), northern and southern populations share more than 20 genera.

The fauna of the NEPR is dominated by the tubeworm *Ridgeia piscesae* (siboglinidae). It forms extremely dense populations in diffuse emission zones [SAR 99c]. This species has distinct morphotypes, which vary according to the physico-chemical conditions of its habitat [URC 03]. The edifices are colonized by a variety of assemblages not found in basaltic microhabitats. Therefore, the high abundance of *Paralvinella palmiformis* is associated with the presence of *Ridgeia* on edifices. Another species, *Paralvinella sulfincola*, colonizes the walls of the newly formed chimney. By creating "isolating" marcasite crusts, this polychaete facilitates the implantation of palm worms (*P. palmiformis*) on hot surfaces. Patelliform gastropods form gastropods of the genus *Lepetodrilus* form populations that can reach densities of more than 400,000 individuals/m^2. Scale worms (polychaete annelids *Polynoidae*), small predators and grazers of microbial mats are also found in large numbers. They are represented by several species similar to those on the EPR and belong to the genera *Branchinotogluma* and *Lepidonotopodium*. Large populations of microphagic pycnogonids are also associated with *Siboglinidae*. There are no large specialized carnivores but a species of spider crab *Macroregonia macrochira* has been observed within and outside hydrothermal sites as well as some *Zoarcidae* fish. Considered as being less extreme and more stable than sulphide edifice microhabitats, the diffuse flow areas, widespread over the basaltic seafloor, support a high species diversity [URC 03, BER 07]. A dozen active sites have been located between 41° and 49° N on the Explorer, Juan de Fuca and Gorda ridges, at depths varying between 1,500 and 3,200 m.

5 NEPR: North East Pacific Ridge.

6.6.3. *Mid-Atlantic Ridge*

The MAR is a slow-spreading ridge (1–3 cm/yr), with a pronounced axial valley. To the north, the Atlantic Ocean opened up during the Jurassic as the Tethys Ocean, then in the South, it widened into the Indian Ocean during the Cretaceous. The depth of the MAR varies greatly according to the proximity to hot spots. Emerging at Iceland, it exceeds a depth of 4,000 m in its tropical zone. The depth and composition of underlying rocks (basaltic or ultrabasic) have a significant influence on the composition of fluids and microbial populations.

In 1985, a camera towed from a surface vessel identified black smokers at a depth of 3,600 m at 26°N. Submersible campaigns carried out later identified at this site (named TAG) and at 23°N (Snake Pit), at 3,480 m, a new type of hydrothermal population, very different to those in the Pacific and dominated by swarms of an unusual shrimp known as the "blind shrimp" and associated with a new species *Rimicaris exoculata*. In the adult stage, this shrimp has no eyes, but instead an unusual dorsal "eye", composed of a smooth cornea and a bilobed photosensitive and hypertrophied retina part. It has been suggested that this organ could be used to detect any possible infrared light emitted from hot sources [VAN 89]. Swarms of *R. exoculata* continually move on the walls of hydrothermal chimneys, and bathe in a diluted fluid with a temperature of less than 30°C. Two other species *Chorocaris chacei* and *Mirocaris fortunata* belonging to the same family (*Alvinocarididae*) are also present at these sites but do not have comparable morphological adaptations, since they are associated with cooler habitats. Mussel assemblages, dominated by *Bathymodiolus puteoserpentis*, grow on hydrothermal edifices. Similar to their cousin on the GSC and EPR, they host the commensal worm *Branchipolynoe seepensis* also described at cold seeps in Florida. Studying the gill of these mussels reveals that they harbor a dual bacterial symbiosis that involves both sulfo-oxidizing bacteria and methanotrophic bacteria found in the same cells in the gill filaments [DUP 09]. The existence of these two symbionts increases the trophic plasticity of this animal. At the edge of these sites, there are filter-feeding and carnivorous species such as sea anemones, abundant populations of the worms *Chaetopteridae* and *Ampharetidae* and grazers of microbial mats such as brittle stars and some necrophages such as the gastropods *Turridae*. Crabs are represented by a single species, *Segonzacia mesatlantica*, much smaller than its Pacific counterparts (40 mm). One fish *Zoarcidae*, *Pachycarathermophilum*, is highly abundant at some sites. Since the first discoveries, many hydrothermal fields were located along the ridge from the north of the Azores to the south of the Equator. The depth of these sites varies from 800 m (Menez Gwen field close to the Azores) to 4,100 m (the Ashadze field situated at 12°58'N).

Hydrothermal sites on the MAR are characterized by differences in depth and geological context, which influence the composition of communities. A continuum ranging from assemblages dominated by shrimps *Rimicaris exoculata* to assemblages dominated by *Bathymodiolus* mussels is present along this ridge. The first assemblages are associated with higher concentrations of metals and chlorinity, whereas the second assemblage is associated with a less extreme environment. Two species of mussels are present along the MAR: *Bathymodiolus azoricus* at shallow sites (< 2,500 m, Menez Gwen, Lucky Strike, Rainbow) and *Bathymodiolus puteoserpensis* at deeper sites (> 3,000 m, Snake Pit, Broken Spur and Logatchev). They are, however, absent from the sites TAG and Ashadze-1.

The hydrothermal site Ashadze-1 can be distinguished from the others by the presence of a high abundance of anemones *Maractis rimicarivora* on hydrothermal edifices, near to diffuse emission zones [FAB 11]. Anemones are usually found on the edge of active hydrothermal sites. Low densities of shrimp *Rimicaris exoculata* may be attributed to the absence of hot emission zones at this site. However, polychaete worms of the family *Chaetopteridae* and necrophage species are very abundant. These observations together with the almost complete absence of engineer species living in symbiosis (with the exception of a few *R. exoculata*) suggest that this site could be in decline, thereby favoring species usually less common on active sites.

Variations in community structure observed along the depth gradient in the MAR may be associated with changes in the toxicity of fluids as well as the abundance of suspended particles [DES 00, DES 01]. In addition, the greater dispersion of non-hydrothermal species at the shallower Menez Gwen site is attributed to the presence of a lower hydrostatic pressure. The low similarity between the different shallow fields on the MAR (< 30% of species shared), and despite the absence of geographic discontinuity, suggests the presence of islands of fauna that have different physico-chemical requirements.

6.6.4. *Sites in the West Pacific*

To the west of the Pacific Ocean, there is a series of front and back-arc basins and active, underwater volcanoes: from north to south, the Okinawa Basin, Bonin ridge (Izu-Ogasawara), Mariana Basin, Manus Basin, Woodlarck Basin, North Fidji Basin, Lau Basin, Havre Basin and the Kermadec volcanic arc. These basins formed over several million years sun-like large oceanic ridges. The Okinawa Basin opened upon the Chinese ridge after 5 million years, the North Fidji Basin after 8 million years and the Lau Basin after 3 million years. These discontinuous structures, which extend from Japan to New Zealand, are separated by shelves and have different biogeographical entities.

The exploration of these basins began in 1986 with a photographic study of the Manus Basin. The images revealed not only active hydrothermal chimneys, but also biological populations dominated by large gastropods. The first dives took place in the Alvin submersible in the Mariana Basin on the axial ridge at a depth of 3,600 m. They revealed an original fauna, different from that of the East Pacific. A new large gastropod *Alviniconcha hessleri* was found to dominate communities concentrated around vents [HES 91]. In this species, the gills can represent up to 40% of the body volume, the circulatory system is hypertrophic and the gut is unusually reduced compared with other species of the same family. Sulfo-oxidizing bacterial symbionts are present in the gills [STE 88]. Isotopic ratios reveal that *A. Hessleri* primarily subsists on symbiosis. The worm *Paralvinella hessleri* colonizes the walls of smokers and forms tangled of encrusted tubes. At the site periphery, there are mats of anemones *Marianactis bythios*, undoubtedly a carnivorous species, abundant populations of primitive barnacles *Neoverruca brachylepadoformis* as well as large populations of limpets *Shinkailepa ssp.* Around these sessile populations and on the basaltic seafloor, swarms of the shrimp *Chorocaris vandoverae* are present. The crab *Austinograea williamsi* and scale worms form the necrophage/carnivore level.

From this model, many variations along the accretion centers of the West Pacific can be observed. The zone currently with the most data is the Lau Basin and North Fidji Basin. These two basins are situated at the boundary between the Pacific and Indo-Australian plates. According to some authors [AUZ 88], these plates formed from the separation of a single ridge between 0.8 and 10 million years ago. The first explorations carried out by French-Japanese teams [DES 94] revealed populations dominated by two species of large gastropods. *Alviniconcha sp.*, genetically similar to those found in the Mariana Basin, form large populations that are distributed close to vents. They are surrounded by populations of *Ifremeria nautilei.* As with *Alviniconcha*, the gills and circulatory system are hypertrophic and the digestive tract is reduced. Two types of symbiotic bacteria were observed in the gills, dominant sulfur-oxidizing bacteria, and those that oxidize methane [BOR 02]. Two bivalve species *Bathymodiolus brevior* and *B. elongatus* are present (*B. brevior* only in the Lau Basin), and these populations are situated on the external margins of the vent community. *Alviniconcha sp.* has a very high growth rate and carbon fixation rate is also one of the greatest measured to date. In *Ifremeria* and *Bathymodiolus* [HEN 08], the carbon fixation rate is an order of magnitude less than that of *Alviniconcha*. Populations of filter-feeders such as barnacles, brisingid starfish and carnivorous sponges were also observed at the edge of these sites.

Here, it is difficult to detail all the differences between the fauna in the different basins of the West Pacific (see [DES 06b]). The Okinawa Basin contains soft and rocky substrata at sites situated at depths varying from 640 to 1,500 m. On hard substrata, mussels dominate (*Bathymodiolus platifrons* and *B. japonicus*) and several species of *Siboglinidae* tubeworms belonging to the genera *Lamellibrachia* and *Alaysia* form populations at the base of chimneys where the worm *Paralvinella hessleri* extends its tubes. Large colonies of the squat lobster *Shinkaia crosnieri* are close to vents in densities exceeding 550 individuals/m^2. Stable isotope studies have shown that this animal feeds off chemoautotrophic filamentous bacteria either from epibionts or microbial mats. Filter feeders on the periphery include diverse species of barnacles. In sedimentary environments, three burrowing species of clam *Calyptogena solidissima*, *C. okutanii* and *C. nankaiensis* dominate the community.

The system of the volcanic arcs Izu-Ogasawara (Bonin) and Mariana is a convergent margin that extends from Tokyo to the Guam island. Many underwater volcanoes are distributed along this system of arcs. The depths, chemistry of fluids and substrata are very heterogeneous at these sites that extend from the photic zone to a depth of 1,500 m. In shallower zones, corals and hydrothermal fauna intermingle. The fauna is dominated by mussels *Bathymodiolus septemdierum* and *Gigantidas horikoshii* associated with the scale worm *Branchipolynoe pettibonae*. The flatfish *Symphurus orientalis* is present in large populations and many species of barnacles (*Neoverruca*) colonize the periphery.

Some populations of *Siboglinidae* and the species *Lamellibrachia satsuma* are found on the volcano Nikko at depths varying from 430 to 600 m. This species is also present in the Kagoshima bay at a depth of around 100 m. Further south, between the Lau Basin and New Zealand, along the Havre Basin emerge a series of active volcanoes (Kermarec arc) whose summits vary from 300 to 1,200 m high. Research on hydrothermal plumes indicates that seven of the 12 volcanoes studied are likely to be active. The populations appear to be dominated by a species of mussel *Gigantigas gladius* whose shell may exceed 30 cm in length.

6.6.5. *Ridges in the Indian Ocean*

Due to their isolation, ridges in the Indian Ocean have barely been explored. During the summer of 2000, the Japanese Kairei vessel deployed the ROV[6] Kaiko on a segment of the Central Indian Ridge[7] situated to the north of the Rodriguez Triple Junction (25°39' S). Two active hydrothermal fields, Kairei and Edmond, separated by 160 km, were discovered. The hydrothermal field Kairei (2,415–2,460 m) is located at 25°19' S, 70°02' E. At least seven active sites including black smokers have been observed, with edifices reaching up 10 m in height. The maximum temperature measured was 360°C, with a pH of 3.4. The hydrothermal field Edmond (23°52' S, 69°35' E; 3,290–3,320 m) was discovered at approximately 160 km from the field Kairei. Hydrothermal fluids (273–382°C) form four sites on the Edmond field. These exhibit a higher chlorinity than that of ambient seawater, making them the hottest brines collected on the ridges. The invertebrate community on the Kairei field is characterized by central populations on black smokers, dominated by dense swarms of shrimp (*Rimicaris kairei*) that colonize fluids at 10–20°C and a cold peripheral zone (1–2°C) dominated by anemones (*Marianacti ssp.*). In intermediate zones, at the base of the shrimp swarms, assemblages of several hundred mussels (*Bathymodiolus aff. brevior*) and gastropods (*Alviniconcha n. sp.* and *scaly foot* vetigastropods) are found. Brachyuran crabs (*Austinograea rodriguezensis*), flat worms of the class turbellarians, nemertean worms and a second species of shrimp (*Chorocaris n. sp*) are often abundant in the transition zone between the anemones and shrimp. Other gastropods (limpets and provannids), polynoid worms (*Branchinotogluma n. sp.*) and

6 ROV: *remotely operated vehicle.*

7 CIR: *Central Indian Ridge.*

archinomids (*Archinome sp.*) as well as populations of cirriped crustaceans of the genus *Neolepas* are also present in this intermediate zone. Vesicomyid shells were collected 1 km from the Kairei field but no living individuals were found so far [HAS 01].

The fauna on the Edmond field is a subgroup of those found in the Kairei field. The biomass is dominated by dense swarms of shrimp *R. kairei* and the peripheral zones are colonized by anemones. Cirriped crustaceans, *scaly foot* gastropods, archinomid polychaetes, nemertians and turbellarians are absent.

The *scaly foot* vetigastropod Peltospiridae, still being described, has many interesting anatomical adaptations [WAR 03] including a reduced or absent opercula and the presence of a foot covered with sclerites. The esophageal gland is between 10 and 1,000 times larger than in other gastropods, whereas the rest of the gut is 10–100 times smaller; the circulatory system is hypertrophic to supply the gills and the esophageal gland and the radula is much smaller. Several studies [GOF 04] reveal that γ-proteobacteria colonize cells of the esophageal gland and their genomes contain all the functions of chemoautotrophic symbionts of hydrothermal organisms. Studying the ratio of natural carbon isotopes confirmed that autotrophy is present in these symbionts.

A Japanese campaign [NAK 12] found two other sites situated between 18 and 20 S on the Central Indian Ridge at a depth of 2,600–2,750 m. The two sites called Dodo and Solitaire have a similar fauna to that described on the Kairei field. Dense populations of shrimp *Rimicaris kairei* and *Mirocaris indica* were observed on active structures. Zonation patterns were observed with populations of worms as well as *Alviniconcha sp.* and *scaly foot* vetigastropods distributed along hydrothermal diffuse emission zones. *Bathymodiolus* mussels with large populations of limpets *Lepetodrilidae* are found further from the chimneys, and at the periphery, *Phymorhynchus* necrophages and at least two species of filter-feeding cirripeds *Neolepas* and *Eochionelasmus* were observed. *A. rodriguezensis* and a new species of shrimp *Chorocaris* were present close to the gastropods and mussels, whereas only some individuals of *R. Kairei* were observed at this site. Macrourid fish, *Marianactis* anemones, *Munidopsis* squat lobsters and holothurians are found outside the zones colonized by hydrothermal fauna. One remarkable discovery is the presence of alvinellid polychaetes. This family was never recorded outside the Pacific Ocean and its presence reinforces the proposed biogeographical link between the fauna in the west Pacific and that in the Indian Ocean. In January 2011, an expedition by RRS

James Cook located four hydrothermal fields on the ultra-slow South West Indian Ridge[8] at 37°47'S–38°28'S at a depth of 2,750–2,840 m[9]. The faunal composition appears quite similar to those at other Indian ridge sites, with the exception of the presence of populations of the *Kiwaidae* "yeti crab", *Kiwan. sp.*, which are very different from species in the West Pacific [ROT 13] as well as white holothuria on the periphery of mussel beds. This suggests that the characteristics of hydrothermal fluids (and therefore their quality and availability in microhabitats) rather than geographic affinities would play a predominant role in the structure of these communities.

6.6.6. *Ridges in the Southern Ocean*

With the development of research ROVs, the exploration of ridges has reached Polar Regions, which up until now have been infrequently visited. Active sites were located in the Arctic Ocean at depths of 700–750 m to the north of the Jan Mayen Island on the Mohn Ridge. There populations are dominated by non-specialized bathyal species despite being at similar depths to the Menez Gwen vent field in the Azores where hydrothermal species dominate. However, it appears that bathyal fauna partly subsists of carbon from chemosynthetic origin [SWE 13]. In the Southern Ocean, sites visited on the East Scotia Ridge[10] at a depth of 2,600 m provide new information about the biogeography of this region and its association with other biogeographical provinces [ROG 12]. The Scotia Sea is surrounded by a series of ridges and islands (Scotia arc), and reach depth of approximately 3,000 m. The ESR has a length of approximately 500 km, and its expansion began approximately 15 million years ago. It has an accretion rate of 70 mm/yr. In 2010, the English ROV Isis located two hydrothermal fields, the first at 60°02'S known as *Devil's Punchbowl* (2,400 m) and the second at 56°5'S called *Mermaid's Purse* (2,600 m). The temperatures of the hydrothermal fluids are very high (352–382°C) and diffuse emission zones, where temperatures vary from 3 to 20°C, are present. Ambient temperatures vary from –0.11 to –1.3°C. Although sample studies are still underway, general information about the population distribution and composition is already available [MAR 12].

8 SWIR: South West Indian Ridge.

9 www.thesearethevoyages.net/jc67/nov30.html.

10 ESR: East Scotia Ridge.

The hottest parts of hydrothermal edifices are covered by anhydrite deposits without fauna and only some microbial mats are visible. The adjacent parts are occupied by large populations of the crab *Kiwan. sp.* [ROT 13] whose size decreases as the distance from vents increases, whereas densities increase from 65 to 4,017 individuals/m^2. The "yeti crabs" closest to the vents have filaments of epibiontic bacteria on the ventral side of their carapace, whereas individuals further away from this fluid do not. The only fauna associated with these populations is a new species of limpet belonging to the genus *Lepetodrilus*. At the base of the structures, a large peltospiroid gastropod forms very dense populations around the "yeti crabs" (> 1,000 individuals/m^2). Severals stalked that barnacles *Vulcanolepasn. sp.* are observed around the periphery and reach densities of approximately 900 individuals/m^2 associated with *Lepetodrilusn. sp.* Several species of actinostolid anemones and pycnogonids (*Sericosura* and *Colossendeis*) are also present. Predators are represented by seven arm starfish and an undetermined octopus.

6.6.7. *Cayman Ridge*

The Mid-Cayman Spreading Center[11] (MCSC) is an ultra-slow ridge with a length of approximately 110 km, situated in the Cayman Basin, in the Caribbean Sea. It spreads at a speed of 1.5 mm/yr and has been active for 49 million years. This is the deepest accretion center known in the world's oceans, the depth of its axis being between 4,200 and 6,000 m. A chemical signal was detected north of the ridge, on a neo-volcanic formation, and active chimneys were located at 18°32.785′N and 81°43.080′W at a depth of 4,960 m. The site was named in honor of William Beebe, the first naturalist to observe a bysso-pelagic fauna *in situ* (*Beebe Vent Field* or BVF). The BVF is a mound of polymetallic sulfides with a diameter of 80 m and height of 50 m on top of which are active chimneys [CON 12]. The fauna is dominated by dense (> 2,000 individuals/m^2) swarms of shrimp *Rimicaris hybisae* [NYE 13] which are found on the walls of the chimneys. These shrimps have the same adaptations and epibioses as *R. exoculata* from the MAR. Populations of undetermined anemones are spread over the peripheral zones of the site.

A second site (Von Damm Vent Field or VDVF) was found and studies were done on the Dent vent, a structure rising 2,700 m above the axis of the ridge at

11 MCSC: Mid-Cayman Spreading Center.

13 km to the west (at a depth of 2,294 m). This site corresponds to a hydrothermal mound with a diameter of 100 m and a height of 30 m. At the summit, hydrothermal vent releases a light fluid with a temperature of around 18°C. *Zoarcidae* fish (see *Pachycara*) as well as squat lobsters have been observed but have not yet been sampled. Beds of *Mytilidae* shells and empty *Siboglinidae* tubes were present at approximately 400 m to the south of the VDVF field, which reveals the presence of recent activity. Some data available for MCSC show that despite the isolation of this zone, it shares many similarities with the fauna found at deep hydrothermal sites at the MAR.

6.6.8. *Biogeography of deep hydrothermal vents*

Hydrothermal communities are composed of a reduced group of species capable of withstanding the harsh environmental conditions and to exploit chemosynthetic microbial production. Most of the species described in this environment have not been reported in the surrounding abyssal area, hence the high level of endemism encountered in these ecosystems. The concept of biogeographic provinces is used to highlight the result of processes such as selection, extinction, dispersion, etc., acting on faunal distribution on long temporal scales. Associations between hydrothermal species that are distributed over linear structures and consisting of endemic taxa would, therefore, be models that are particularly adapted to biogeographic studies, provided that hydrothermal fauna evolved independently of other ecosystems based on microbial chemosynthesis, which is the least controversial hypothesis.

Moreover, for the conclusions to be unequivocal, a set of consistent taxonomic data and sampling that covers all geographic zones must be obtained, which is not yet the case. Zones at high latitudes have been less explored and may provide corridors for the dispersion of many vent taxa. In addition, data from the Indian Ocean are still patchy (including data from isolated basins such as the Andaman basin, still unexplored), those in the North Atlantic are limited to zones between 45°N and 4.48°S with the exception of the recent studies on the Mohn ridge and the ESR. Taxonomic information, whether it is based on morphological and/or genetic information, is still incomplete and heterogeneous. Therefore, the information available on the biogeography of hydrothermal communities is still incomplete to draw an accurate large-scale distribution portrait of vent fauna.

The most recent global study [MOA 12], based on the analysis of faunal distance[12] (Jaccard distances) of a genus/site matrix using the network theory, outlines five robust groups of hydrothermal fields:

– EPR, GSC, PAR[13];

– Northeast Pacific;

– West Pacific basins;

– Indian ridges;

– MAR.

Interestingly, results on the connectivity between provinces reveal the central position of East Pacific basins. The fauna of the MAR has no direct relationship with EPR but is linked to the Pacific via the Indian Ocean. This central position is reinforced if we focus on the main taxa (annelids, mollusks and crustaceans). In their recent publication on ESR sites, Rogers *et al.* [ROG 12] propose the existence of a sixth biogeographical province that groups the Antarctic ridge sites.

6.7. Faunal microdistribution and interactions

A hydrothermal site (also known as a field) is a zone with more or less localized hydrothermal emissions. It includes several active hydrothermal structures and also diffuse emission zones. Fauna assemblages colonize narrow mixing gradients between hydrothermal fluids and seawater, on hard substrataor in thick sediment layers. These organisms subsist at this physico-chemical interface due to their capacity to withstand the harshness of the environment but also to sustain their trophic requirements.. The characterizations of physico-chemical conditions in the habitats of hydrothermal species are still relatively limited, primarily due to the difficulty in carrying out *in situ* measurements close to these organisms. Only some factors such as temperature and oxygen are relatively "easy" to measure in the deep sea.

12 The most recent information for the Indian Ocean (CIR and SWIR), as well as those for ESR, the Mohn ridge and the MCSC,is not taken into account in this study.

13 *Pacific Antarctic Ridge.*

Initial studies outlining the concept of zonation date back to the 1980s. The studies describe that species are distributed around active hydrothermal structures depending on their distances from vent emissions [HES 83]. EPR communities are distributed along an environmental gradient: siboglinid polychaete worms are found in high diffuse flow zones (≤ 25°C), bivalves in moderate flow zones (≤ 10°C), whereas filter-feeding species colonize low flow peripheral zones. A similar zonation is also observed at other sites on the Pacific Ridge and on the MAR, where less tolerant species colonize the periphery, whereas the most active zones are colonized by more tolerant pioneer species. Less extreme zones also provide a refuge for larvae, juveniles as well as spawning females [TYL 99]. A recent observation rather showed that for the shrimp *Rimicaris exoculata*, males instead of females appear to colonize the less active zones.

6.7.1. *Environmental conditions*

Environmental conditions affect the distribution of organisms, influencing the diversity and modifying the structure of communities. A species becomes absent from a zone where the abiotic factors exceed its tolerance limit (e.g. temperature) and from an environment that cannot meet its requirements (e.g. nutrients). Even if all factors, but one, are found within the tolerance limits, the species will be eliminated or absent. Environmental conditions also influence the dispersion, recruitment and fixation of larvae. The frequency and intensity of environmental fluctuations may exceed the tolerance limits, affect recruitment, modify the quantity and quality of food available and even initiate succession. Recent studies on the EPR show that the species distribution at a site is complex and would result from the combination between the temporal variability in larval installation [MUL 98], competition between the first colonizers [MUL 00], predator–prey relationships [MIC 02], physiological limits and nutrient gradients [LEN 08].

6.7.1.1. *Temperature*

Temperature fluctuations at scales varying from seconds to hours and sudden changes in chemical conditions result from the turbulent mixing of hot fluids with the surrounding cold seawater as well as from the influence of biotic and abiotic processes on this mixing. The community structure is greatly influenced by temperature, which, locally, is a good “proxy” for the chemical composition of hydrothermal fluids [SAR 99a].

The hottest zones are exploited by species that can tolerate high temperatures and significant environmental fluctuations, whereas habitats at lower temperatures, characterized by a greater stability, are colonized by less tolerant species. Several studies highlight the reduced diversity at in the hottest areas of the ecosystem. Some species have a broader niche [CUV 11] which reflects their capacity to move between different microhabitats as suggested for the shrimp *Rimicaris exoculata* [RAV 03] or for gastropods in the Lau basin [POD 09].

Several examples support the fact that species colonize zones where temperatures are below their tolerance limits which allow them to survive to significant and sudden fluctuations in environmental conditions [BAT 10]. Thus, the polychaete worm *Paralvinella sulfincola*, which occupies newly formed chimney surfaces, has a much greater thermal preferendum than the temperatures measured in its habitat. In addition, the result of one study showed that the spatial distribution of species in the Lau basin was influenced by different thermal tolerances [POD 09]. The gastropods *Alviniconcha sp.* and *Ifremeria nautilei* are present close to fluid sources, whereas the less tolerant mussel *Bathymodiolus brevior* colonizes peripheral zones.

Physiological adaptations play an important role in faunal distribution patterns [FIS 90]. The distribution of *Riftia pachyptila* and *Calyptogena magnifica* on the EPR would be determined by their physiological tolerance as well as their strict requirements for hydrogen sulfide to sustain their endosymbiotic bacteria [CHI 92]. Furthermore, behavioral adaptations also participate to the adaptation of vent species by allowing organisms to survive in fluctuating environmental conditions. For example, alvinellid polychaetes regulate the temperature in their habitats by moving their gills [CHE 93].

6.7.1.2. *Sulfides*

Hydrogen sulfide (H_2S) is toxic to aerobic organisms as it inhibits mitochondrial respiration and the binding of oxygen to most hemoglobins [FEL 85]. However, the presence of H_2S in fluids is also linked to the availability of food. The fauna associated with these emissions has developed different detoxication mechanisms that allow them to carry out aerobic respiration regardless of whether H_2S is present or not [MAR 97]. Some species with sulfur-oxidizing symbionts such as *Riftia pachyptila* have a particular type of hemoglobin that allows the transport of sulfides to their symbionts by reversibly binding them [ZAL 98]. Furthermore, the presence of many mitochondria and lysosomes in the gills of some polynoid species and alvinellids allow them to withstand high concentrations of sulfide in their environment [HOU 00].

This ability to tolerate H_2S influences the composition and spatial distribution of hydrothermal communities [LUT 01, SAR 99c]. Whereas some species appear to be adapted to habitats rich in H_2S, others are limited to diffuse emission zones. For example, the pioneer species, *Tevnia jerichonana*, on the EPR colonizes habitats with higher concentrations of H_2S than in the case of *Riftia pachyptila*, the species that follows in the proposed biological succession model [LUT 12, SHA 98]. The speciation of sulfides, and therefore their bioavailability, varies significantly from one habitat to another. In the Lau basin, the gastropod *Alviniconcha spp.* colonizes habitats with higher concentrations of H_2S than the other two pioneer species at this site (*Ifremeria nautilei* and *Bathymodiolus brevior* [POD 09]). The presence of high concentrations of iron may reduce the availability of H_2S [LUT 01] and constrain the distribution of some species such as mussels and siboglinids [GOV 05]. pH favors H_2S over HS increasing its ability to diffuse through membranes and therefore, its toxicity [VIS 91].

The distribution of nematodes and copepods, which usually dominate the meiofauna, varies according to temperature and/or concentrations of sulfides and their diversity is higher in zones where sulfide concentrations are lower [GOL 10b]. Dirivultid copepods are present in a wide range of environmental conditions [GOL 10a], whereas tisbids are restricted to stricter conditions [IVA 11]. On the Juan de Fuca Ridge, only the copepod *Stygiopontius quadrispinosus* is associated with the higher temperature environments [TSU 03a]. Moreover, it was recently shown that adult and larval copepods may not be found in the same microhabitats [SAR in press].

6.7.1.3. *Oxygen*

Variations in oxygen are extreme in the hydrothermal environment and organisms colonize the interface between anoxic (hydrothermal fluid) and oxic (seawater) conditions. Some taxa are physiologically well adapted to frequent fluctuations in oxygen [ARN 98] whereas others are less so, hence its influence on species distribution. A study at 9°N (EPR) reveals that oxygen concentrations were less than 120 μM within the habitats and that above 10°C, the fluids were totally anoxic [LUT 12]. Oxygen concentrations in the habitat of *Tevnia jerichonana*, following a series of volcanic eruptions, reveal that this pioneer species can colonize habitats characterized by low oxygen concentrations, while *Riftia pachyptila* and *Bathymodiolus thermophilus* are found in more oxygenated environments. Similarly, the mussel *Bathymodiolus brevior* in the Lau basin lives in more oxygenated conditions whereas species associated with warmer zones, *Alviniconcha spp.* and *Ifremeria nautilei*, colonize less oxygenated zones [LUT 12]. Several physiological adaptations increase the capacity of species to withstand low concentrations of oxygen [CHA 04, HOU 07].

6.7.1.4. *Metals and radioactivity*

The hydrothermal environment harbors natural concentrations of heavy metals and toxic radioelements, which may influence species distribution. The accumulation of metals by organisms depends on their bioavailability in surrounding waters and the capacity of organisms to manage the presence of these metals. Therefore, some species appear to have developed a high tolerance to high metal concentrations and possess high concentrations in their tissues [COS 08]. With regard to radioelements, the results show different trends depending on the site, position of organisms in the mixing zone and the trophic regime of the species. Significant accumulation, much higher than in species from other marine habitats, can be found in their tissues [CHA 09]. The tolerance of hydrothermal species to these abiotic constraints probably has a role to play in their realized spatial distribution.

6.7.1.5. *Hydrothermal fluid flow*

Fluid flow is a complex variable that integrates variations in temperature, availability of hydrogen sulfide and other chemical elements and particles. Changes in fluid flow can also affect the overall metabolism of species, as well as growth, reproduction and colonization processes. Fluid flow also affects the availability of food resources by altering the input of particulate matter and the uptake of chemicals by microorganisms. The lower abundance of siboglinid communities south of the EPR (13°N) was correlated with lower hydrothermal activity [FUS 87] whereas on the Juan de Fuca Ridge (NEPR), an increase in activity caused visible changes in the distribution of fauna [SAR 97]. The development of dense aggregations of shrimp on the MAR was associated with the presence of high emission zones [COP 97] and the distribution of fauna assemblages on the Eiffel Tower edifice (MAR) is associated with the proximity to fluid emissions

[CUV 09b]. On the Juan de Fuca ridge, some species of copepods are dominant in high flow zones, whereas low abundances are found in senescent zones [TSU 03a]. At the community level, high flow zones are more associated with low diversity, whereas low flow zones, which are less restrictive, are home to the highest diversity [SAR 97, SAR in press].

Results of a study carried out by Kelly *et al.* [KEL 07] on the Juan de Fuca ridge suggest that the properties of fluids influence the installation and colonization of hydrothermal species. The dispersion of larvae over great distances despite their reduced swimming capacity questions the role of hydrothermal emissions and hydrodynamic processes on their distribution [KIM 98]. Changes in fluid flow may even initiate species succession in some hydrothermal communities [SAR 97, SHA 98]. While species succession in diffuse flow areas is relatively easy to predict, the growth of hydrothermal communities after reactivation of a site is still poorly understood. An increase in fluid flow may cause a variety of different assemblages, reflecting the patchiness of emissions across a site [LEB 06].

6.7.1.6. *pH*

pH influences calcification/decalcification processes in marine organisms and has severe consequences on metabolism and reproduction. It also influences the acquisition of nutrients, speciation and the bioavailability of metals. Changes in pH can have significant impacts on metabolism and a too acidic pH can even be lethal to most organisms. Hydrothermal organisms in warmer zones may experience acidic pH, as in the case of *Alvinella pompejana* (pH of approximately 4) on the SEPR. Few studies have focused on the direct effects of pH on the distribution of hydrothermal fauna since this variable is correlated to most other environmental variables. pH significantly influences the distribution of some taxonomic groups, such as copepods [IVA 11] and nematodes [SAR in press], and a linear relationship between pH and diversity has been observed in meiofauna on the EPR [GOL 10b]: a higher pH favors diversity.

6.7.1.7. *Hydrodynamic and tidal processes*

Variations in hydrodynamic processes affect the structure of benthic communities at different spatial and temporal scales. On a small spatial scale, these processes have an influence on competition and predation rates, the transport and deposition of particles and larvae as well as oxygenation. Ocean currents play an important role in the dispersion of pelagic larvae and juveniles [MUL 95] as well as nutrients. Cuvelier *et al.* [CUV 11] suggest that the distribution of small mussels on the Eiffel Tower edifice (MAR) was significantly influenced by the orientation of this structure, the side most exposed to currents being colonized by a greater abundance of small individuals. Thus, the presence of relief generates currents that create microhabitats and increase the spatial heterogeneity [KIM 98] favoring the installation of filter-feeding communities. The role that the currents play on the structure of hydrothermal communities is not fully understood. As larvae cannot swim great distances, their dispersion would be insured by deep currents [TYL 99]. In fact, tidal currents and hydrothermal plumes may help transport larvae several kilometers [MUL 95]. Moreover, spatio-temporal variations in currents may influence the recolonization of new sites after an eruption [MUL 10].

The flow of inorganic and organic particles close to hydrothermal vents is an important factor both in terms of toxicity and nutrition. Some benthic organisms are particularly sensitive to water turbidity. High concentrations of particles may have adverse effects by clogging the filtration system and the surfaces involved in gaseous exchanges. Thus, the high particle loads found at the Rainbow hydrothermal site were considered as a factor that may explain the low abundance of Bathymodiolus mussels [DES 00].

Several studies reveal that emissions of hydrothermal fluids are affected by tidal signals [CHE 91, TIV 02]. This suggests an impact on the distribution of fauna by modifying the nature and composition of hydrothermal fluids. Recently, a study on the activity of siboglinids on the Juan de Fuca ridge shows that the rhythm at which they enter and exit their tubes is directly linked with tidal rhythms [CUV 14].

6.7.1.8. *Depth*

The effect of pressure on the spatial distribution of marine communities is significant. Since biochemical reactions are usually sensitive to differences in pressure, animals at great depths have developed metabolic and physiological adaptations to survive [SOM 92]. At hydrothermal vents, the composition of communities varies from one depth to another. For example, on the MAR, *Bathymodiolus* mussel assemblages dominate shallower sites, which are also home to a greater abundance of abyssal species. The deepest sites are rather dominated by shrimp assemblages [DES 01]. Although the influence of photosynthesis is virtually absent in the trophic network of deep hydrothermal environments, its influence may be notable at some shallower sites.

6.7.1.9. *Type of substratum*

The nature, texture, orientation, position, slope and heterogeneity of substrata influence the structure of benthic communities. The succession model proposed by [SAR 97] for edifice communities on the Juan de Fuca ridge takes into account the evolution of the substratum. Therefore, pioneer communities colonize newly formed substrata, characterized by a high permeability and the presence of more elevated fluid emissions, whereas climax communities are found on less porous and more stable substrata. Substrata also play a role in species distribution and density. Thus, the abundance of meiofauna in siboglinid assemblages on the EPR is associated with the amount of sediment trapped in the tubworm bushes [GOL 07]. The distribution of dirivultid copepods would also be influenced by the type of substratum since these organisms are preferentially found on hard ones such as sulfide or basalt edifices [GOL 10a]. However, the diversity of nematodes, associated with *Bathymodiolus* mussel assemblages on the SEPR, is greater in soft substratum communities [ZEK 06].

6.7.1.10. *Nutrients*

The quantity and quality of available food may influence the reproduction and recruitment of benthic invertebrates and modify the composition, density and abundance of communities on hard substrata. This observation is particularly true for hydrothermal ecosystems where high concentrations of food support an exceptionally large biomass for the deep-sea environment [SAR 99b]. Several authors have shown that the physico-chemical conditions of the hydrothermal environment influence the food available to organisms and that the composition of particulate organic matter varies according to time and space [LEV 05]. Nutrients are distributed heterogeneously and biofilms partially cover the different surfaces available. A study along the thermal gradient revealed a higher abundance of microorganisms closer to emissions [SIE 99]. Thus, spatial variations in flow intensity may influence the resources used by the fauna [LIM 07]. This spatialization of resources certainly has an impact on the distribution of vent species. An increase in the density of abyssal megafauna around active sites suggests that their distribution and abundance are strongly linked to the increased availability of food.

6.7.1.11. *Habitat heterogeneity*

Assemblages of siboglinids, alvinellids and bathymodiolin mussels provide a protected habitat for larvae, juveniles and small species. These engineer species create three-dimensional habitats with gradients in environmental conditions, thereby reducing environmental stress and favoring the local and regional diversities [KEL 08]. These structuring species also contribute to the concentration of organic particles [SAR 02]. In siboglinid assemblages of the Pacific, species richness is positively correlated with the total area provided by the tubes of the worms [GOV 05]. On a smaller scale, colonies of cilia and microbial mats also form three-dimensional microstructures that favor the colonization of juveniles and meiofauna in high abundances [KEL 08, KOU 07].

6.7.1.12. *Natural disturbances*

Catastrophic events may severely affect the structure of benthic communities, depending on the severity and frequency of the disturbance. These disturbances may cause partial or total destruction of a community and provide new spaces to be colonized. Major events such as volcanic eruptions and disturbances related to plate tectonics regularly affect hydrothermal ecosystems. On fast-spreading ridges, such as the EPR, significant tectonic and magmatic events occur [HAY 91] whereas on slow-spreading ridges, these events are rarer [DZI 04]. They may completely destroy a habitat or cause the opening or closing of new conduits. On a local scale, the collapse of hydrothermal walls, changes in fluid composition and modifications of fluid flow represent sources of disruption for these communities. These more or less regular disruptions may explain the high productivity and low diversity of hydrothermal ecosystems [KIM 98].

6.7.2. *Biotic interactions*

Benthic species are often restricted to only part of their potential niche due to the constraints imposed by other organisms. Thus, the environmental conditions determine the succession of species capable of colonizing the substratum but their success and the final structure of communities are also governed by biological interactions. The level of species interactions will vary according to the degree of physical stress [CON 75]. Results from a transplant experiment of hydrothermal fauna between active and inactive areas on the EPR reveal that species distribution was not only limited by environmental factors but also by the food available to species in inactive areas and by biotic interactions in active areas [MUL 09]. Competition, predation and grazing are the main biological interactions that control the structure and diversity of benthic communities [GAG 91].

6.7.2.1. *Competition*

Competition for space in the redox transition zone is high due to the dependence of organisms on the resources brought by hydrothermal fluids. The mixing zone, where both hydrogen sulphide and oxygen necessary to the fauna are present, is characterized by high density of fauna favoring competition. A decrease in hydrothermal activity in the GSC caused significant community changes. Bathymodiolin mussels were favored to the detriment of siboglinids probably due to their greater capacity to exploit low sulfide concentrations [HES 88]. Mixotrophy and their ability to move would also give the mussels a competitive advantage against other species since they can adapt, at least temporarily, to disruption in vent fluids and even senescence. This may partly explain their dominance at the end of succession on the EPR.

On the MAR, the greater requirement for chemical compounds of large mussels may explain their migration toward areas of higher fluid flux and the size segregation observed on edifices [COM 98]. The superposition of isotopic signatures of two species of shrimp (*Mirocaris fortunata* and *Chorocaris chacei*) on the Eiffel Tower edifice on the MAR may indicate that they are in competition for nutrients [DEB 09]. Nevertheless, the habitats colonized by *M. fortunata* have greater temperature ranges and iron concentrations than *C. chacei*, which is restricted to habitats with slightly higher temperatures [SAR 15].

On the Juan de Fuca ridge, several accompanying species have similar isotopic signatures, which suggest that they use the same food sources [BER 07]. The success of *Lepetodrilus fucensis* that may be present in extremely high densities near the emissions is probably linked to its capacity to exploit different food sources through different feeding modes. Recent experiments suggest that this species may exert a high pressure on these resources, causing the decrease in abundance of other species [LEL, thesis in progress]. A behavioral study of *Paralvinella sulfincola* reveals that its aggressive behavior ensures that it has privileged access to nutrients [GRE 06] and would explain its dominance in the most extreme colonized areas on the

Juan de Fuca ridge. Other studies show that competition seems to drive less competitive species toward peripheral zones [MAR 02]. Competition may play a role in the exclusion of some gastropod species by leading them toward poorer habitats [BAT 05]. Finally, the distribution of the three dominant species of megafauna in the Lau basin (*Alviniconcha spp.*, *I. nautilei* and *B. brevior*) may be linked to interspecific interactions [POD 09].

The influence of competition on the structure of communities is not limited to adult stages. On the SEPR and MAR, Bathymodiolin mussels appear to prevent recruitment of invertebrates. Thus, an inverse relationship was observed between mussel density and that of larvae of other species [COM 98, LEN 08].

6.7.2.2. *Predation*

Some experiments show that large mobile predators influence the structure of communities on the SEPR by reducing the abundance of limpets through grazing, as well as by eradicating some invertebrate recruits [MIC 02]. Thus, the fish *Thermarces cerberus* may play a key role by feeding on organisms that control the recruitment of several sessile species [SAN 05]. A camera deployed on the axial segment of the Juan de Fuca ridge revealed that predation by fish and polynoids could be the main source of death in siboglinid tubeworms *Ridgeia piscesae* [TUN 90a]. Polynoids dominate the guild of predators by feeding not only siboglinids but also on many bacterivore and detritivore species [BER 07]. On the MAR, isotopic signatures and lipid analyses show that the crab *Segonzacia mesatlantica* and the shrimp *Mirocaris fortunata* would be predators, feeding on bacterivorous species [COL 02]. More recent studies (isotopic analyses and behavioral study) show that, even though they are at the top of the food chain, both crustacean taxa would be scavengers [DEB 09, MAT 15].

6.8. Temporal dynamics of hydrothermal ecosystems

Most temporal studies carried out in the hydrothermal environment are not based on continuous temporal monitoring but on regular, annual or multi-annual visits by manned or unmanned submersibles. Many attempts have been made to deploy deep-sea cameras since the discovery of hydrothermal vents in 1977, but due to the large number of technical failures, very few have succeeded. There are still considerable technological limitations associated with the deployment of cameras for long periods of time in this harsh environment. Nevertheless, since about 10–15 years, recent technological developments have facilitated the deployment of deep-sea observatories and temporal data are starting to accumulate.

Most temporal studies published to date are restricted to a few known sites and are located on the EPR. There are only few studies on the MAR [COP 07, CUV 11] despite several campaigns in this region. Temporal studies may be divided into two categories: those that began during or following an eruption (section 6.8.1) and those that were monitored over several years at active sites (section 6.8.2). The first are all located in the Pacific Ocean due to the frequency of seismic events and volcanic activity on ridges with high accretion rates. In fact, only one major event has been recorded on the MAR for 40 years [DZI 04] compared with several dozen in the Pacific. Finally, the first articles based on data from deep-sea observatories are currently being published, providing a perspective on the temporal variability of these ecosystems and associated fauna [CUV 14, SAR 14, MAT in press].

6.8.1. *Temporal studies associated with an eruption*

The chronology is as follows: the initial faunal communities are buried under basaltic lava and, several years later, the sites are recolonized by new communities. New emission sites are first covered by extensive white microbial mats that cover fresh lava flows [TSU 01]. The surrounding water may be milky-colored due to the presence of microorganism flocculates. Then, mobile species are observed, feeding on this new biological production [DES 98, MAR 09]. Several months later, the microbial cover decreases and hydrothermal emissions concentrate in more restricted zones.

The recovery of a community after a disturbance varies according to the pool of species and depends on the larvae available in the region as well as the size of the zone affected [KIM 96]. In the hydrothermal ecosystem, the recolonization of a new site can be rapid, from several days to several weeks, if the conditions are favorable. However, it takes several years before all species from the initial community recolonize a site. Disturbances caused by an eruption appear to directly affect the "pool" of available larvae and consequently, the structure of new colonizing communities [MUL 10].

6.8.1.1. *Sites at GSC and EPR*

Discovered in 1977 on the Galapagos ridge, the Rose Garden site is the first site to be discovered [LON 77]. In the 1980s, many studies were carried out and in 2002, this site was revisited. The initial flourishing hydrothermal communities were buried under fresh basaltic larva [SHA 98] and a new diffuse emission site, with low temperatures, was discovered approximately 300 m to the northwest. This site was called "Rosebud" and is characterized by the presence of assemblages of small siboglinids (< 6 cm), Bathymodiolin mussels, anemones and vesicomyid bivalves colonizing fissures in the fresh basaltic seafloor.

In April 1991, a series of underwater volcanic eruptions led to the formation of several hydrothermal sites between 9°45' and 9°52' N on the EPR. A large proportion of the ridge summit was covered, wiping out almost all faunal communities present [DES 98]. The sites were buried under fresh lava and new sites, devoid of megafauna, were created. A microbial cover attracted large populations of crabs *Bythograea thermydron* and other mobile species. Nine months later, this microbial cover was greatly reduced and fissures were occupied by dense populations of the tubeworm *Tevnia jerichonana*. After 2.5 years, large colonies of *Riftia pachyptila* were present in all preexisting colonies of *Tevnia*. After 3.5 years, small mytilids *Bathymodiolus thermophilus* were observed close to these assemblages and approximately 1 year later, they were attached to siboglinid tubes [SHA 98]. More than 4 years later, *Bathymodiolus thermophilus* surrounded most living siboglinid assemblages, even replacing some of them. In fact, due to their ability to divert fluids, Bathymodiolin mussels caused a significant decrease in the H_2S available [LUT 08]. In this sequence, *Tevnia jerichonana* would be the pioneer species, gradually replaced by *Riftia pachyptila*, which would then be excluded by *B. thermophilus*. Another large eruption occurred in January 2006, destroying an entire hydrothermal site at 9.5° N. While most of the hydrothermal systems were dominated by Bathymodiolin mussels before the eruption, *Tevnia jerichonana* rapidly colonized the new emission sites, characterized by an increase in temperature and sulfide concentrations [LUT 08].

6.8.1.2. *NEPR sites*

The colonization of new hydrothermal sites was monitored for 2 years following an eruption on the CoAxial segment in 1993 [TUN 97]. The eruption was caused by an intrusion of lava [DZI 07] and characterized by a high abundance of microorganisms from subsurface [JUN 95]. This microbial population seems to have played an important role in the recruitment of macrofauna since sites with high concentrations of microorganisms had a higher density of fauna [TUN 97]. Seven months after the eruption, several species had colonized the active sites. The most abundant species were siboglinids *Ridgeia piscesae*, alvinellids *Paralvinella pandorae* as well as nemerteans. Within 2 years, a third of the regional pool of species was present at the site. This colonization was linked to an increase in temperature and concentrations of H_2S. However, after 2 years, some sites were already in decline. The succession of species at these senescent sites differs from that observed at active sites, since alvinellids disappeared and many predators (crabs and octopus) were observed.

In 1986, a hot water plume rose 800 m above the sea floor and a second mega plume was observed in 1987, indicating a volcanic eruption on the North Cleft segment. The faunal succession patterns were observed between 1988 and 1994 [TSU 01]. In 1988, 2 years after the eruption, large areas of microbial mats, thick clusters of *Ridgeia piscesae*, high densities of *Paralvinella pandorae* and the presence of fissures in the basalt, emitting "microbial flocs", were observed. More than half of the species in the segment were present. From 1988 to 1990, the microbial coverage and extent of colonization by *P. pandorae* decreased, siboglinids increased and the "floc" disappeared. Five years after the eruption, most low temperature sites were gone and, by 1994, they had all disappeared. High temperature emissions, however, remained. The similarity of the fauna between 1988 and 1994 did not appear to be linked to the region, nor to the year but instead to local habitat characteristics (sulfide input, food availability and competition) which would represent key factors structuring these communities.

In 1998, intense earthquakes were detected on the Axial Volcano and a new lava flow, covering the faunal communities in place, was found a few months later. New active sites, where temperatures reached 80°C, were sampled for 4 years after the eruption [MAR 09]. Just after the eruption, the layer of basalt, covered with microbial mats, attracted alvinellids and polynoids. In June 1999, siboglinids and other species began to colonize the site [LEV 02]. Contrary to what had been observed on the CoAxial segment, polychaetes took more than 3 years to colonize all the new active sites despite the presence of a source population [MAR 09]. *Paralvinella pandorae* dominated all siboglinid assemblages in the first year. During the second year, limpets (*Lepetodrilus fucensis*) increased in abundance, exceeding the densities of *P. pandorae* in the third year. Two and half years after the eruption, all new active sites were colonized by siboglinids and over 64% of known species of this volcano were present. The causes of this "delay" in siboglinid colonization are unknown, but the H_2S/heat ratio was higher at the sites before the worms arrived.

6.8.2. *Temporal studies of active sites*

Hydrothermal vents are extremely dynamic, despite the absence of catastrophic events such as volcanic eruptions. The distribution of emission zones at the same site may vary. The geological structures themselves are constantly changing, some grow, others are eroded and the substratum itself is getting modified over time. Finally, the physico-chemical characteristics of the habitats colonized by fauna are also changing, causing variations in the structure of communities.

6.8.2.1. *GSC and EPR sites*

After its first visit in 1979, the GSC was revisited in 1985. During this period, major changes to the fauna were observed, despite the presence of stable emissions [HES 88]. In 1979, siboglinid polychaete worms *Riftia pachyptila* were abundant and dominated almost all active sites, while in 1985 their abundance was reduced to 10%, being replaced by mytilid mussels *Bathymodiolus thermophilus*, which then dominated. The abundance of squat lobsters *Munidopsis subsquamosa* had greatly increased. Even the density of the whelk *Phymorhynchus sp.*, which lives on the periphery and at senescent sites, increased. In 1985, the population of clams *Calyptogena magnifica* increased, and by 1988, it was surrounded by assemblages of mussels, likely due to a decrease in emissions. In fact, clams can insert their feet into cracks to obtain sulfides. A decrease in filter-feeding organisms was also observed. Between 1985 and 1990, the community barely changed, despite an increase in the abundance of fauna at the periphery of the active sites [DES 98].

Between 1984 and 1987, changes in subsurface circulation on the hydrothermal field at 13°N resulted in the cessation of the activity of two active sites and the reactivation of a previously non-active area [DES 98]. In 1984, the temperature of the Genesis edifice was that of seawater and only one assemblage of small Bathymodiolin mussels and a few empty tubes of *Tevnia jerichonana* were present. In 1987, the Genesis site appeared to have reactivated and warm fluids (< 23°C) as well as microbial mats were visible close to emission zones. The fauna was dominated by the crab *Bythograea thermydron* well as large populations of *Tevnia jerichonana*, together with small *Riftia pachyptila* [DES 98]. Three years later, populations of *Tevnia* were replaced by small *Riftia*. Active edifices greater than 5 m in height were created and colonized by large populations of alvinellids dominated by *A. pompejana*. In 1991 and 1992, hydrothermal emissions seem to have stabilized and the communities described remained unchanged. Four years later, several edifices had collapsed, illustrating the short lifetime (~ 5 years) of the active sites in this field.

One site, called "Clam Acres" located at 21°N, was visited in 1979, 1982 and 1990 [DES 98]. Large clusters of *Riftia* were found close to hydrothermal emissions, their tubes densely colonized by limpets. The summit of the edifices was covered by mineralized tubes of *Alvinella pompejana*. *Bathymodiolus* were absent [HES 88]. Peripheral areas were sparsely covered by dead vesicomyid shells. The emissions and distribution of communities appeared to remain constant between 1979 and 1990, indicating a decadal stability.

6.8.2.2. *NEPR sites*

Hydrothermal species of the Axial volcano were distributed along a horizontal gradient from the center to the periphery [MAR 02]. The zonation changes as the distance to the emission sites increases. Along this gradient *Ridgeia piscesae*, *Paralvinella pandorae* and *P. palmiformis* were found near sources and tended to disappear toward peripheral areas, while gastropods (*Lepetodrilus fucensis* and *Depressigyra globulus*) occurred in the same proportions along this gradient. In 1986, a video camera was deployed on the "mushroom" site on this volcano. The most remarkable phenomenon was the rapid collapse of small anhydrite chimneys which resulted in the death of 44% *Ridgeia piscesae* worms within the first 26 days of the study [TUN 90b], confirming the significant potential impact of these natural disturbances on the structure of hydrothermal communities.

Six assemblages of fauna were identified on S&M, a large hydrothermal edifice on the Main Endeavour Field. Assemblages were distributed heterogeneously in relation with changes in environmental conditions [SAR 97, SAR 99a]. Four assemblages include *Ridgeiapiscesae* and two others are dominated byalvinellids (*Paralvinella sulfincola* and *P. palmiformis*). Community dynamics was monitored between 1991 and 1995 during repeated visits. The percentage of coverage by different assemblages varied over time and several bare areas were colonized. Assemblages of *P. palmiformis*, *Lepetodrilus fucensis* and gastropods (*Provanna variabilis* and *Depressigyra globulus*) as well as assemblages of senescent siboglinids increased, indicating a decrease in fluid input. The assemblages identified may reflect various stages of succession initiated by variations in fluid flow and porosity. The progressive mineralization of edifices causes changes in flow on small spatial scales and the succession occurs at the same time as the substratum evolves [SAR 02]. Gradually, as the mineralization and succession of the edifice progress, biotic factors such as predation and competition play an increasing significant role.

6.8.2.3. *MAR sites*

The hydrothermal vent TAG, located at a depth of 3,600 m, was the first hydrothermal vent field discovered on this ridge. The hydrothermal community here has been long known and a temporal study of this fauna was conducted between 1994 and 2004 [COP 07]. This study showed that there were no significant differences in coverage and abundance of the shrimp *Rimicaris exoculata*, one of the dominant assemblage of this site. This stability coincides with that of the geochemistry of high temperature fluids on the same time scale.

The hydrothermal vent field Lucky Strike (1,700 m) is undoubtedly the most studied on the MAR. With an area of 1 km^2, it is one of the largest known active vent field and is home to over 20 active edifices [OND 09]. It has been visited during many cruises and the images collected by different submersibles between 1994 and 2008 were analyzed [CUV 11]. Four assemblages of fauna, associated with the vent mussel *Bathymodiolus azoricus* and shrimp *Mirocaris fortunata* and two types of substrata were identified on the Eiffel Tower edifice [CUV 09b]. The temporal dynamics was assessed on the edifice, between its different sides and also on a small scale, which was a first for a MAR edifice. The percentage of colonization and mussel coverage remained stable throughout the 14-year study. However, on a shorter time scale and smaller spatial scale, significant differences in microbial coverage and distribution of fauna were observed. These fluctuations were linked to changes in the local hydrothermal activity and a negative correlation between the percentage of colonization of the edifice and the number of active emissions on the structure (black smoker, flanges and diffusion zones) was detected. The orientation of the edifice with regard to the hydrodynamic process was suggested to play an important role in the distribution of fauna [CUV 11].

6.8.3. *Deep-sea observatories*

Deep-sea observatories, autonomous (uncabled) or wired with fiber optic cables, have experienced a recent boom. These observatories are used to continuously monitor the ecosystems studied using several sensors. In total, two deep-sea observatories are currently in operation on hydrothermal sites. The EMSO-Azores observatory, located on the Lucky Strike field on the MAR, has been in operation since 2010. It has two autonomous nodes of energy that provide to the various sensors control and energy [COL 11]. These nodes communicate acoustically with a surface buoy, which sends the data via satellite to the archive center at the Ifremer center in Brest (France, Figure 6.7). The Ocean Network Canada observatory is located on the Juan de Fuca tectonic plate in the northeast of the Pacific Ocean. It is traversed by an 800 km long fiber optic cable that supplies six energy nodes. One of the nodes is deployed at the Endeavour segment hydrothermal vents at the base of the Grotto active edifice (2,200 m). In operation since 2009, this network allows 24 h access to data via the Internet and enables the remote control of instruments. Data are archived in Port Alberoni, located on the Vancouver Island (British Columbia, Canada).

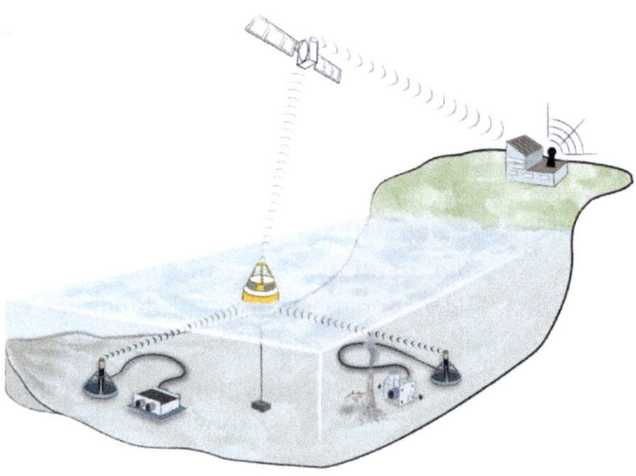

Figure 6.7. *The EMSO-Azores observatory has been in operation since 2010 on the Lucky Strike vent field (1,700 m) on the Mid-Atlantic Ridge. It was implemented by French interdisciplinary teams – Institut de Physique du Globe, University of Toulouse and Ifremer (Capsule Graphik 2014)*

The first ecological studies from these two observatories show the relative annual stability of vent communities, both on the MAR [SAR 14] and the Juan de Fuca Ridge [CUV 14]. Few environmental variations were recorded during the periods studied so far (48 days on the MAR and 28 days on NEPR). Temperature fluctuations were closely linked to those of the tides, with a potential impact on the behavior and physiology of the surrounding fauna. A correlation between the activity of siboglinids coming in and out their tubes and the tidal cycle supports this hypothesis [CUV 14]. The EMSO-Azores observatory was also used to analyze the behavior of two species of crustaceans, the shrimp *Mirocaris fortunata* and the crab *Segonzacia mesatlantica* [MAT 15] in relation with abiotic factors. Further studies are currently underway, in particular attempting to associate variations in faunal abundance with larger scale phenomena such as currents, vent fluid temperature and earthquake intensity.

6.8.4. *Dispersion of larvae and recruitment*

Invertebrates that colonize vent sites maintain their populations due to the dispersion of larvae that must travel long distances to find a substratum and settle successfully. The geographical isolation of breeding groups and the genetic pool explain the inverse relationship observed between the distance of the sites and the similarity of fauna. Thus, the nearest hydrothermal fields, located on adjacent ridge segments, share pools of genes and species [SHA 98], whereas remote sites share few species [TUN 88].

Most hydrothermal species are sedentary and produce planktonic larvae that are dispersed by the currents [ADA 12]. Larval dispersion maintains populations and ensures genetic connectivity between sites. The larvae can colonize new habitats or even an existing community, influencing the structure and diversity. Understanding dispersion between active sites is essential to understand community dynamics and identify the processes that help maintain the diversity of hydrothermal ecosystems.

Larvae of hydrothermal species have morphological and physiological characteristics that differ from those of adults. They survive in the water column until they arrive at an optimal habitat. Studying the larvae in deep-sea ecosystems is still a challenge, mainly because of the difficulty in sampling. It is not known where these larvae end up once released since larval densities, even near active sites, are extremely low [MUL 05]. Several indicators suggest that the larvae could be found higher in the water column. For example, some experiments have shown that the larvae of several hydrothermal species such as those of *Riftia pachyptila* and *Alvinella pompejana* could not tolerate the higher temperature conditions found in the habitats of adults [PRA 07].

The dispersion process is closely linked to the reproduction of hydrothermal species. While reproductive mechanisms are relatively well known for most of the dominant species, there is little information about their lifecycles due to sampling constraints. Indeed, these sites are revisited by research vessels at the same time of the year, which does not allow access to the intra-annual variability in reproduction. Only a few studies have directly observed spawning events. Otherwise, they are often deduced from observations of species breeding cycles (gonadal maturation, presence of spawning females or juveniles, etc.). Reproduction is continuous in several hydrothermal species [TYL 03], which would explain the rapid colonization of new active sites. However, for others such as *Bathymodiolus azoricus* on the Menez Gwen field (MAR) and the crab *Bythograea thermydron* on the SEPR, reproduction may be influenced by surface production [TEN 06] and therefore be discontinuous. The French campaign BICOSE, conducted in early 2014, revealed the presence of a high percentage of spawning females of the shrimp *Rimicaris exoculata* and large populations of juveniles, absent from previous campaigns, indicating the possible temporality in the reproduction of this species. Other factors, such as tides, seem to influence reproduction in some alvinellids.

It is still difficult to estimate the development time of larval hydrothermal species, particularly due to difficulties in raising them in pressurized containers. However, some experiments have been carried out in required conditions to observe larval development at their natural pressure. But, none of them has yet made it possible to observe the whole developmental sequence [PRA 01]. The lifetime of larvae in the water column will depend on their ability to feed themselves. Non-planktotrophic larvae have a shorter planktonic stage and disperse over shorter distances than those that feed. However, the special conditions of deep-sea environments could favor a longer residence time [TYL 03] and some species, such as *Alvinella pompejana*, have the ability to halt their development until they encounter a favorable habitat [PRA 01].

The coupling between the temporal monitoring, larval culture and the study of currents have led to an increased understanding of larval dispersal of hydrothermal species. Identification and modeling hydrodynamic processes are key elements to explain these dispersal mechanisms [ADA 12]. Thus, currents, directly related to seabed topography, will influence the trajectory and the distance traveled by benthic species larvae before their recruitment. Larval behavior, meanwhile, is beginning to be studied in some species, providing basic knowledge about their dispersal capacities. Other additional approaches such as colonization experiments provide an insight into the larval installation process, the species pool available over time and the role of environmental conditions in this colonization [CUV 14, MUL 98]. Recent technological developments such as seafloor observatories and sequential larvae pumps allow intra-annual temporal variations to be observed and therefore collect basic data on the lifecycle of hydrothermal species.

6.9. Mineral resources and exploitation

The global demand for minerals is increasing and the risk of shortages of strategic metals, are evident, particularly in Europe [FOU 12]. It is in this context that several countries are currently seeking alternatives, which include the exploitation of deep-sea mineral resources. The resources of the deep sea could become vital to meet the global demand for energy and raw materials. Like oil, mineral resources are non-renewable and their formation is slower than their rate of consumption [FOU 12].

Scientific research over the last 40 years in the deep-sea has helped highlight several geological and geochemical processes that promote the concentration of metals in certain types of ecosystems (nodules, crusts and hydrothermal sulfides) and the creation of energy resources (oil, methane clathrate and hydrogen). These resources include hydrothermal minerals, rich in several base metals (copper, zinc, lead, silver, gold, etc.) as well as rare metals (indium, germanium, cadmium, selenium, etc.). These strategic metals are those whose demand is the greatest [FOU 12].

Hydrothermal activity contributes to the concentration of metals that accumulate in the form of massive deposits of several hundred meters in diameter. A large proportion of copper, zinc, silver and gold exploited ashore are also extracted from these types of deposits. Sulfide edifices may account for several millions to several tens of million tons of ore. While current technologies can easily localize new active hydrothermal sites, no systematic method is recognized to search for inactive sites, for exploitation.

So far, only operations carried out near the seabed can locate them and drilling is essential to accurately characterize the content of these deposits and assess their economic potential. Current knowledge estimates that approximately 100 hydrothermal vents could be exploitable. Technological developments to explore deep-sea environments and associated ecosystems are required in order to assess the resources and the potential impacts of such an operation on their associated communities.

To date, there are 26 applications to explore deep-sea resources in the "Zone[14]". While in exclusive economic zones[15], where sovereign states regulate access to marine mineral resources (exploration and exploitation of the continental shelf), beyond the continental shelf, in international waters, this mandate falls under the International Seabed Authority[16] (ISA). The most active countries in the exploration of these resources are China, Korea, Russia, India and France. France currently has two exploration permits: one for nodules in the Clarion-Clipperton zone (granted in 2001), located in the center of the Pacific Ocean and more recently a permit for polymetallic sulfides on the MAR (signed in 2014).

The exploitation of resources contained in hydrothermal deposits is imminent. Thus, Nautilus Minerals Inc. is the first company in the world to have a license to explore polymetallic sulfide deposits on the seabed of the EEZ of Papua New Guinea (Solwara I site). Having fulfilled all obligations associated with this license, Nautilus proceeded to the second phase and received its first license to exploit these resources. Despite the fact that many technical details are still unknown, extraction scenarios are similar (see Figure 6.8). A device is used to extract the ore base by scraping and grinding the surface; the ore is raised to the surface by pumping, with discharge of part of the waste into the water column in the form of fine particles. The ore is then accumulated on barges plying up a terrestrial deposit. The operating time of a deposit is estimated to be short, between 2 and 5 years.

Our knowledge of the fauna colonizing inactive hydrothermal deposits is scarce. However, the complexity of the topography and the presence of hydrodynamic conditions and heterogeneous substrata suggest an increased

14 The international seabed (called the "Zone") is composed of the seabeds and their subsoil beyond the limits of the national court.

15 EEZ: exclusive economic zone.

16 ISA: *International Seabed Authority*.

biodiversity [FOU 12]. While it is difficult to assess with certainty what the consequences of such an activity may be on marine communities, several studies suggest a range of potential impacts, including both direct and indirect impacts. The first direct impact is the complete destruction of the habitat in the mined areas, decimating the entire associated fauna. The spread of sediment plumes, the physico-chemical changes in the environment, noise and light are some of the indirect factors that may affect benthic and surrounding pelagic communities. A recent study compared the potential impact of the exploitation of resources to those that occur naturally during an underwater volcanic eruption [VAN 14]. However, our current knowledge is still insufficient to establish reasonable plans for the protection of the environment and biodiversity associated with deep-sea ecosystems. Acquiring basic knowledge is essential to better understand their functioning, biodiversity, natural variability and resilience. This knowledge is essential for the development of management tools that will assess the potential impacts of such an operation and, if applicable, to ensure the total or partial restoration of these original ecosystems following an operation. Establishing protected areas, to ensure the conservation of biodiversity at the regional level, is essential [FOU 12]. However, only mobilization will allow scientists to make decisions with the knowledge necessary for the establishment of these areas, for the sustainability and respect of the environment.

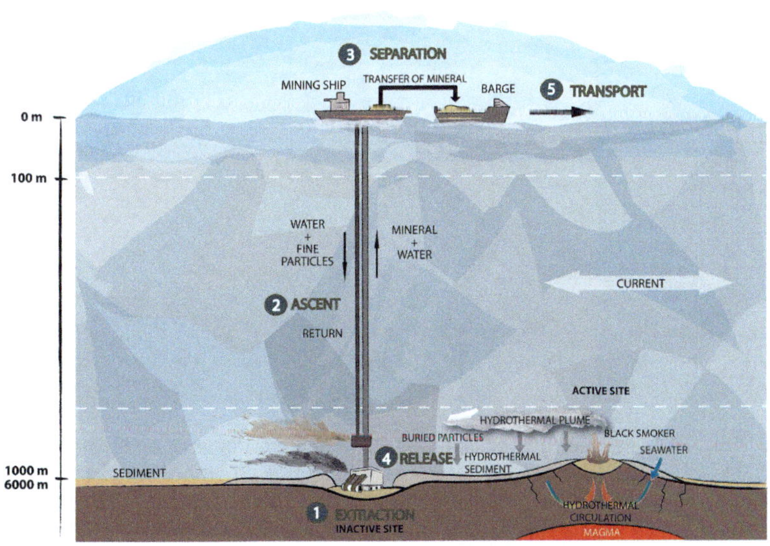

Figure 6.8. *Diagram showing the potential exploitation of sulfide deposits from the deep sea (Capsule Graphik 2014)*

Figure 6.9. *The main engineer species of the hydrothermal ecosystem. a) Riftia pachyptila, EPR; b) Alvinella pompejana, EPR; c) Bathymodiolus azoricus, MAR; d) Rimicaris exoculata, MAR; e) Alviniconcha spp., Lau basin; f) Vulcanolepas sp., Lau basin. (Photos a, b, c and d belong to Ifremer and photos eandfarea courtesy of C. Fisher, Penn State University). For a color version of the figure, see color section*

6.10. Bibliography

[ADA 12] ADAMS D.K., ARELLANO S.M., GOVENAR B., "Larval dispersal: vent life in the water column", *Oceanography*, vol. 25, pp. 256–268, 2012.

[ARN 98] ARNDT C., SCHIEDEK D., FELBECK H., "Metabolic responses of the hydrothermal vent tube worm Riftia pachyptila to severe hypoxia", *Marine Ecology Progress Series*, vol. 174, pp. 151–158, 1998.

[ARP 84] ARP A.J., CHILDRESS J.J., FISHER C.R., "Metabolic and blood-gas transport characteristics of the hydrothermal vent bivalve Calyptogena magnifica", *Physiological Zoology*, vol. 57, pp. 648–662, 1984.

[AUZ 88] AUZENDE J.M., YVES L., MARSEET B., "Recent geodynamic evolution of the North Fiji Basin, *Southwest Pacific*", *Geology*, vol. 16, pp. 925–929, 1988.

[BAT 05] BATES A.E., TUNNICLIFFE V., LEE R.W., "Role of thermal conditions in habitat selection by hydrothermal vent gastropods", *Marine Ecology Progress Series*, vol. 305, pp. 1–15, 2005.

[BAT 10] BATESA.E., LEE R.W., TUNNICLIFFE V. *et al.*, "Deep-sea hydrothermal vent animals seek cool fluids in a highly variable thermal environment", *Nature Communications*, vol. 1, pp. 12, 2010.

[BER 07] BERGQUIST D.C., ECKNER J.T., URCUYO I.A. *et al.*, "Using stable isotopes and quantitative community characteristics to determine a local hydrothermal vent food web", *Marine Ecology Progress Series*, vol. 330, pp. 49–65, 2007.

[BOE 00] BOETIUS A., RAVENSCHLAG K., SCHUBERT C.J. *et al.*, "A marine microbial consortium apparently mediating anaerobic oxidation of methane", *Nature*, vol. 407, pp. 623–626, 2000.

[BOR 02] BOROWSKI C.G.O., KRIEGER J., AMANN R. *et al.*, "New aspects of the symbiosis in the provannid snail Ifremeria nautilei from the North Fiji Back Arc Basin", *Cahiers de Biologie Marine*, vol. 43, pp. 321–324, 2002.

[CHA 00] CHARLOU J.L.D.J., DOUVILLE E., JEAN-BAPTISTE P. *et al.*, "Compared geochemical signatures and the evolution of Menez Gwen (37°50'N) and Lucky Strike (37°17'N) hydrothermal fluids, South of the Azores triple junction on the Mid-Atlantic Ridge", *Chemical Geology*, vol. 171, pp. 49–75, 2000.

[CHA 04] CHAUSSON F., SANGLIER S., LEIZE E. *et al.*, "Respiratory adaptations of a deep-sea hydrothermal vent crab", *Micron*, vol. 35, pp. 27–29, 2004.

[CHA 09] CHARMASSON S., SARRADIN P.-M., LE FAOUDER A. *et al.*, "High levels of natural radioactivity in biota from deep-sea hydrothermal vents: a preliminary communication", *Journal of Environmental Radioactivity*, vol. 100, pp. 522–526, 2009.

[CHE 91] CHEVALDONNE P., DESBRUYERES D., LEHAITRE M., "Time-series of temperature from three deep-sea hydrothermal vent sites", *Deep-Sea Research Part I – Oceanographic Research Papers*, vol. 38, pp. 1417–1430, 1991.

[CHE 93] CHEVALDONNE P., JOLLIVET D., "Videoscopic study of deep-sea hydrothermal vent alvinellid polychaete populations-biomass estimation and behavior", *Marine Ecology Progress Series*, vol. 95, pp. 251–262, 1993.

[CHI 91] CHILDRESS J.J., FISHER C.R., FAVUZZI J.A. *et al.*, "Sulfide and carbon dioxide uptake by the hydrothermal vent clam Calyptegena magnifica and its chemoautotrophic symbionts", *Physiological Zoology*, vol. 64, pp. 1444–1470, 1991.

[CHI 92] CHILDRESS J.J., FISHER C.R., "The biology of hydrothermal vent animals–physiology, biochemistry and autotrophic symbioses", *Oceanography and Marine Biology*, vol. 30, pp. 337–441, 1992.

[COL 02] COLACO A., DEHAIRS F., DESBRUYERES D., "Nutritional relations of deep-sea hydrothermal fields at the Mid-Atlantic Ridge: a stable isotope approach", *Deep-Sea Research Part I – Oceanographic Research Papers*, vol. 49, pp. 395–412, 2002.

[COL 07] COLACO A., DESBRUYERES D., GUEZENNEC J., "Polar lipid fatty acids as indicators of trophic associations in a deep-sea vent system community", *Marine Ecology – An Evolutionary Perspective*, vol. 28, pp. 15–24, 2007.

[COL 11] COLACO A., BLANDIN J., CANNAT M. *et al.*, "MoMAR-D: a technological challenge to monitor the dynamics of the Lucky Strike vent ecosystem", *ICES Journal of Marine Science*, vol. 68, pp. 416–424, 2011.

[COM 98] COMTET T., DESBRUYERES D., "Population structure and recruitment in mytilid bivalves from the Lucky Strike and Menez Gwen hydrothermal vent fields (37 degrees 17'N and 37 degrees 50'N on the Mid-Atlantic Ridge)", *Marine Ecology Progress Series*, vol. 163, pp. 165–177, 1998.

[CON 75] CONNELL J.H., "Some mechanisms producing structure in natural communities: a model and evidence from field experiments", *Ecology and Evolution of Communities*, Belknap Press of Harvard University, MA, 1975.

[CON 12] CONNELLYD.P., COPLEY J.T., MURTON B.J. *et al.*, "Hydrothermal vent fields and chemosynthetic biota on the world's deepest seafloor spreading centre", *Nature Communications*, vol. 3, 2012.

[COP 97] COPLEY J.T.P., TYLER P.A., MURTON B.J. *et al.*, "Spatial and interannual variation in the faunal distribution at Broken Spur vent field (29 degrees N, Mid-Atlantic Ridge)", *Marine Biology*, vol. 129, pp. 723–733, 1997.

[COP 07] COPLEY J.T.P., FLINT H.C., FERRERO T.J. *et al.*, "Diversity of melofauna and free-living nematodes in hydrothermal vent mussel beds on the northern and southern East Pacific Rise", *Journal of the Marine Biological Association of the United Kingdom*, vol. 87, pp. 1141–1152, 2007.

[COS 08] COSSON R.P., THIEBAUT E., COMPANY R. *et al.*, "Spatial variation of metal bioaccumulation in the hydrothermal vent mussel Bathymodiolus azoricus", *Marine Environmental Research*, vol. 65, pp. 405–415, 2008.

[CUV 09a] CUVELIER D., SARRAZIN J., COLACO A. *et al.*, "Distribution and spatial variation of hydrothermal faunal assemblages at Lucky Strike (Mid-Atlantic Ridge) revealed by high-resolution video image analysis", *Deep-Sea Research Part I – Oceanographic Research Papers*, vol. 56, pp. 2026–2040, 2009.

[CUV 09b] CUVELIER D., SARRADIN P.M., SARRAZIN J. *et al.*, "Hydrothermal faunal assemblages and habitat characterisation at the Eiffel Tower edifice (Lucky Strike, Mid-Atlantic Ridge)", *Marine Ecology – An Evolutionary Perspective*, vol. 32, pp. 243–255, 2009.

[CUV 11] CUVELIER D., SARRAZIN J., COLACO A. *et al.*, "Community dynamics over 14 years at the Eiffel Tower hydrothermal edifice on the Mid-Atlantic Ridge", *Limnology and Oceanography*, vol. 56, pp. 1624–1640, 2011.

[CUV 14] CUVELIER D., LEGENDRE P., LAES A. *et al.*, "Rhythms and community dynamics of a hydrothermal tubeworm assemblage at main endeavour field – a multidisciplinary deep-sea observatory approach", *Plos One*, vol. 9, no. 5, 2014.

[DEB 09] DE BUSSEROLLES F., SARRAZIN J., GAUTHIER O. *et al.*, "Are spatial variations in the diets of hydrothermal fauna linked to local environmental conditions?", *Deep-Sea Research Part II – Topical Studies in Oceanography*, vol. 56, pp. 1649–1664, 2009.

[DES 94] DESBRUYÈRES D., ALAYSE-DANET A.-M., OHTA S. *et al.*, "Deep-sea hydrothermal communities in Southwestern Pacific back-arc basins (the North Fiji and Lau Basins): composition, microdistribution and food web", *Marine Geology*, vol. 116, pp. 227–242, 1994.

[DES 98] DESBRUYÈRES D., CHEVALDONNE P., ALAYSE A.M. *et al.*, "Biology and ecology of the 'Pompeii worm' (Alvinella pompejana Desbruyeres and Laubier), a normal dweller of an extreme deep-sea environment: a synthesis of current knowledge and recent developments", *Deep-Sea Research Part II – Topical Studies in Oceanography*, vol. 45, pp. 383–422, 1998.

[DES 00] DESBRUYÈRES D., ALMEIDA A., BISCOITO M. *et al.*, "A review of the distribution of hydrothermal vent communities along the northern Mid-Atlantic Ridge: dispersal vs. environmental controls", *Hydrobiologia*, vol. 440, pp. 201–216, 2000.

[DES 01] DESBRUYÈRES D., BISCOITO M., CAPRAIS J.C. *et al.*, "Variations in deep-sea hydrothermal vent communities on the Mid-Atlantic Ridge near the Azores plateau", *Deep-Sea Research Part I – Oceanographic Research Papers*, vol. 48, pp. 1325–1346, 2001.

[DES 06] DESBRUYERES D., HASHIMOTO J., FABRI M.-C., "Composition and biogeography of hydrothermal vent communities in western pacific back-arc basins", *Back-Arc Spreading Systems: Geological, Biological, Chemical, and Physical Interactions*, Wiley Ed., 2006.

[DIX 06] DIXON D.R., LOWE D.M., MILLER P.I. *et al.*, "Evidence of seasonal reproduction in the Atlantic vent mussel Bathymodiolus azoricus, and an apparent link with the timing of photosynthetic primary production", *Journal of the Marine Biological Association of the United Kingdom*, vol. 86, pp. 1363–1371, 2006.

[DUP 09] DUPERRON S., LORION J., SAMADI S. *et al.*, "Symbioses between deep-sea mussels (Mytilidae: Bathymodiolinae) and chemosynthetic bacteria: diversity, function and evolution", *Comptes Rendus Biologies*, vol. 332, pp. 298–310, 2009.

[DZI 04] DZIAK R.P., SMITH D.K., BOHNENSTIEHL D.R. *et al.*, "Evidence of a recent magma dike intrusion at the slow spreading Lucky Strike segment, Mid-Atlantic Ridge", *Journal of Geophysical Research-Solid Earth*, vol. 109, B12102, 2004.

[DZI 07] DZIAK R.P., BOHNENSTIEHL D.R., COWEN J.P. *et al.*, "Rapid dike emplacement leads to eruptions and hydrothermal plume release during seafloor spreading events", *Geology*, vol. 35, pp. 579–582, 2007.

[FAB 11] FABRI M.C., BARGAIN A., BRIAND P. *et al.*, "The hydrothermal vent community of a new deep-sea field, Ashadze-1, 12 degrees 58 ' N on the Mid-Atlantic Ridge", *Journal of the Marine Biological Association of the United Kingdom*, vol. 91, pp. 1–13, 2011.

[FEL 85] FELBECK H., "CO_2 fixation in the hydrothermal vent tube worm Riftia pachyptial (Jones)", *Physiological Zoology*, vol. 58, pp. 272–281, 1985.

[FIS 90] FISHER C.R., KENNICUTT M.C., BROOKS J.M., "Stable carbon isotopic evidence for carbon limitation in hydrothermal vent vestimentiferans", *Science*, vol. 247, pp. 1094–1096, 1990.

[FOU 12] FOUQUET Y., LACROIX, D., *Les ressources minérales marines profondes*, Editions Quae,Versailles, 2012.

[FUS 87] FUSTEC A., DESBRUYERES D., JUNIPER K.S., "Deep-sea hydrothermal communities at 13°N on the East Pacific Rise: microdistribution and temporal variations", *Biological Oceanography*, vol. 4, pp. 121–164, 1987.

[GAG 91] GAGE J.D., "Biological rates in the deep-sea – a perspective from studies on processes in the benthic boundary layer", *Reviews in Aquatic Sciences*, vol. 5, pp. 49–100, 1991.

[GIR 06] GIRGUIS P.R., LEE R.W., "Thermal preference and tolerance of alvinellids", *Science*, vol. 312, pp. 231–231, 2006.

[GOF 04] GOFFREDI S.K., WAREN A., ORPHAN V.J. *et al.*, "Novel forms of structural integration between microbes and a hydrothermal vent gastropod from the Indian Ocean", *Applied and Environmental Microbiology*, vol. 70, pp. 3082–3090, 2004.

[GOL 07] GOLLNER S., ZEKELY J., GOVENAR B. *et al.*, "Tubeworm-associated permanent meiobenthic communities from two chemically different hydrothermal vent sites on the East Pacific Rise", *Marine Ecology Progress Series*, vol. 337, pp. 39–49, 2007.

[GOL 10a] GOLLNER S., IVANENKO V.N., ARBIZU P.M. *et al.*, "Advances in taxonomy, ecology, and biogeography of dirivultidae (Copepoda) associated with chemosynthetic environments in the deep sea", *Plos One*, vol. 5, no. 8, 2010.

[GOL 10b] GOLLNER S., RIEMER B., MARTINEZ ARBIZU P. *et al.*, "Diversity of meiofauna from the 9Â°50'N East Pacific rise across a gradient of hydrothermal fluid emissions", *Plos One*, vol. 5, no. 8, 2010.

[GOV 05] GOVENAR B., LE BRIS N., GOLLNER S. *et al.*, "Epifaunal community structure associated with Riftia pachyptila aggregations in chemically different hydrothermal vent habitats", *Marine Ecology Progress Series*, vol. 305, pp. 67–77, 2005.

[GRA 92] GRASSLE J.F., MACIOLEK N.J., "Deep-sea species richness -regional and local diversity estimates from quantitative bottom samples", *American Naturalist*, vol. 139, pp. 313–341, 1992.

[GRE 06] GRELON D., MORINEAUX M., DESROSIERS G. *et al.*, "Feeding and territorial behavior of Paralvinella sulfincola, a polychaete worm at deep-sea hydrothermal vents of the Northeast Pacific Ocean", *Journal of Experimental Marine Biology and Ecology*, vol. 329, pp. 174–186, 2006.

[HAL 08] HALARY S., RIOU V., GAILL F. *et al.*, "3D FISH for the quantification of methane- and sulphur-oxidizing endosymbionts in bacteriocytes of the hydrothermal vent mussel Bathymodiolus azoricus", *ISME Journal*, vol. 2, pp. 284–292, 2008.

[HAS 01] HASHIMOTO J., OHTA S., GAMO T. *et al.*, "First hydrothermal vent communities from the Indian Ocean discovered", *Zoological Science*, vol. 18, pp. 717–721, 2001.

[HAY 91] HAYMON R.M., FORNARI D.J., EDWARDS M.H. *et al.*, "Hydrothermal vent distribution along the East Pacific Rise Crest (9°09'-54'N) and its relationship to magmatic and tectonic processes on fast-spreading midocean ridges", *Earth and Planetary Science Letters*, vol. 104, pp. 513–534, 1991.

[HEN 08] HENRY M.S., CHILDRESS J.J., FIGUEROA D., "Metabolic rates and thermal tolerances of chemoautotrophic symbioses from Lau Basin hydrothermal vents and their implications for species distributions", *Deep-Sea Research Part I – Oceanographic Research Papers*, vol. 55, pp. 679–695, 2008.

[HES 83] HESSLER R., SMITHEY W.M., "The distribution and community structure of megafauna at the Galapagos Rift hydrothermal vents", *Hydrothermal Processes at Seafloor Spreading Centers*, Plenum Press, New York, 1983.

[HES 88] HESSLER R., SMITHEY W.M., BOUDRIAS M.A. *et al.*, "Temporal change in megafauna at the rose garden hydrothermal vent (Galapagos Rift – Eastern Tropical Pacific)", *Deep-Sea Research Part I – Oceanographic Research Papers*, vol. 35, pp. 1681–1709, 1988.

[HES 91] HESSLER R., LONSDALE P.F., "Biogeography of Mariana Trough hydrothermal vent communities", *Deep Sea Research Part I – Oceanographic Research Papers*, vol. 38, pp. 185–199, 1991.

[HOL 01] HOLM N.G., CHARLOU J.L., "Initial indications of abiotic formation of hydrocarbons in the rainbow ultramafic hydrothermal system, Mid-Atlantic Ridge", *Earth and Planetary Science Letters*, vol. 191, pp. 1–8, 2001.

[HOU 00] HOURDEZ S., LALLIER F.H., DE CIAN M.C., *et al.*, "Gas transfer system in Alvinella pompejana (Annelida Polychaeta, Terebellida): functional properties of intracellular and extracellular hemoglobins", *Physiological and Biochemical Zoology*, vol. 73, pp. 365–373, 2000.

[HOU 07] HOURDEZ S., LALLIER F.H., "Adaptations to hypoxia in hydrothermal-vent and cold-seep invertebrates", *Reviews in Environmental Science and Bio/Technology*, vol. 6, pp. 143–159, 2007.

[HUE 11] HUEGLER M., PETERSEN J.M., DUBILIER N. *et al.*, "Pathways of carbon and energy metabolism of the epibiotic community associated with the deep-sea hydrothermal vent shrimp Rimicaris exoculata", *Plos One*, vol. 6, no. 1, 2011.

[IVA 11] IVANENKO V.N., FERRARI F.D., DEFAYE D. *et al.*, "Description, distribution and microhabitats of a new species of Tisbe (Copepoda: Harpacticoida: Tisbidae) from a deep-sea hydrothermal vent field at the Mid-Atlantic Ridge (37 degrees N, Lucky Strike)", *Cahiers de Biologie Marine*, vol. 52, pp. 89–106, 2011.

[JAN 79] JANNASCH H.W., WIRSEN C.O., "Chemosynthetic primary production at East Pacific sea floor spreading centers", *BioScience*, vol. 29, pp. 592–598, 1979.

[JAN 95] JANNASCH H.W., Microbial Interactions with Hydrothermal Fluids, American Geophysical Union, Washington, DC, 1995.

[JOH 94] JOHNSON K.S., CHILDRESS J.J., BEEHLER C.L. *et al.*, "Biogeochemistry of hydrothermal vent mussel communities: the deep-sea analogue to the intertidal zone", *Deep Sea Research Part I – Oceanographic Research Papers*, vol. 41, pp. 993–1011, 1994.

[JUN 95] JUNIPER S.K., MARTINEU P., SARRAZIN J. *et al.*, "Microbial-mineral floc associated with nascent hydrothermal activity on Coaxial segment, Juan de Fuca ridge", *Geophysical Research Letters*, vol. 22, pp. 179–182, 1995.

[KAR 95] KARL D.M., "Ecology of free-living, hydrothermal vent microbial communities", *The Microbiology of Deep-Sea Hydrothermal Vents*, CRC Press, New York, pp. 35–124, 1995.

[KEL 07] KELLY N., METAXAS A., BUTTERFIELD D., "Spatial and temporal patterns of colonization by deep-sea hydrothermal vent invertebrates on the Juan de Fuca Ridge, NE Pacific", *Aquatic Biology*, vol. 1, pp. 1–16, 2007.

[KEL 08] KELLY N., METAXAS A., "Diversity of invertebrate colonists on simple and complex substrates at hydrothermal vents on the Juan de Fuca Ridge", *Aquatic Biology*, vol. 3, pp. 271–281, 2008.

[KIM 96] KIM J.H., DEWREEDE R.E., "Effects of size and season of disturbance on algal patch recovery in a rocky intertidal community", *Marine Ecology Progress Series*, vol. 133, pp. 217–228, 1996.

[KIM 98] KIM S.L., MULLINEAUX L.S., "Distribution and near-bottom transport of larvae and other plankton at hydrothermal vents", *Deep-Sea Research Part II – Topical Studies in Oceanography*, vol. 45, pp. 423–440, 1998.

[KOU 07] KOURIS A., JUNIPER S.K., FREBOURG G. *et al.*, "Protozoan-bacterial symbiosis in a deep-sea hydrothermal vent folliculinid ciliate (Folliculinopsis sp.) from the Juan de Fuca Ridge", *Marine Ecology – An Evolutionary Perspective*, vol. 28, pp. 63–71, 2007.

[LEB 06] LE BRIS N., GOVENAR B., LE GALL C. *et al.*, "Variability of physico-chemical conditions in 9 degrees 50'N EPR diffuse flow vent habitats", *Marine Chemistry*, vol. 98, pp. 167–182, 2006.

[LEL XX] LELIÈVRE Y., Characterization of the factors controlling the dynamics of the macrofauna associated with hydrothermal vents using deep-sea observatory data, PhD Thesis, University of Montréal, in progress.

[LEN 08] LENIHAN H.S., MILLS S.W., MULLINEAUX L.S. *et al.*, "Biotic interactions at hydrothermal vents: recruitment inhibition by the mussel Bathymodiolus thermophilus", *Deep-Sea Research Part I – Oceanographic Research Papers*, vol. 55, pp. 1707–1717, 2008.

[LEV 02] LEVESQUE C., JUNIPER S.K., "Particulate matter as a food source at a nascent hydrothermal vent on the Juan de Fuca Ridge", *Cahiers de Biologie Marine*, vol. 43, pp. 289–292, 2002.

[LEV 05] LEVESQUE C., LIMEN H., JUNIPER S.K., "Origin, composition and nutritional quality of particulate matter at deep-sea hydrothermal vents on Axial Volcano, NE Pacific", *Marine Ecology Progress Series*, vol. 289, pp. 43–52, 2005.

[LEV 06] LEVESQUE C., JUNIPER S.K., LIMEN H., "Spatial organization of food webs along habitat gradients at deep-sea hydrothermal vents on Axial Volcano, Northeast Pacific", *Deep-Sea Research Part I – Oceanographic Research Papers*, vol. 53, pp. 726–739, 2006.

[LIM 07] LIMEN H., LEVESQUE C., JUNIPER S.K., "POM in macro-/meiofaunal food webs associated with three flow regimes at deep-sea hydrothermal vents on Axial Volcano, Juan de Fuca Ridge", *Marine Biology*, vol. 153, pp. 129–139, 2007.

[LON 77] LONSDALE P., "Deep-tow observations at mounds abyssal hydrothermal field, Galapagos rift", *Earth and Planetary Science Letters*, vol. 36, pp. 92–110, 1977.

[LUT 01] LUTHER G.W., ROZAN T.F., TAILLEFERT M. *et al.*, "Chemical speciation drives hydrothermal vent ecology", *Nature*, vol. 410, pp. 813–816, 2001.

[LUT 08] LUTZ R.A., SHANK T.M., LUTHER G.W. *et al.*, "Interrelationships between vent fluid chemistry, temperature, seismic activity, and biological community structure at a mussel-dominated, deep-sea hydrothermal vent along the East Pacific Rise", *Journal of Shellfish Research*, vol. 27, pp. 177–190, 2008.

[LUT 12] LUTHER G.W., GARTMAN A., YUECEL M. *et al.*, "Chemistry, temperature, and faunal distributions at diffuse-flow hydrothermal vents comparison of two geologically distinct ridge systems", *Oceanography*, vol. 25, pp. 234–245, 2012.

[MAR 97] MARTINEU P., JUNIPER S.K., FISHER C.R. *et al.*, "Sulfide binding in the body fluids of hydrothermal vent alvinellid polychaetes", *Physiological Zoology*, vol. 70, pp. 578–588, 1997.

[MAR 02] MARCUS J., TUNNICLIFFE V., "Living on the edges of diffuse vents on the Juan de Fuca Ridge", *Cahiers de Biologie Marine*, vol. 43, pp. 263–266, 2002.

[MAR 09] MARCUS J., TUNNICLIFFE V., BUTTERFIELD D.A., "Post-eruption succession of macrofaunal communities at diffuse flow hydrothermal vents on Axial Volcano, Juan de Fuca Ridge, Northeast Pacific", *Deep-Sea Research Part II – Topical Studies in Oceanography*, vol. 56, pp. 1586–1598, 2009.

[MAR 12] MARSH L., COPLEY J.T., HUVENNE V.A.I. *et al.*, "Microdistribution of faunal assemblages at deep-sea hydrothermal vents in the Southern Ocean", *Plos One*, vol. 7, no. 10, 2012.

[MAT 15] MATABOS M.C.D., BROUARD J., SHILLITO B. *et al.*, "Behavioural study of two hydrothermal crustacean decapods: Mirocaris fortunata and Segonzacia mesatlantica, from the Lucky Strike vent field (Mid Atlantic Ridge)", *Deep Sea Research, Part II,* forthcoming 2015 (online 04.2015).

[MIC 02] MICHELI F., PETERSON C.H., MULLINEAUX L.S., *et al.*, "Predation structures communities at deep-sea hydrothermal vents", *Ecological Monographs*, vol. 72, pp. 365–382, 2002.

[MOA 12] MOALIC Y., DESBRUYERES D., DUARTE C.M. *et al.*, "Biogeography revisited with network theory: retracing the history of hydrothermal vent communities", *Systematic Biology*, vol. 61, pp. 127–137, 2012.

[MUL 95] MULLINEAUX L.S., WIEBE P.H., BAKER E.T., "Larvae of benthic invertebrates in hydrothermal vent plumes over Juan de Fuca ridge", *Marine Biology*, vol. 122, pp. 585–596, 1995.

[MUL 98] MULLINEAUX L.S., MILLS S.W., GOLDMAN E., "Recruitment variation during a pilot colonization study of hydrothermal vents (9°50'N, East Pacific Rise)", *Deep-Sea Research Part II – Topical Studies in Oceanography*, vol. 45, pp. 441–464, 1998.

[MUL 00] MULLINEAUX L.S., FISHER C.R., PETERSON C.H. *et al.*, "Tubeworm succession at hydrothermal vents: use of biogenic cues to reduce habitat selection error?", *Oecologia*, vol. 123, pp. 275–284, 2000.

[MUL 05] MULLINEAUX L.S., MILLS S.W., SWEETMAN A.K. *et al.*, "Vertical, lateral and temporal structure in larval distributions at hydrothermal vents", *Marine Ecology Progress Series*, vol. 293, pp. 1–16, 2005.

[MUL 09] MULLINEAUX L.S., MICHELI F., PETERSON C.H., *et al.*, "Imprint of past environmental regimes on structure and succession of a deep-sea hydrothermal vent community", *Oecologia*, vol. 161, pp. 387–400, 2009.

[MUL 10] MULLINEAUX L.S., ADAMS D.K., MILLS S.W. *et al.*, "Larvae from afar colonize deep-sea hydrothermal vents after a catastrophic eruption", *Proceedings of the National Academy of Sciences of the United States of America*, vol. 107, pp. 7829–7834, 2010.

[NAK 12] NAKAMURA K., WATANABE H., MIYAZAKI J. *et al.*, "Discovery of new hydrothermal activity and chemosynthetic fauna on the Central Indian Ridge at 18°-20° S", *Plos One*, vol. 7, no. 3, 2012.

[NYE 13] NYE V., COPLEY J.T., TYLER P.A., "Spatial variation in the population structure and reproductive biology of Rimicaris hybisae (Caridea: Alvinocarididae) at hydrothermal vents on the Mid-Cayman spreading centre", *Plos One*, vol. 8, pp. e60319–e60319, 2013.

[OND 09] ONDREAS H., CANNAT M., FOUQUET Y. *et al.*, "Recent volcanic events and the distribution of hydrothermal venting at the Lucky Strike hydrothermal field, Mid-Atlantic Ridge", *Geochemistry Geophysics Geosystems*, vol. 10, no. 2, 2009.

[POD 09] PODOWSKI E.L., MOORE T.S., ZELNIO K.A. *et al.*, "Distribution of diffuse flow megafauna in two sites on the Eastern Lau Spreading Center, Tonga", *Deep-Sea Research Part I – Oceanographic Research Papers*, vol. 56, pp. 2041–2056, 2009.

[PON 98] POND D.W., BELL M.V., DIXON D.R. *et al.*, "Stable-carbon-isotope composition of fatty acids in hydrothermal vent mussels containing methanotrophic and thiotrophic bacterial endosymbionts", *Applied and Environmental Microbiology*, vol. 64, pp. 370–375, 1998.

[PON 13] PONSARD J., CAMBON-BONAVITA M.-A., ZBINDEN M. *et al.*, "Inorganic carbon fixation by chemosynthetic ectosymbionts and nutritional transfers to the hydrothermal vent host-shrimp Rimicaris exoculata", *ISME Journal*, vol. 7, pp. 96–109, 2013.

[POR 15] PORTAIL M., OLU K., ESCOBAR-BRIONES E., *et al.*, "Comparative study of vent and seep macrofaunal communities in the Guaymas basin", *Biogeosciences*, forthcoming 2015.

[PRA 01] PRADILLON F., SHILLITO B., YOUNG C.M. *et al.*, "Deep-seaecology – developmental arrest in vent worm embryos", *Nature*, vol. 413, pp. 698–699, 2001.

[PRA 07] PRADILLON F., GAILL F., "Adaptation to deep-sea hydrothermal vents: some molecular and developmental aspects", *Journal of Marine Science and Technology-Taiwan*, vol. 15, pp. 37–53, 2007.

[RAU 79] RAU G.H., HEDGES J.I., "C_{13} Depletion in a hydrothermal vent mussel – suggestion of a chemo-synthetic food source", *Science*, vol. 203, pp. 648–649, 1979.

[RAV 03] RAVAUX J., GAILL F., LE BRIS N. *et al.*, "Heat-shock response and temperature resistance in the deep-sea vent shrimp Rimicaris exoculata", *Journal of Experimental Biology*, vol. 206, pp. 2345–2354, 2003.

[ROG 12] ROGERS A.D., TYLER P.A., CONNELLY D.P. *et al.*, "The discovery of new deep-sea hydrothermal vent communities in the Southern Ocean and implications for biogeography", *Plos Biology*, vol. 10, no. 1, 2012.

[ROT 13] ROTERMAN C.N., COPLEY J.T., LINSE K.T. *et al.*, "The biogeography of the yeti crabs (Kiwaidae) with notes on the phylogeny of the Chirostyloidea (Decapoda: Anomura)", *Proceedings of the Royal Society B – Biological Sciences*, vol. 280, 2013.

[SAN 05] SANCHO G., FISHER C.R., MILLS S. *et al.*, "Selective predation by the zoarcid fish Thermarces cerberus at hydrothermal vents", *Deep-Sea Research Part I – Oceanographic Research Papers*, vol. 52, pp. 837–844, 2005.

[SAR 97] SARRAZIN J., ROBIGOU V., JUNIPER S.K. *et al.*, "Biological and geological dynamics over four years on a high-temperature sulfide structure at the Juan de Fuca Ridge hydrothermal observatory", *Marine Ecology Progress Series*, vol. 153, pp. 5–24, 1997.

[SAR 99a] SARRADIN P.M., CAPRAIS J.C., RISO R. *et al.*, "Chemical environment of the hydrothermal mussel communities in the Lucky Strike and Menez Gwen vent fields, Mid Atlantic ridge", *Cahiers de Biologie Marine*, vol. 40, pp. 93–104, 1999.

[SAR 99b] SARRAZIN J., JUNIPER S.K., "Biological characteristics of a hydrothermal edifice mosaic community", *Marine Ecology Progress Series*, vol. 185, pp. 1–19, 1999.

[SAR 99c] SARRAZIN J., JUNIPER S.K., MASSOTH G. *et al.*, "Physical and chemical factors influencing species distributions on hydrothermal sulfide edifices of the Juan de Fuca Ridge, northeast Pacific", *Marine Ecology Progress Series*, vol. 190, pp. 89–112, 1999.

[SAR 02] SARRAZIN J., LEVESQUE C., JUNIPER S.K. *et al.*, "Mosaic community dynamics on Juan de Fuca Ridge sulphide edifices: substratum, temperature and implications for trophic structure", *Cahiers de Biologie Marine*, vol. 43, pp. 275–279, 2002.

[SAR 14] SARRAZIN J., CUVELIER D., PETON L. *et al.*, "High-resolution dynamics of a deep-sea hydrothermal mussel assemblage monitored by the EMSO-Acores MoMAR observatory", *Deep-Sea Research Part I – Oceanographic Research Papers*, vol. 90, pp. 62–75, 2014.

[SAR 15] SARRAZIN J., LEGENDRE P., DE BUSSEROLLES F. *et al.*, "Biodiversity patterns, environmental drivers and indicator species on a high temperature hydrothermal edifice, Mid-Atlantic Ridge", *Deep Sea Research*, forthcoming 2015 (online 04.2015).

[SCH 08] SCHMIDT C., VUILLEMIN R., LE GALL C. *et al.*, "Geochemical energy sources for microbial primary production in the environment of hydrothermal vent shrimps", *Marine Chemistry*, vol. 108, pp. 18–31, 2008.

[SHA 98] SHANK T.M., FORNARI D.J., VON DAMM K.L. *et al.*, "Temporal and spatial patterns of biological community development at nascent deep-sea hydrothermal vents (9 degrees 50'N, East Pacific Rise)", *Deep-Sea Research Part II – Topical Studies in Oceanography*, vol. 45, pp. 465–515, 1998.

[SIB 02] SIBUET M., ROY O.-L., *Cold Seep Communities on Continental Margins: Structure and Quantitative Distribution Relative to Geological and Fluid Venting Patterns. Ocean Margin Systems*, Springer Verlag, Berlin, 2002.

[SIE 99] SIEVERT S.M., BRINKHOFF T., MUYZER G. *et al.*, "Spatial heterogeneity of bacterial populations along an environmental gradient at a shallow submarine hydrothermal vent near Milos Island (Greece)", *Applied and Environmental Microbiology*, vol. 65, pp. 3834–3842, 1999.

[SOM 92] SOMERO G.N., "Adaptations to high hydrostatic pressure", *Annual Review of Physiology*, vol. 54, pp. 557–577, 1992.

[STE 88] STEIN J.L., CARY S.C., HESSLER R.R. *et al.*, "Chemoautotrophic symbiosis in a hydrothermal vent gastropod", *Biological Bulletin*, vol. 174, pp. 373–378, 1988.

[SWE 13] SWEETMAN A.K., LEVIN L.A., RAPP H.T. *et al.*, "Faunal trophic structure at hydrothermal vents on the southern Mohn's Ridge, Arctic Ocean", *Marine Ecology Progress Series*, vol. 473, pp. 115–131, 2013.

[TSU 01] TSURUMI M., TUNNICLIFFE V., "Characteristics of a hydrothermal vent assemblage on a volcanically active segment of Juan de Fuca Ridge, northeast Pacific", *Canadian Journal of Fisheries and Aquatic Sciences*, vol. 58, pp. 530–542, 2001.

[TSU 03a] TSURUMI M., DE GRAAF R.C., TUNNICLIFFE V., "Distributional and biological aspects of copepods at hydrothermal vents on the Juan de Fuca Ridge, north-east Pacific ocean", *Journal of the Marine Biological Association of the United Kingdom*, vol. 83, pp. 469–477, 2003.

[TSU 03b] TSURUMI M., TUNNICLIFFE V., "Tubeworm-associated communities at hydrothermal vents on the Juan de Fuca Ridge, northeast Pacific", *Deep-Sea Research Part I – Oceanographic Research Papers*, vol. 50, pp. 611–629, 2003.

[TUN 85] TUNNICLIFFE V., JUNIPER S.K., DE BURGH M., "The hydrothermal vent community on axial seamount, Juan de Fuca Ridge", *Bulletin of the Biological Society of Washington*, vol. 6, pp. 453–464, 1985.

[TUN 88] TUNNICLIFFE V., "Biogeography and evolution of hydrothermal vent fauna in the Eastern Pacific Ocean", *Proceedings of the Royal Society B – Biological Sciences*, vol. 233, pp. 347–366, 1988.

[TUN 90a] TUNNICLIFFE V., GARRETT J.F., JOHNSON H.P., "Physical and biological factors affecting the behavior and mortality of hydrothermal vent tubeworms (Vestimentiferans)", *Deep-Sea Research Part I – Oceanographic Research Papers*, vol. 37, pp. 103–125, 1990.

[TUN 90b] TUNNICLIFFE V., JUNIPER S.K., "Dynamic character of the hydrothermal vent habitat and the nature of sulfide chimney fauna", *Progress in Oceanography*, vol. 24, pp. 1–13, 1990.

[TUN 96] TUNNICLIFFE V., FOWLER C.M.R., "Influence of sea-floor spreading on the global hydrothermal vent fauna", *Nature*, vol. 379, pp. 531–533, 1996.

[TUN 97] TUNNICLIFFE V., EMBLEY R.W., HOLDEN J.F., *et al.*, "Biological colonization of new hydrothermal vents following an eruption on Juan de Fuca Ridge", *Deep-Sea Research Part I – Oceanographic Research Papers*, vol. 44, pp. 1627–1644, 1997.

[TYL 99] TYLER P.A., YOUNG C.M., "Reproduction and dispersal at vents and cold seeps", *Journal of the Marine Biological Association of the United Kingdom*, vol. 79, pp. 193–208, 1999.

[TYL 03] TYLER P.A., YOUNG C.M., "Dispersal at hydrothermal vents: a summary of recent progress", *Hydrobiologia*, vol. 503, pp. 9–19, 2003.

[URC 03] URCUYO I.A., MASSOTH G.J., JULIAN D. *et al.*, "Habitat, growth and physiological ecology of a basaltic community of Ridgeia piscesae from the Juan de Fuca Ridge", *Deep-Sea Research Part I – Oceanographic Research Papers*, vol. 50, pp. 763–780, 2003.

[VAN 89] VAN DOVER C., FRY B., "Stable isotopic compositions of hydrothermal vent organisms", *Marine Biology*, vol. 102, pp. 257–263, 1989.

[VAN 94] VAN DOVER C., FRY B., "Microorganisms as food resources at deep-sea hydrothermal vents", *Limnology and Oceanography*, vol. 39, pp. 51–57, 1994.

[VAN 00] VAN DOVER C., *The Ecology of Deep-Sea Hydrothermal Vents*, Princeton University Press, Princeton, 2000.

[VAN 02] VAN DOVER C., GERMAN C.R., SPEER K.G. *et al.*, "Marine biology – evolution and biogeography of deep-sea vent and seep invertebrates", *Science*, vol. 295, pp. 1253–1257, 2002.

[VAN 14] VAN DOVER C., "Impacts of anthropogenic disturbances at deep-sea hydrothermal vent ecosystems: a review", *Marine Environmental Research*, vol. 102, pp. 59–72, 2014.

[VIS 91] VISMANN B., "Sulfide tolerance – physiological mechanisms and ecological implications", *Ophelia*, vol. 34, pp. 1–27, 1991.

[WAR 03] WAREN A., BENGTSON S., GOFFREDI S.K. *et al.*, "A hot-vent gastropod with iron sulfide dermal sclerites", *Science*, vol. 302, pp. 1007, 2003.

[ZAL 98] ZAL F., LEIZE E., LALLIER F.H. *et al.*, "S-sulfohemoglobin and disulfide exchange: the mechanisms of sulfide binding by Riftia pachyptila hemoglobins", *Proceedings of the National Academy of Sciences of the United States of America*, vol. 95, pp. 8997–9002, 1998.

[ZEK 06] ZEKELY J., VAN DOVER C.L., NEMESCHKAL H.L. *et al.*, "Hydrothermal vent meiobenthos associated with mytilid mussel aggregations from the Mid-Atlantic Ridge and the East Pacific Rise", *Deep-Sea Research Part I – Oceanographic Research Papers*, vol. 53, pp. 1363–1378, 2006.

List of Authors

Mehdi ADJEROUD
IRD – CoReUs – UPVD
Perpignan
France

Cédric BACHER
Ifremer – DYNECO
Brest
France

Luc BEAUFORT
CEREGE
University of Aix-Marseille
France

Laure CARASSOU
Department of Marine Science &
Fisheries
Muscat
Oman

Pascale CHABANET
IRD – CoReUs
La Réunion
France

Daniel DESBRUYÈRES
Ifremer – LEP
Brest
France

Tania HERNÁNDEZ-FARIÑAS
Ifremer – DYNECO
Nantes
France

Christian HILY
CNRS – IUEM
Brest
France

Fanny HOULBREQUE
IRD – CoReUs
Nouméa
France

Michel KULBICKI
IRD – CoReUs – UPVD
Perpignan
France

André MARIOTTI
UPMC – IUF
Paris
France

André MONACO
CNRS – Cefrem – UPVD
Perpignan
France

Valeriano PARRAVICINI
EPHE – Criobe – UPVD
Perpignan
France

Jean-Charles POMEROL
UPMC – INSIS/CNRS
Paris
France

Dominique PONTON
IRD – CoReUs
La Réunion
France

Patrick PROUZET
Ifremer – DS
Issy-les-Moulineaux
Paris
France

Jozée SARRAZIN
Ifremer – LEP
Brest
France

Laurent VIGLIOLA
IRD – CoReUs
Nouméa
France

Index

A, B, C, D

E, F, G

H, I, L, M, N

O, P, R

S, T, V, Z

Other titles from

ISTE

in

Earth Systems – Environmental Engineering

2015

MERCIER-LAURENT Eunika
The Innovation Biosphere

2014

DE LARMINAT Philippe
Climate Change: Identification and Projections

MONACO André, PROUZET Patrick
Seas and Oceans Set
Volume 1 – Ocean in the Earth System
Volume 2 – The Land-Sea Interactions
Volume 3 – Ecosystem Sustainability and Global Change
Volume 4 – Vulnerability of Coastal Ecosystems and Adaptation
Volume 5 – Development of Marine Resources
Volume 6 – Value and Economy of Marine Resources

MUTTIN Frédéric
Marine Coastal and Water Pollutions: Oil Spill Studies

2013

LÉCUYER Christophe
Water on Earth: Physicochemical and Biological Properties

LEGCHENKO Anatoly
Magnetic Resonance Imaging for Groundwater

2012

LE MENN Marc
Instrumentation and Metrology in Oceanography

PAVÉ Alain
Modeling of Living Systems: From Cell to Ecosystem

2010

TANGUY Jean-Michel
Physical Processess and Measurement Devices – EHS volume 1

TANGUY Jean-Michel
Mathematical Models – EHS volume 2

TANGUY Jean-Michel
Numerical Methods – EHS volume 3

TANGUY Jean-Michel
Practical Applications in Engineering – EHS volume 4

TANGUY Jean-Michel
Modeling Software – EHS volume 5